Signal Processing with Python

A practical approach

Online at: https://doi.org/10.1088/978-0-7503-5929-0

Signal Processing with Python

A practical approach

Edited by
Irshad Ahmad Ansari
ABV-Indian Institute of Information Technology and Management Gwalior, Gwalior, India

Varun Bajaj
Maulana Azad National Institute of Technology, Bhopal, India

IOP Publishing, Bristol, UK

ISBN 978-0-7503-5929-0 (ebook)
ISBN 978-0-7503-5927-6 (print)
ISBN 978-0-7503-5930-6 (myPrint)
ISBN 978-0-7503-5928-3 (mobi)

DOI 10.1088/978-0-7503-5929-0

Version: 20240301

IOP ebooks

British Library Cataloguing-in-Publication Data: A catalogue record for this book is available from the British Library.

Published by IOP Publishing, wholly owned by The Institute of Physics, London

IOP Publishing, No.2 The Distillery, Glassfields, Avon Street, Bristol, BS2 0GR, UK

US Office: IOP Publishing, Inc., 190 North Independence Mall West, Suite 601, Philadelphia, PA 19106, USA

Dedicated to my lovely daughter Eimaan

—Irshad Ahmad Ansari

Dedicated to my father, the late Mahendra Bajaj, and family members

—Varun Bajaj

Contents

Preface

Python, being an open source and popular language, has attracted a lot of application development. Signal processing has also seen tremendous growth in the recent past. The majority of signal processing work revolves around simulation and testing before the final hardware implementation. Many times, code-based implementation itself is sufficient. These implementations make use of a cloud processing unit for a faster and more efficient user experience. Python has been developed as a tool for simulation, visualization, understanding, and manipulation of signals. The same has attracted a lot of new development in the field of signal processing. Python and signal processing both go hand in hand for many applications developments. The present book is an attempt to explore the domain of signal processing with the help of working examples of the Python language.

There are many signal processing applications such as acquisition, sensing, representation, understanding, feature extraction, classification, compression etc. There are standard ways to implement different signal processing methods. Signal processing has become a backbone of modern technical systems. It is almost impossible to develop any systems without using signal processing concepts. Python has helped developers to club together different concepts of signals to make use of them for practical systems. The proposed book is an attempt to introduce the concepts of Python via signal processing with hands-on examples for the readers. The book is planned to include code-led examples that will help readers to get an in-depth understanding of the subject.

Furthermore, the book first focuses on the basics of Python and signal processing, then moves forward to more advanced applications like machine learning (ML) and research avenues in the domain. This field of academia and new system development is rapidly changing. Python is a multifaceted area of development including the concepts of signal processing, ML etc. Therefore, the aim of this book is to bring forward professionals from academia and industry, presenting their recent work with Python support. The chapter-wise description of this book is as follows:

Chapter 1 presents a generalized model of automatic feature extraction from raw I/Q samples of received signals, which is further used for intelligent wireless systems built using automatic modulation classification and emerges very efficiently in modern software-defined radio. **Chapter 2** provides a two-step process for hypothesis testing to compare individuals with intellectual and developmental disorder and typically developing controls using electroencephalogram (EEG) signals. **Chapter 3** demonstrates filter design, and denoising techniques for electrocardiogram (ECG) signals are explored using time domain filters and frequency selective filters. **Chapter 4** presents Python libraries mne, numpy, scipy and pandas to preprocess and make data usable for further ML algorithms and models. **Chapter 5** demonstrates the technique of data augmentation as applied in synthetic data augmentation for tabular data and appropriate features; it is a crucial preprocessing step in classification that involves removing attributes that are unnecessary, repeated, and intrusive. **Chapter 6** discusses techniques such as normalization and standardization of electrooculogram (EOG) signals, wavelet decomposition of EOG signals, and estimation of signal power at different levels. **Chapter 7** develops artificial neural network models using the Python-based deep learning framework TensorFlow and Keras libraries, allowing for hypertuning, efficient training, implementation, and easy integration into existing workflows. **Chapter 8** discusses a distinctive analysis of various CNN-based approaches for prior breast cancer detection using histopathological image data. **Chapter 9** provides an overview of various MPPT techniques and their applications, usages, and code-based implementation. A detailed analysis of maximum power tracking is provided. **Chapter 10** presents an approach of leveraging Python scripts to analyze Monte Carlo simulation data in an Anaconda environment.

Acknowledgements

Dr Ansari expresses his gratitude and sincere thanks to his wife, family members and teachers for their constant support and motivation.

Dr Bajaj expresses his heartfelt appreciation to his mother Prabha, wife Anuja, and daughters Avadhi and Avyana for their wonderful support and encouragement throughout the completion of this important book on signal processing with Python. His deepest gratitude goes to his mother-in-law and father-in-law for their constant motivation. This book is the outcome of sincere efforts that could be given to the book only due to the great support of his family.

We sincerely thank Professor S N Singh, Director of the ABV-Indian Institute of Information Technology and Management (ABV-IIITM), Gwalior, and Professor Karunesh Kumar Shukla, Director of Maulana Azad National Institute of Technology (MANIT) Bhopal for their support and encouragement. We would like to thank all our friends, well-wishers and all those who keep us motivated to do more and more, better and better. We sincerely thank all contributors for writing relevant the theoretical background for and applications of this book.

We humbly thank Dr John and all editorial staff of IOP Publishing for their excellent support, necessary help, appreciation, and quick responses. We also wish to thank IOP Publishing for giving us this opportunity to contribute to a relevant topic with a reputable publisher. Finally, we want to thank everyone, in one way or another, who helped us edit this book.

Dr Bajaj especially thanks his family, who encouraged him throughout the time of the editing of this book. This book is heartily dedicated to his father, who took the lead to heaven before the completion of this book.

Last but not least, we would also like to thank God for showering us with his blessings and the strength to do this type of novel and quality work.

Irshad Ahmad Ansari
Varun Bajaj

Editor biographies

Irshad Ahmad Ansari

Irshad Ahmad Ansari (PhD, SMIEEE20) has been working as an Assistant Professor Grade I in the Department of Electrical and Electronics Engineering at ABVIIITM, Gwalior, India, since June 2023. He received his Bachelor of Technology degree in Electronics and Communication Engineering from Gautam Buddh Technical University (formally Uttar Pradesh Technical University), Lucknow, India, in 2010 and his Master's of Technology (M.Tech.) degree in Control and Instrumentation from Dr B R Ambedkar National Institute of Technology (NIT) Jalandhar, Punjab, India in 2012. He completed his PhD at the Indian Institute of Technology Roorkee with Ministry of Human Resource Development (MHRD) teaching assistantship in 2017 and subsequently joined the Gwangju Institute of Science and Technology, South Korea as a postdoctoral fellow. Afterward, he joined PDPM IIITDMJ as Assistant Professor Grade II. His major research interests include signal and image processing, electronic design, ML, biomedical signal processing, computer vision etc. He is contributing as an active technical reviewer of leading international publishers such as IEEE, IOP Publishing, Elsevier, and Springer. He has more than 70 publications, which include 29 SCI/SCIE journal papers, 28 international conference papers, 4 edited books, and 6 book chapters. He has more than 70 publications, which include 29 SCI/ SCIE journal papers, 28 international conference papers, 6 edited books and 6 book chapters. The citation impact of his publications is around 1200 citations, with an h-index of 19, and an i10 index of 27 (Google Scholar December 2023). He has guided three (3 awarded) PhD scholars and 14 M. Tech scholars. He has been listed as the world's top 2% of researchers/scientists by Stanford University, USA (October, 2023).

Varun Bajaj

Varun Bajaj (PhD, SMIEEE20) is working as an Associate Professor in the discipline of Electronics and Communication Engineering, at Maulana Azad National Institute of Technology Bhopal, India from Jan 2024. He served as Associate Professor in the discipline of Electronics and Communication Engineering, Indian Institute of Information Technology, Design and Manufacturing (IIITDM) Jabalpur, India since July 2021 to Jan 2024. He worked as an Assistant Professor at IIITDMJ from March 2014 to July 2021. He also worked as a visiting faculty in IIITDMJ from September 2013 to March 2014. He worked as Assistant Professor at the Department of Electronics and Instrumentation, Shri Vaishnav Institute of Technology and Science, Indore, India from 2009–2010. He received his PhD degree in the Discipline of Electrical Engineering at Indian Institute of Technology Indore, India, in 2014. He received his M.Tech. degree with honours in microelectronics and very large scale integration (VLSI) design from Shri Govindram Seksaria Institute of Technology and Science, Indore, India in 2009, Bachelor of Engineering (B.E.) degree in Electronics and Communication Engineering from Rajiv Gandhi Technological University, Bhopal, India in 2006.

He is an Associate Editor of the *IEEE Sensors Journal* and Biomedical Signal Processing for *Frontiers in Signal Processing* and Subject Editor-in-Chief of *Electronics Letters*. He served as a Subject Editor of *Electronics Letters* from November 2018 to June 2020 and as a Guest Editor in Elsevier and IET Journals. He is an IEEE Senior Member June 2020, MIEEE 16–20, and contributes as an active technical reviewer of leading international journals of IEEE, IET, and Elsevier etc. He has served as a member of the review boards for around 80 scientific journals. He has also served on the scientific committees of various international conferences. He has delivered more than 50 expert talks and lectures in conferences, workshops, and short-term courses organized by various institutes. He has received several awards including Achievement Award and Best Paper Award (ICHIT conference 2012 and the 2nd International Conference on Computational Electronics for Wireless Communication (Springer) 2022) and the Award for Excellence in Research by the 7th South Asian Education Awards Summit-22. He has been listed as being within the world's top 2% of researchers/ scientists by Stanford University, USA (October 2020, October 2021, and October 2022).

He has 170 publications, which include journal papers (104), conference papers (35), books (13), and book chapters (18). He has also granted two international patents and published three Indian patents. The citation impact of his publications is around 5 758 citations, with an h-index of 42 and i10 index of 108 (Google Scholar March 2023). He has guided nine (six completed and three in progress) PhD scholars and eight M.Tech. scholars. He has worked on five research projects funded by India's Department of Science and Technology (DST) and Council of Scientific and Industrial Research and three consultancy projects. His research interests include biomedical signal processing, artificial intelligence (AI) in healthcare, brain–computer interface, pattern recognition, and ECG signal processing.

List of contributors

Mosabber Uddin Ahmed
Department of Electrical and Electronic Engineering, University of Dhaka, Dhaka, Bangladesh

Meena Anandan
Indian Institute of Information Technology, Design and Manufacturing, Kancheepuram, India

Varun Bajaj
Maulana Azad National Institute of Technology Bhopal, India

Zahra Ghanbari
Biomedical Engineering Department, Amirkabir University of Technology, Tehran, Iran

Aakash Kumar Jain
Indian Institute of Information Technology, Design and Manufacturing, Kancheepuram, India

K P Madhavan
PDPM Indian Institute of Information Technology, Design and Manufacturing, Jabalpur, India

Manas Kumar Mishra
Indian Institute of Information Technology, Design and Manufacturing, Kancheepuram, India

Ishrat Jahan Mohima
Department of Information and Communication Technology, Bangladesh University of Professionals, Dhaka, Bangladesh

Mohammad Hassan Moradi
Biomedical Engineering Department, Amirkabir University of Technology, Tehran, Iran

Poorya Moradi
Department of Industrial Engineering, Engineering Faculty, Khatam University, Iran

Rohini Palanisamy
Indian Institute of Information Technology, Design and Manufacturing, Kancheepuram, India

Nakul Pathak
PDPM Indian Institute of Information Technology, Design and Manufacturing, Jabalpur, India

K K Mujeeb Rahman
Ajman University, United Arab Emirates

Mozhdeh Saghalaini
Biomedical Engineering Department, Amirkabir University of Technology, Tehran, Iran

Deepika Sainani
Banasthali Vidyapith, Rajasthan, India

Urvashi Prakash Shukla
Banasthali Vidyapith, Rajasthan, India

Abhishek Kumar Singh
Indian Institute of Information Technology, Design and Manufacturing, Kancheepuram, India

Mahdi Taghaddossi
Biomedical Engineering Department, Amirkabir University of Technology, Tehran, Iran

Pandiyarasan Veluswamy
Indian Institute of Information Technology, Design and Manufacturing, Kancheepuram, India

Contributor biographies

Mosabber Uddin Ahmed

Mosabber Uddin Ahmed is currently a professor in the Department of Electrical and Electronic Engineering (EEE) at the University of Dhaka. He received his Bachelor of Science (B.Sc.; honours) and Master's of Science (M.Sc.) degrees from the Department of Applied Physics and Electronics, University of Dhaka, in 1998 and 1999, respectively. He received a second M.Sci. degree in Communications and Signal Processing and a PhD degree in Signal Processing from the Imperial College London (ICL), United Kingdom, in 2006 and 2012, respectively. From 2015–2016, he was a postdoctoral fellow in the EEE Department of ICL. To date, he has authored more than 60 research articles that are published in refereed journals, books, and conference proceedings. Dr Ahmed, a senior member of IEEE, has also co-edited a book titled *Signal Processing Techniques for Computational Health Informatics*, which has been published by Springer Nature. He has won several fellowships and research grants including the Commonwealth Academic Fellowship, the Commonwealth Scholarship, the Charles Wallace Bangladesh Trust grant, the Mirza Samsul Huda and Mahmuda Begum Memorial Trust Gold Medal, the Roushan Innas Ali Welfare Trust Gold Medal and the Taylor & Francis Commonwealth Scholar Highly Commended Journal Article Prize. His research interests include signal processing, time-frequency analysis, multivariate time series analysis, complexity analysis, multiscale entropy analysis, embedded system design, Internet of things (IoT), information security and ML.

Meena Anandan

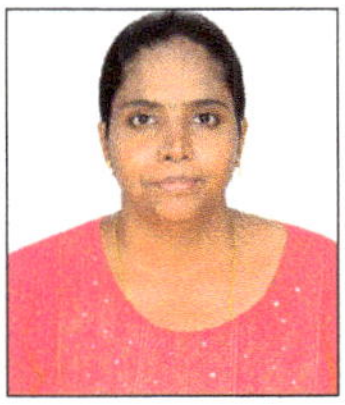

Meena Anandan received her M.Tech. degree in Embedded Systems and Technologies from Vel Tech Rangarajan Dr Sagunthala R&D Institute of Science and Technology and is currently pursuing a PhD at IIITDM, Kancheepuram, India. Her research interests include biomedical signals and systems, wearable devices and AI. She received the Savithri Jyotirao Phule Single Girl Child fellowship from the University Grants Commission to pursue her PhD studies.

Zahra Ghanbari

Zahra Ghanbari received her B.Sc. in electronics engineering from Ferdowsi University of Mashhad, Iran, in 2005. She received her M.Sc. in biomedical engineering from Sharif University of Technology, Tehran, Iran, in 2011. Zahra received her PhD in biomedical engineering from Amirkabir University of Technology, Tehran, Iran, in 2020. She was an adjunct professor at Sadjad University of Technology, Imam Reza International University and the Kayyam Institute for Higher Education. Furthermore, she taught at the electrical engineering and computer engineering departments of Alzahra University, Tehran, Iran (2016–2021). Now she is an adjunct professor at Ferdowsi University, Mashhad, Iran, Shahid Beheshti University, Tehran, Iran, and Amirkabir University of Technology, Tehran, Iran. She has authored more than 14 journal and conference papers in addition to 8 book chapters. Her research and teaching interests include biomedical signal and image processing, neuroscience, cognitive science, and analogue and digital electronics.

Aakash Kumar Jain

Aakash Kumar Jain is an Assistant Professor at IIITDM Kancheepuram, India, and a respected member of IEEE. He obtained his M.Tech. degree in Microelectronics and VLSI stream from IIT Bombay, Mumbai, India, in 2014, and later achieved his PhD degree in electrical engineering from IIT Delhi, New Delhi, India, in 2020. Previously, he served as a Postdoctoral Research Fellow in the Faculty of ECE at the National University of Singapore, Singapore. Dr Jain has made significant contributions in the field of nanoscale devices, junctionless field-effect transistors, and various nanoelectronics phenomena, as evidenced by his multiple publications in esteemed journals and conferences. His primary research interests revolve around novel nano-scale device design, modeling of emerging non-volatile memory device technology, reconfigurable and neuromorphic computing, and low-power device design.

K P Madhavan

K P Madhavan is currently working as a power electronics engineer at ESAB India Ltd. He completed his M. Tech. at IIITDMJ in Electronics and Communication (power and Control). His research interests include digital control of power electronics and power electronics applications for renewable energy sources.

Manas Kumar Mishra

Manas Kumar Mishra received his Bachelor of Technology (B.Tech.) with honours and M.Tech. degrees in Communication System Design from the IIIDTM, Kancheepuram, India in 2023. He is currently working at Nokia Solutions and Networks Pvt Ltd as an associate engineer. His research interests include biomedical signals and systems, digital filter design, 5G/6G communication systems, and statistical signal processing. He received academic proficiency awards for three consecutive years during his study.

He has been involved in multiple papers submitted to IEEE and one patent regarding wireless textile-based health monitoring. He has successfully published one paper at the IEEE International Conference on Emerging Electronics.

Ishrat Jahan Mohima

Ishrat Jahan Mohima is currently a student in the Department of Information and Communication Technology (ICT) at Bangladesh University of Professionals. She has completed her B.Sc. (Hons.) from the same university in Information and Communication Engineering. She is currently studying for her M.Sc. in the same subject. In recognition of her academic achievements, she received a Dean's Appreciation Letter and departmental merit scholarship. She held leadership position as Vice Chair in the IEEE BUP Student Branch (Robotics and Automation Society). Her research interests include image processing, signal processing, IoT, AI and ML.

Mohammad Hassan Moradi

Professor Mohammad Hassan Moradi, PhD received his B.Sc. and M.Sc. degrees in Electronic Engineering from University of Tehran, Iran, in 1988 and 1990, respectively. He received his PhD degree from Tarbiat Modarres University, Tehran, Iran, in 1995. He has been with the faculty of biomedical engineering, Amirkabir University of Technology, Tehran, Iran, since 1995. His primary research and teaching interests include design, manufacturing and application of medical instrumentation, biomedical signal processing, cognitive neuroscience, neuro-marketing, wavelet system design, time-frequency transforms and deep fuzzy neural systems. He has published over 160 technical papers in various high impact factor peer-reviewed journals. Moreover, he presented more than 300 papers in international conferences. Professor Moradi translated a book on wavelet signal processing into Persian. He has written 12 book chapters as well.

Poorya Moradi

Poorya Moradi, M.Sc. received his B.Sc. degrees in Industrial Engineering from Islamic Azad University, North Tehran branch, in 2020 and M.Sc. degrees in Engineering Management and Industrial Engineering from Khatam University, Tehran, in 2023. He is currently a foreign trade expert in an international company. His research interests include engineering management, neuromarketing and data analysis.

Rohini Palanisamy

Rohini Palanisamy received her undergraduate degree in Biomedical Engineering from Department of Electronics and Communication Engineering, College of Engineering Guindy, Anna University. Furthermore, she pursued her graduate studies in Biomedical Engineering offered by the Department of Applied Mechanics in Indian IIT Madras. In 2016, she joined as a research scholar in the Non-Invasive Imaging and Diagnostics Laboratory to pursue a PhD in Biomedical Engineering at the IIT Madras. Dr Palanisamy is currently serving as an Assistant Professor in the Department of Electronics and Communication Engineering, IIITDM, Kancheepuram, India, an institute of national importance. Her research interests include biomedical signal processing, medical image analysis, pattern recognition, early diagnosis, and non-invasive disease prognosis.

Nakul Kishor Pathak

Nakul Kishor Pathak received his B.E. degree in Electronics and Communication Engineering from Sant Gadge Baba Amravati University, Amravati, India, in 2010, his Master's of Engineering (M.E.) degree in Communication Engineering from the Birla Institute of Technology and Science, Pilani, India, in 2013. He worked as an Assistant Professor with the Sir M Visvesvaraya Institute of Technology, Bengaluru, India, from 2013–2017, and with the Shri Ramdeobaba College of Engineering and Management, Nagpur, India, from 2017–2019. Currently, he is working as scientist/ engineer 'SC' at the Space Application Centre, Indian Space Research Organisation, Ahmedabad, India, and is also pursuing a PhD (external) from PDPM IIITDMJ, India. He has published papers and a patent (patent number 201821006225). His areas of interest are wireless communication, satellite communication, deep learning and its applications to communication systems.

K K Mujeeb Rahman

K K Mujeeb Rahman, PhD, received his B.Tech. in Electronics Engineering from the Institution of Engineers, Kolkata, India, in 1999, his M.Tech. in Biomedical Instrumentation Engineering from Visvesaraya Technological University, Karnataka, in 2002, with a gold medal, and a PhD degree in Biomedical Engineering from Vellore Institute of Technology, Vellore, India, in 2023. Since August 2004, he has been with the College of Engineering and Information Technology, Ajman University, Ajman, United Arab Emirates, where he is currently a Senior Lecturer in the Department of Biomedical Engineering. His research focuses on ML applications in the field of biomedical engineering, specifically developing ML solutions for screening autism in children using objective biomarkers such as eye-movement anomalies, facial dysmorphia, and EEG energy. Dr Mujeeb has authored/co-authored 19 publications in reputed journals and presented 10 papers at international conferences. He is a senior member of IEEE and an associate member of IE (India).

Mozhdeh Saghalaini

Mozhdeh Saghalaini has been a Master's student in biomedical engineering (bioelectric) at Amirkabir University of Technology, Tehran, Iran since 2021, where she is currently working on her Master's thesis, titled "Estimation of the Severity of Obsession with Contamination using Electroencephalogram Signal," under the supervision of Professor Mohammad Hassan Moradi. She also received her Bachelor degree in electrical engineering with a focus on electronics from Isfahan University of Technology, Isfahan, Iran in 2020, where she was working in the power-electronic laboratory as a research project assistant and completed her bachelor's thesis on 'analysis and review of multi-input switching converters' under the supervision of Dr Hosein Farzanehfard. Mozhdeh also attended the managerial functions of leadership, finance, marketing, and entrepreneurship virtual internship, which was held by Upkey. Meanwhile, she succeeded in passing several courses, including Arduino, robot mechanics, Python programming etc.

Deepika Sainani

Deepika Sainani is pursuing her PhD at Banasthali Vidyapith, Rajasthan, India. She completed her M.Tech. (computer science) in 2014 from Rajasthan Technical University, Kota and B.E. (information technology) in 2009 from the University of Rajasthan, Jaipur. She has been part of various national and international conferences for her research contributions. Her areas of research interest are ML, data analytics and data mining.

Urvashi Prakash Shukla

Urvashi Prakash Shukla is at present working as an Assistant Professor in the Department of Computer Science at Banasthali Vidyapith, Rajasthan, India. She received her bachelor's degree in Electronics and Communication Engineering in 2009. She pursued her M.Tech. in 2013 in the field of compressive sensing. She completed her PhD from NIT, Jaipur, and Rajasthan, India on 'Clustering Algorithms Based on Social Spider Optimization for Hyper Spectral Image Analysis'. She has published many articles in reputed journals and conferences and is an active reviewer of *IEEE Access* and Elsevier journals. Her research interests involve obtaining solutions for problems associated with medical signal processing, hyperspectral Images, data mining, and AI.

Abhishek Kumar Singh

Abhishek Kumar Singh received his integrated Master's degree in Electronics and Communication Engineering with a specialization in VLSI design from IIITDM Kancheepuram, India. He was a part of the analog computer-aided design (CAD) team at STMicroelectronics, Greater Noida. Currently he works as a Hardware Description Language (HDL) Code Generation Engineer at MathWorks, Hyderabad, India. His major interest lies in register transfer level (RTL) design, computer architecture and hardware accelerators.

Mahdi Taghaddossi

Mahdi Taghaddossi, M.Sc., received his B.Sc. degree in Electrical Engineering from Islamic Azad University, Yadegare-Imam branch, Iran, in 2017 and his M.Sc. degree in Biomedical Engineering from Amirkabir University of Technology (polytechnic Tehran), Tehran, Iran, in 2021. He is currently an EEG signal processing and registration specialist at the National Brain Mapping Laboratory, Tehran University, Tehran, Iran. His research interests include biomedical signal processing, biomedical data mining, fuzzy system, wavelet and time-frequency analysis, and neuroscience.

Pandiyarasan Veluswamy

Pandiyarasan Veluswamy has received his B.E. and M.E. degrees from Anna University, Chennai, India. He received his PhD in Engineering from National University Corporation Shizuoka University, Hamamatsu, Japan. He also received a Postdoctoral Research Fellowship at the Nanoelectronics and Energy Device Laboratory, Korea Advanced Institute of Science and Technology, South Korea. Dr Pandiyarasan is currently working as an Assistant Professor at IIITDM, Kancheepuram, India. He was a DST INSPIRE Faculty Fellow from October 2018 to January 2021. He has contributed to three international book chapters and over 60 publications. He is a visiting faculty member at various universities across the globe.

Chapter 1

Automatic feature extraction using deep learning for automatic modulation classification implemented with Python

Nakul Kishor Pathak and Varun Bajaj

The radio spectrum has become extremely packed due to an immense spike in user terminals as wireless technology has grown rapidly. Thus, giving reliable and quality services is greatly challenging due to system complexities at various system levels. Current intelligent wireless communication system needs to provide cutting edge technology with improving quality of service and proper spectrum usage to provide efficient resource management. Intelligent wireless systems built using automatic modulation classification (AMC) are emerging as very efficient in modern software-defined radio. Recently, AMC based on deep learning (DL) paved a new way of tackling the problem of extracting various features automatically where machine learning and conventional likelihood approaches failed. This chapter covers various DL-based automatic feature extraction methods from the raw I/Q (in phase/quadrature) samples of received communication systems. The DL framework includes automatic feature extraction layers followed by classification layers, where raw I/Q samples are directly fed to the model. For the learning of model weights, models go through training and validation and testing phases. DL model building, data preparation, training the model, and prediction were all implemented in Python over Keras and TensorFlow libraries.

1.1 Introduction

With its applicability spanning across military and civilian realms, AMC has taken the spotlight, captivating the interest of researchers worldwide. The goal of AMC is to reduce the overheads on the preamble data and improve the system's overall throughput. When adaptive modulation and coding schemes were implemented, as receive unaware about the modulation and coding scheme used by the transmitter, there was a need for the transmitter to transmit these data along with the preamble.

doi:10.1088/978-0-7503-5929-0ch1

But as the communication system moves towards 5G and beyond, due to the higher frequencies, to tackle fading effect and improve system reliability, preamble data needs to be transmitted more frequently, which indeed reduces the system's throughput [1–4]. To mitigate this, researchers proposed that detection of the modulation scheme take place at the receiver instead of transmitting it over the network that in turn improves system throughput.

According to the existing literature [1], *researchers* have put forth various methodologies for AMC. Among these, two widely utilized approaches include likelihood-based (LB) and feature-based (FB) methods. As a probability-based solution, the LB method utilizes various techniques for extraction of unknown parameters in its operation such as the likelihood ratio test and its variants [5–9]. But due to the computational complexity of the LB approach in real time, research proposed the FB approach with various features such as instantaneous values, transform based, higher order moments and cumulants, entropy features etc [8–15]. As part of this process, the extracted features from the wireless communication data are fed into diverse classifiers like multilayer perceptron and support vector machines, as discussed in previous works [15–18]. In the realm of wireless communication, researchers have increasingly turned their attention to DL-based approaches, driven by the remarkable success these techniques have achieved in computer vision. O'Shea *et al* [19], pioneers in the field of AMC using DL approaches, have already proven the effectiveness of the convolutional neural network (CNN), which outperforms other statistics-based methodologies. Even other research utilizes various DL-based frameworks for extraction of various high level and low level features from incoming raw I/Q samples of received signals [20–25]. These featurs are generally spatial features extracted from convolutional layers (CLs) in CNN or may be temporal features between consecutive samples in long short-term memory (LSTM) networks or both, using integration of CNN and LSTM into a single DL framework [26–34]. The primary constituents of a general DL-based approach are the feature extraction block (FEB) and the classifier block (CB), which work in tandem. The FEB consists of feature extraction layers, which automatically extract various features such as spatial features, temporal features etc from incoming signals. The incoming signal may either directly give to FEB, or the signal may go through a pre-processing unit, which may extract different features such as bispectrum [22], smooth pseudo-Wigner-Ville distributions [24] that in turn feed into the FEB for further extraction of features. Extracted features then further pass to the CB for mapping into hyperplanes for efficient classification of various modulation schemes.

The FEB and CB are both packed into a single DL framework and implemented using open source libraries such as Keras and TensorFlow. These libraries provide various functional blocks that can directly be incorporated as FEB's and CB's various layers. They also provide various functions for training, testing, prediction etc. In the sections below, this chapter discusses various DL frameworks for AMC that extract features automatically along with their implementation in Python, various obtained results and conclusion.

1.2 Proposed AMC based on DL framework

Various tools are available for the implementation of DL frameworks for the classification of different modulation schemes. This chapter is based on [35], which includes two main parts.

(a) In the first part, the literature consists of a communication system model implemented in MATLAB for the dataset generation. This can also be accomplished using GNU radio implementation in Python.
(b) The second part focuses on realization of DL frameworks (state-of-the-art and proposed model). DL models can also be realized in many ways. To the authors' knowledge, two approaches are widely used: MATLAB based and Python based using Keras/TensorFlow libraries. The literature [35] follows a Python-based approach.

Complete DL training and prediction take place on the Google Colab Pro platform, which supports accessing the GPU through infrastructure as a service platform. Diffferent runtimes were used for simulation purposes. A total of 256 were used as a batch size for accessing the data during training. Various callback methods have been used, such as reduction in learning rate if there was no improvement in validation loss performace for 5 consecutive epochs, saving the best model weights with best results, and stopping training early if validation loss did not improve for 50 consecutive epochs. A learning rate of 0.001 was used when starting training. THe dataset has been split into three segments with 60% for training, 20% for validation and 20% for testing.

Subsequent subsections discuss various parts of the implementation of the DL framework as snippets of Python code.

1.3 System model for dataset generation

The article discussed in reference [35] explores the incorporation of a system model to generate datasets, as depicted in figure 1.1. The modulation schemes

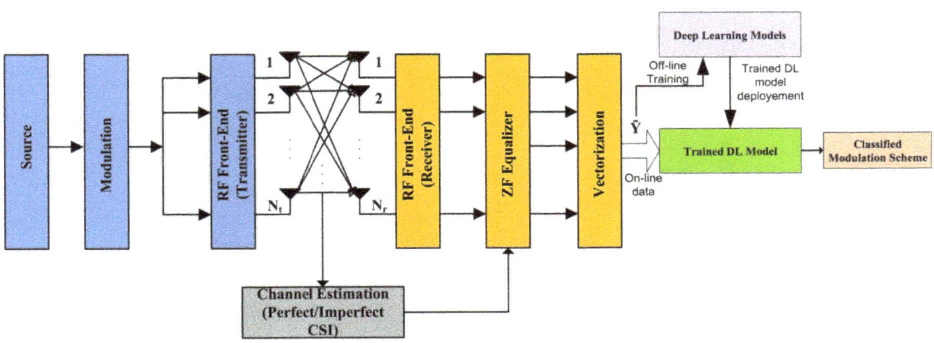

Figure 1.1. The AMC system framework for the dataset generation and online deployment of the trained model for classification of schemes. N_r represents number of receiver antennas, while N_t represents the number of transmitter antennas.

investigated in the article encompass binary phase-shift keying (BPSK), quadrature phase-shift keying (QPSK), 8 phase-shift keying (8PSK), and 16 quadrature-amplitude modulation (16QAM). The length of the modulated symbols considered in the article is 128, but this may be variable as per requirements and classifications of accuracy improvement. The signal is then transmitted via the multiple input-multiple output (MIMO) system (N_r x N_t). The receiver signal passes through a zero-forcing (ZF) equalizer for reducing the channel effect that needs channel state information which can either be perfect or impact in nature. The output of the ZF equalizer is then reshaped to store the data in an M x (2xL) matrix form where each signal has length L while M is the total number of signals in the dataset for training for all the signal-to-noise ratio (SNR) and modulation schemes. The individual element of L consists of two real parts: I is the real part of the equalized data, and Q is the imaginary part of the equalized data. The data for each signal are stored as a 2 X L matrix form. Once the DL model gets trained offline, it is then deployed in the online system for the modulation scheme classification task.

1.4 Generalized feature extraction system using DL

Figure 1.2 describes the generalized model of automatic feature extraction from raw I/Q samples of received signals. This feature extraction block can be any state-of-the-art DL framework layer such as a CL, LSTM layers, auto-encoders (AEs), their combinations (series, parallel, multichannel) etc. The final classification block is generally built using a neural network (NN) that maps extracted features into a hyperplane for efficient classification of modulation schemes. Multiple CLs along with NNs form CNNs that are used as baseline models for this chapter. In this way, various DL frameworks are developed such as LSTMs, convolutional LSTM deep (CLDNNs) [36, 37], multichannel CLDNNs (MCLDNN) [34], generalized adversarial networks, AEs, bimodal multichannel configurable CLDNNs (BMCCLDNNs) [35] etc.

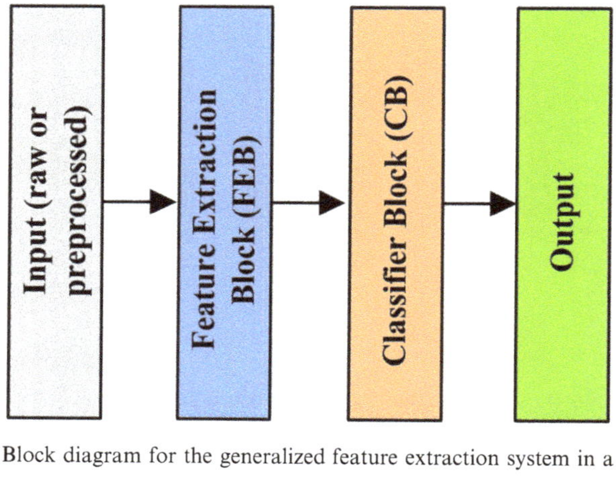

Figure 1.2. Block diagram for the generalized feature extraction system in a DL model.

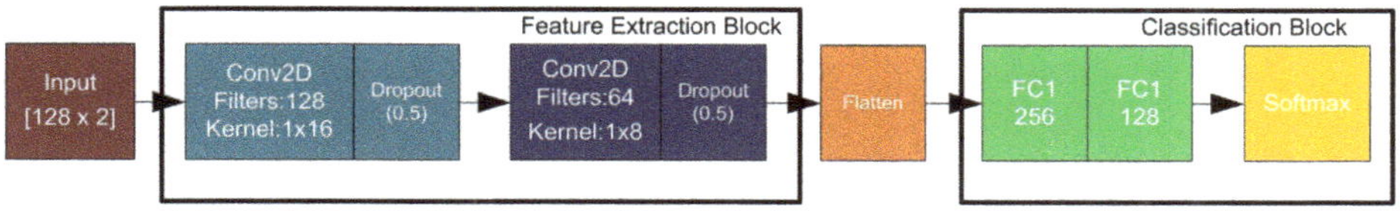

Figure 1.3. Block diagram for a CNN.

1.4.1 Spatial feature extraction using CNNs

CNNs have already proved their effectiveness in the extraction of spatial features such as edges, shapes etc from input images. CNNs consist of multiple CLs that operate convolutional operation on the input. CNNs employed in the field of AMC may accept raw I/Q samples or pre-processed data as inputs (smoothed pseudo-Wigner-Villey distribution images, spectrograms, wavelet transforms, constellation images etc). Spatial features are extracted from one of these inputs or combinations of these inputs as multimodes using CLs. Extracted features then pass through the flattened layer if not present in $1 \times N$ array format, which then feeds into the NN for non-linear mapping into the hyperplanes for classification. CNNs can be constructed in many ways, but figure 1.3 shows the way implemented in [35].

1.4.1.1 Method

1.4.1.2 Python code for CNN model

```
#Convolutional Neural Network function
def CNN(weights=None, input_shape1=[2,128], input_shape2=[128,1], classes =
4, **kwargs):

 dr=0.5
 input1=Input(input_shape1+[1],name='I/Q__channel')

#Feature Extraction Block - (Two Convolutional Layers)
    x1=Conv2D(128, kernel_size=(1,16), padding='valid', activation="relu",
 name="Convol1", kernel_initializer="glorot_uniform", data_format =
 "channels_last")(input1)
   x1 = (Dropout(dr))(x1)
   x1=BatchNormalization()(x1)

   x2=Conv2D(64, kernel_size=(1,8), padding='valid', ,activation="relu",
 name="Convol2", kernel_initializer="glorot_uniform" data_format = "channels_last"
 )(x1)
   x2 = (Dropout(dr))(x2)
   x2=BatchNormalization()(x2)
```

```
#Flatten
    x2 = (Flatten())(x2)

#CB
    x3=Dense(256,activation="selu",name="FC1")(x2) #also try with 1020
    x3=Dropout(dr)(x3)
    x3=BatchNormalization()(x3)
    x3=Dense(128,activation="selu",name="FC1")(x2) #also try with 1020
    x3=Dropout(dr)(x3)
    x3=BatchNormalization()(x3)
    x4=Dense(classes,activation="softmax",name="Softmax")(x3)
    my_model=Model(inputs=[input1],outputs=x4)

    return my_model

#from tf.keras.optimizers import adam
if __name__ == '__main__':

   my_model = CNN(classes=4)
   my_model.compile( metrics=['accuracy'], loss='categorical_crossentropy',
 optimizer='adam')
   my_model.summary()
```

1.4.2 CLDNN-based approach

When it comes to feature extraction, CNNs demonstrate strength in spatial domains, while LSTMs specialize in temporal domains, and deep neural network (DNN)s efficiently map the extracted features onto non-linear hyperplanes [36–38]. Aforementioned research proved effectiveness of CLDNN models over CNNs, LSTMs, and Residual Network (Resnet).

1.4.2.1 Method

1.4.2.2 Python code for CLDNN model

```
#CLDNN model function
def ConvUnit(input_fun,size_kernel = 8,indx = 0):
    x = Conv1D(80, kernel_size= size_kernel, kernel_initializer='glorot_uniform',
 activation='relu,' padding='same' ,name='CL{}'.format(indx + 1))(input_fun)
    x = MaxPool1D(2, 2, name='MXpl{}'.format(indx + 1))(x)
 return x


 rt_dr = 0.5
 tap_val =8

 def my_CLDNN_model(input_sh=[1024,2], classes=4, **kwargs):

    input_val = Input(input_sh, name='IN')


#Feature Extraction Block
   #Convolutional Block
```

```
    C1 = 4
    for CLindx in range(C1):
       in_cnn = ConvUnit(input_val, tap_val, CLindx)
   #LSTM Unit
    x_lstm = LSTM(80, return_sequences = True)(in_cnn)
    x_lstm = LSTM(80)(x_lstm)
   #Classification Block (DNN)
    in_dnn = Dense(256,activation='selu',name='fullconnect1')(x_lstm)
   in_dnn = Dropout(rt_dr)( in _dnn)
   in_dnn = BatchNormalization()(in_dnn)
   in _dnn = Dense(128, activation='selu', name='fullconnect2')( in _dnn)
   in _dnn = Dropout(rt_dr)( in _dnn)
   in_dnn = BatchNormalization()(in_dnn)
   in _dnn = Dense(classes,activation='softmax',name='softmax')( in _dnn)

    archi = Model(inputs=[input_val], outputs=[in_dnn])
    return archi

my_archi = my_CLDNN_model((1024,2),classes=4)
my_archi.compile(metrics=['accuracy'], loss='categorical_crossentropy',
optimizer='adam')
```

1.4.3 MCLDNN based approach

In [34], the authors propose a multichannel structure of CLDNNs to improve accuracy further by extracting features from a multichannel structure accepting inputs from three channels: I/Q, I and Q simultaneously.

1.4.3.1 Method

For complete model details, kindly refer to article [34].

1.4.3.2 Python code for MCCLDNN models

For complete model details, kindly refer to [39].

1.4.4 Complement of multichannel structure from bimodal information using BMCCLDNN models

Inspired by the framework in [34], authors of article [35] extracted features from two different modalities: I/Q and amplitude/phase. This modal information fed into two parallel multichannel structures with CLs and combined their output features. It further fed to LSTM layers for temporal feature extraction and finally deep NNs employed for furnishing classification tasks as shown in figure 1.4. Part 1 and 2 form the FEB. For detailed information about the structure (figure 1.5), kindly refer to the model in [35].

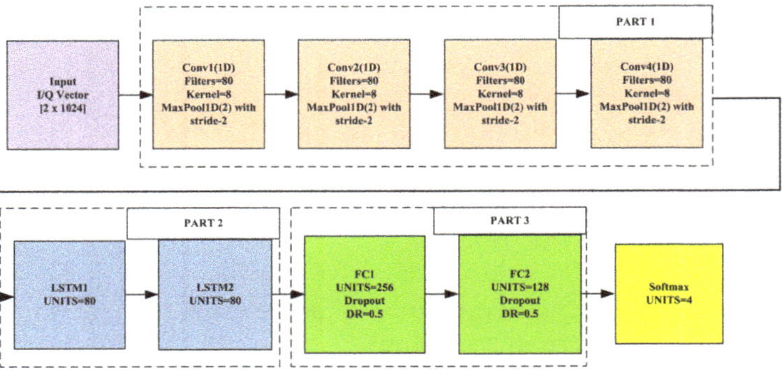

Figure 1.4. Block diagram for a CLDNN.

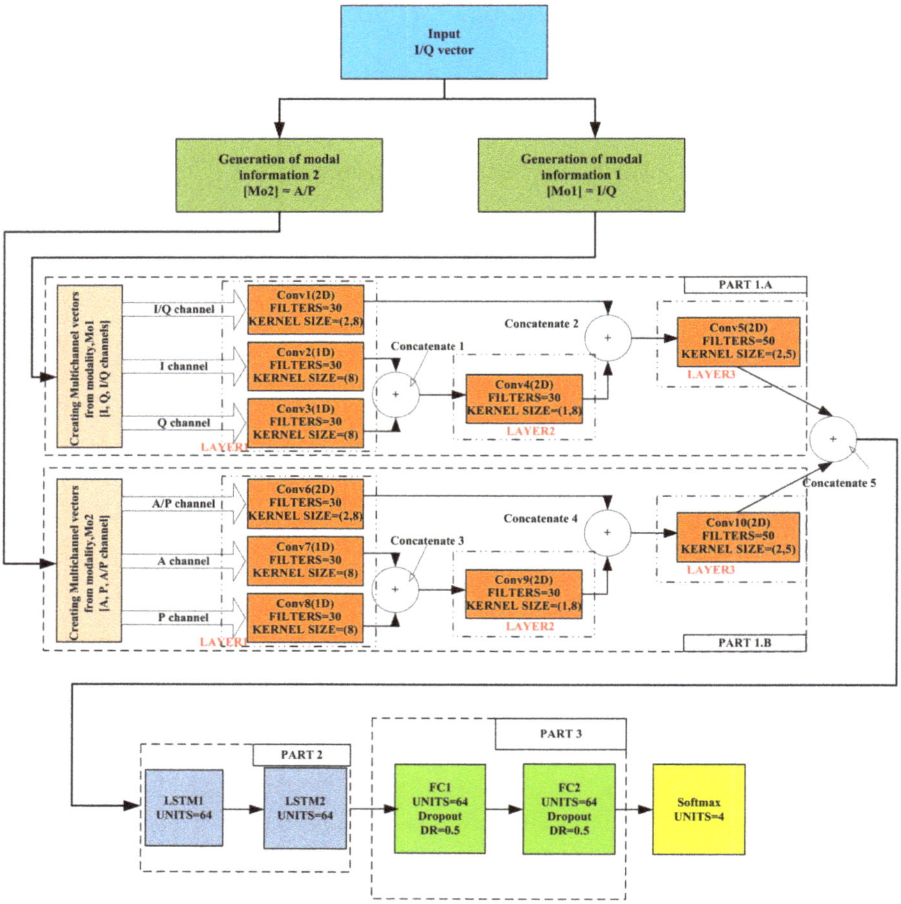

Figure 1.5. Detailed structure of a BMCCLDNN for feature extraction and classification in AMC [35].

1.4.4.1 Method

1.4.4.2 Python code for a BMCCLDNN model

```
#BMCCLDNN Model

def my_BMCCLDNN_archi( input_shape1=[2,128], input_shape2=[128,1], classes=4,
**kwargs):
  rt_dr =0.5

   input1=Input(input_shape1+[1],name='I/Qchannel')
   input2=Input(input_shape2,name='Ichannel')
   input3=Input(input_shape2,name='Qchannel')

   input4=Input(input_shape1+[1],name='A/Pchannel')
   input5=Input(input_shape2,name='Achannel')
   input6=Input(input_shape2,name='Pchannel')

# Feature Extraction Block
   # Part-1.A: Multi-channel Inputs and Spatial Characteristics Mapping
Section
   x1=Conv2D(30,(2,8), activation= "relukernel_initializer= "glorot_uniform"",
name= "Convol1", padding='same',)(input1)
 x2=Conv1D(30,8,padding='causal',activation="relu",name="Convol2",kernel_initiali
zer="glorot_uniform")(input2)
   x2_reshape=Reshape([-1,128,30])(x2)
 x3=Conv1D(30,8,padding='causal',activation="relu",name="Convol3",kernel_initiali
zer="glorot_uniform")(input3)
   x3_reshape=Reshape([-1,128,30],name="reshap2")(x3)
   x=concatenate([x2_reshape,x3_reshape],axis=1,name='Concatenate1')
   x=Conv2D(30,(1,8), activation="relu",name="Convol4", padding='same',
kernel_initializer="glorot_uniform")(x)
   x=concatenate([x1,x],name="Concatenate2")
 x=Conv2D(50,(2,5),padding="valid",activation="relu",name="Convol5",kernel_initi
alizer="glorot_uniform")(x)

   # Part-1.B: Multi-channel Inputs and Spatial Characteristics Mapping
Section

xx1=Conv2D(30,(2,8),activation="relu",name="Convol6",kernel_initializer="glorot_u
niform",padding='same')(input4)
xx2=Conv1D(30,8,padding='causal',activation="relu",name="Convol7",kernel_initial
izer="glorot_uniform")(input5)
   xx2_reshape=Reshape([-1,128,30])(xx2)

xx3=Conv1D(30,8,padding='causal',activation="relu",name="Convol8",kernel_initial
izer="glorot_uniform")(input6)
```

```
    xx3_reshape=Reshape([-1,128,30],name="reshap22")(xx3)
    xx=concatenate([xx2_reshape,xx3_reshape],axis=1,name='Concatenate3')

xx=Conv2D(30,(1,8),activation="relu",name="Convol9",kernel_initializer="glorot_u
niform", padding='same')(xx)
    xx=concatenate([xx1,xx],name="Concatenate4")

xx=Conv2D(50,(2,5),padding="valid",activation="relu",name="Convol10",kernel_ini
tializer="glorot_uniform")(xx)
   # output shape
   ## Concatenation of two diff multichannel output and pass through conv
layer
    x_A=concatenate([x,xx],name="Concatenate5")

   # Part-2 LSTM
    x=Reshape(target_shape=((124,200)))(x_A)
    x=LSTM(units=128,return_sequences=True,name="LSTM1")(x)
    x_lstm=LSTM(units=128,name="LSTM2")(x)
    n_FC = 128
  # Part-3: CB
    in_dnn=Dense(n_FC, ,name="FullyConnect1" activation="selu")(x_lstm)
    in_dnn =Dropout(rt_dr)( in_dnn)
    in_dnn = BatchNormalization()(in_dnn)
   in_dnn =Dense(n_FC, name="FullyConnect2", activation="selu")( in_dnn)
   in_dnn =Dropout(rt_dr)( in_dnn)
    in_dnn = BatchNormalization()(in_dnn)
    in_dnn=Dense(classes,activation="softmax",name="Softmax")( in_dnn)
model=Model(inputs=[input1,input2,input3,input4,input5,input6],outputs=in_dnn)
return model


  my_archi = my_BMCCLDNN_archi(classes=4)
   my_archi.compile(metrics=['accuracy'], loss='categorical_crossentropy',
optimizer='adam')
   my_archi.summary()
```

1.5 Data preparation

Some parts of the code are adapted from the code (https://github.com/leena201818/radioml) contributed by leena201818.

1.5.1 Python code (libraries, file paths)

Generally, the code mentioned below exercises various libraries used, the path for extraction of the dataset in Google Drive, preparing training, testing, and validation input sets and labels.

```
#Libraries
 import pickle
 import keras
 import matplotlib
 import matplotlib.pyplot as plt
 import os
 import tensorflow as tf
 from tensorflow import keras
 from keras.models import Model
 from keras.layers import Input, Dense, Conv1D, Dropout, concatenate, Reshape,
 LSTM, Flatten, ZeroPadding2D, BatchNormalization
 from keras.layers.convolutional import Conv2D
 import pickle
 import pandas as pd
 import numpy as obj_numpy
 import random
 os.environ["KERAS_BACKEND"] = "tensorflow"
 import keras.backend as K
 from keras.callbacks import LearningRateScheduler
 from keras.regularizers import *
 #from keras.optimizers import adam
 from keras.models import model_from_json,Model
 import argparse


#Code to connect from Google Colab to personal Google-drive
 from google.colab import drive
 drive.mount('/content/drive')

#Defining variable for data received from datasets in Google Drive as .csv files
 filepath = '/content/drive/MyDrive/Database/... /weights_ver1_exp0.h5'
 lblpath='/content/drive/MyDrive/Database/.../Label.csv'
 modtypespath='/content/drive/MyDrive/Database/MIMO.../MOD.csv'
 snrlevelpath='/content/drive/MyDrive/Database/.../SNR.csv'
 signalpath_real = '/content/drive/MyDrive/Database/.../Signal_real.csv'
 signalpath_imag = '/content/drive/MyDrive/Database/.../Signal_imag.csv'
 signalpath_amp = '/content/drive/MyDrive/Database/.../Signal_amp.csv'
 signalpath_phase = '/content/drive/MyDrive/Database/.../Signal_phase.csv'
```

1.5.2 Python code (random shuffling of data)

```
below function data_load() accepts
  my_x:  (I/Q) values
  my_xx: (A/P) values
  my_mods: modulation schemes under consideration
  my_snrs: SNR values
  my_lbl: labels as (Mod, SNR) format. E.g. (QPSK, 10)
  data: 0 (const)

Finally returns:
  (modulation-mods, SNR--snrs, Label- lbl),
  (Input for I/Q, Label): (X_train,Y_train), (X_val,Y_val), (X_test,Y_test)
  Indexes:(train_idx,val_idx,test_idx),
  Inpurts for A/P : (XX_train,XX_test,XX_val)


################# code for shuffling of the  ##################
  def load_data(my_x,my_mods,my_snrs,my_lbl,data,my_xx):

     mods = my_mods
     snrs = my_snrs
     X = my_x
     lbl = my_lbl
     train_idx=[]
     val_idx=[]
    obj_numpy.random.seed(2016)
     a=0

     for n_moduations in mods:
        for n_signal_to_noise_ratio in snrs:
           if data==0:
              train_idx+=list(obj_numpy.random.choice(range(a*7000,(a+1)*7000),
  size=4200, replace=False))
              val_idx+=list(obj_numpy.random.choice(list(set(range(a*7000,(a+1)*7000))-
  set(train_idx)), size=1400, replace=False))
           elif data==1:
              train_idx+=list(obj_numpy.random.choice(range(a*6000,(a+1)*6000),
  size=3600, replace=False))
              val_idx+=list(obj_numpy.random.choice(list(set(range(a*6000,(a+1)*6000))-
  set(train_idx)), size=1200, replace=False))
           a+=1

#Genaration of various datasets: "X_train"—Training, "X_val"--- validation
and "X_test"—testing
    n_examples=X.shape[0]
# below are the indexes for test data
    test_idx=list(set(range(0,n_examples))-set(train_idx)-set(val_idx))
   obj_numpy.random.shuffle(test_idx)
   X_test =X[test_idx]
   obj_numpy.random.shuffle(val_idx)
```

```
 X_val=X[val_idx]
obj_numpy.random.shuffle(train_idx)
X_train =X[train_idx]

# transfor the label form to one-hot
 def onehot_to(yy_org):
    yy_val=obj_numpy.zeros([len(yy_org), len(mods)])
    yy_val[obj_numpy.arange(len(yy_org)), yy_org]=1
    return yy_val
Y_test= onehot_to (list(map(lambda x: mods.index(lbl[x][0]),test_idx)))
Y_val= onehot_to (list(map(lambda x: mods.index(lbl[x][0]),val_idx)))
Y_train= onehot_to (list(map(lambda x: mods.index(lbl[x][0]),train_idx)))

############ modified #############
 XX = my_xx
  XX_test =XX[test_idx]
  XX_val=XX[val_idx]
 XX_train =XX[train_idx]
####################################


  return (mods,snrs,lbl), (X_train,Y_train), (X_val,Y_val), (X_test,Y_test),
(train_idx,val_idx,test_idx), (XX_train,XX_test,XX_val)
```

1.5.3 Python code (data preparation for training)

```
Obtaining training sample for I/Q and A/P inputs
 x_train, x_test, x_val: training, testing, validation data from I/Q inputs
 xx_train, xx_test, xx_val: training, testing, validation data from A/P inputs
 Y_train, Y_test, y_val: labels for training, testing, validation

##################### Code #####################
#######   I/Q channel processing #####################
 dbsignal_real = pd.read_csv(signalpath_real,header=None)
 dbsignal_imag = pd.read_csv(signalpath_imag,header=None)
 L = len(dbsignal_real); # length of total signal vectors for all snr all mod

 lbl_ms = pd.read_csv(lblpath,header=None)
 modtypes = pd.read_csv(modtypespath,header=None)
 snrlevels = pd.read_csv(snrlevelspath,header=None)

 my_lbl = [[[0 for k in range(1)] for j in range(2)] for i in range(L)]
 my_x_real = dbsignal_real.values
 my_x_imag = dbsignal_imag.values
```

```
my_mods = ['bpsk','qpsk','8psk','16qam']

indx=0
my_snrs=[ 0 for i in range(11)]
for snr in range(-10,12,2): ### for -10 to 10 dB==total 11
  my_snrs[indx] = snr
  indx = indx+1

type(my_mods)
index = 0

for i in range(0,len(my_mods)):
  for snr in range(-10,12,2):  ### for -10 to 10 dB
    for j in range(0,7000):   #  for 7000 signals per snr per mod
      my_lbl[index][0] = my_mods[i]
      my_lbl[index][1] = snr
      index = index+1

## creating input signal matrix (my_x)  in dim: [308000,2,128]
x = [[[0 for k in range(128)] for j in range(2)] for i in range(L)]
for i in range(0,L):
  x[:][i][0]=my_x_real[i]
  x[:][i][1]=my_x_imag[i]

my_x = obj_numpy.array(x)

del my_x_real, my_x_imag, dbsignal_real, dbsignal_imag, x
##################################################
#######   A/P channel processing #################
dbsignal_amp = pd.read_csv(signalpath_amp,header=None)

dbsignal_phase = pd.read_csv(signalpath_phase,header=None)
my_x_amp = dbsignal_amp.values
my_x_phase = dbsignal_phase.values

## creating input signal matrix (my_xx)  in dim: [308000,2,128]
xx = [[[0 for kk in range(128)] for jj in range(2)] for ii in range(L)]
for ii in range(0,L):
  xx[:][ii][0]=my_x_amp[ii]
  xx[:][ii][1]=my_x_phase[ii]
my_xx = obj_numpy.array(xx)

del xx, dbsignal_amp, dbsignal_phase, my_x_amp, my_x_phase
####################################################

(mods,snrs,lbl),(X_train,Y_train),(X_val,Y_val),(X_test,Y_test),(train_idx,val_idx,test_
idx),(XX_train,XX_test,XX_val) = \
   load_data(my_x,my_mods,my_snrs,my_lbl ,0,my_xx)
##########################
```

```
in_shp = list(X_train.shape[1:])
in_shp1 = list(XX_train.shape[1:])
classes = mods
#########################

######## data expand dimension function #######

    def change_extent(input_val, index, in_axis):

        output_val= obj_numpy.expand_dims(input_val [:,index,:], axis = in_axis)

    return output_val


    def change_extent_all(input_val, in_axis):

        output_val= obj_numpy.expand_dims(input_val, axis = in_axis)

    return output_val

#########################################
```

1.5.4 Python code (data reshaping)

```
# Select in-phase (I), quadrature-phase (Q), and in-phase/quadrature-phase (I/Q)
separately and expand the data set dimension

    X1_train_data= change_extent (X_train, 0, 2)
    X1_test_data = change_extent (X_test, 0, 2)
    X1_val_data = change_extent (X_val, 0, 2)

    X2_train_data= change_extent (X_train, 1, 2)
    X2_test_data = change_extent (X_test, 1, 2)
    X2_val_data = change_extent (X_val, 1, 2)
    X_train_data= change_extent_all (X_train, 3)
    X_test_data = change_extent_all (X_test, 3)
    X_val_data = change_extent_all (X_val, 3)

    #################### modified #######################
# Select amplitude (A), phase (P), and amplitude/ phase (A/P) separately and
expand the data set dimension

    XX1_train_data= change_extent (XX_train, 0, 2)
    XX1_test_data = change_extent (XX_test, 0, 2)
    XX1_val_data = change_extent (XX_val, 0, 2)

    XX2_train_data= change_extent (XX_train, 1, 2)
    XX2_test_data = change_extent (XX_test, 1, 2)
    XX2_val_data = change_extent (XX_val, 1, 2)

    XX_train_data= change_extent_all (XX_train, 3)
    XX_test_data = change_extent_all (XX_test, 3)
    XX_val_data = change_extent_all (XX_val, 3)
```

1.5.5 Python code (model training)

```
my_archi = my_BMCCLDNN_archi(classes=4)
 my_archi.compile(metrics=['accuracy'], loss='categorical_crossentropy',
optimizer='adam')

history =
my_archi.fit([X_train_data,X1_train_data,X2_train_data,XX_train_data,XX1_train_
data,XX2_train_data],
Y_train_data,
  batch_size=batch_size,
  epochs=epoch,
  verbose=1,
 validation_data=([X_val_data,X1_val_data,X2_val_data,XX_val_data,XX1_val_data,
XX2_val_data],Y_val_data),
callbacks = [
keras.callbacks.ModelCheckpoint(filepath, monitor='val_loss', verbose=1,
save_best_only=True, mode='auto'),
keras.callbacks.ReduceLROnPlateau(monitor='val_loss',factor=0.8,verbose=1,patince
=5,min_lr=0.0000001),
keras.callbacks.EarlyStopping(monitor='val_loss', patience=50, verbose=1,
mode='auto) ]
       )
```

1.5.6 Python code (model evaluation)

```
# We re-load the best weights once training is finished
  my_archi.load_weights(filepath)

  # Show simple version of performance
   score = my_archi.evaluate([X_test_data, X1_test_data, X2_test_data, XX_test_data,
XX1_test_data, XX2_test_data],  Y_test_data, verbose=1, batch_size=batch_size)
   print(score)
```

1.5.7 Python code (model prediction)

```
predict_bmccldnn(model)

def predict_bmccldnn(model):

(mods,snrs,lbl),(X_train,Y_train),(X_val,Y_val),(X_test,Y_test),(train_idx,val_idx,test_
idx),(XX_train,XX_test,XX_val) = \
   load_data(my_x,my_mods,my_snrs,my_lbl ,0,my_xx)
   in_shp = list(X_train.shape[1:])
   classes = mods
     # Plot confusion matrix
  ############ modified ###################
   test_Y_hat = model.predict([X_test,X1_test,X2_test,XX_test,XX1_test,XX2_test],
batch_size=batch_size)
```

```
    ##############################################
     confnorm,_,_ = calculate_confusion_matrix(Y_test,test_Y_hat,classes)

     plot_confusion_matrix(confnorm, labels=classes)

    #calculate_accuracy_each_snr(Y_test,test_Y_hat,classes)
  ###############################

  ######## data expand dimension function #######

     def get_test_data(input_val, test_SNRs, snr):

          output_val= input_val [obj_numpy.where(obj_numpy.array(test_SNRs) ==
          snr)]

     return output_val

    # Plot confusion matrix
     acc = {}
     acc_mod_snr = obj_numpy.zeros( (len(classes),len(snrs)) )
     i = 0
     for snr in snrs:
        test_SNRs = [lbl[x][1] for x in test_idx]
    test_X_i = get_test_data(X_test, test_SNRs, snr)
    test_X1_i = get_test_data(X_test1, test_SNRs, snr)
    test_X2_i = get_test_data(X_test2, test_SNRs, snr)
    test_Y_i = get_test_data(Y_test, test_SNRs, snr)
       #################### modified #######################
    test_XX_i = get_test_data(XX_test, test_SNRs, snr)
    test_XX1_i = get_test_data(XX_test1, test_SNRs, snr)
    test_XX2_i = get_test_data(XX_test2, test_SNRs, snr)

       ############# modified ####################


    test_Y_i_hat =
model.predict([test_X_i,test_X1_i,test_X2_i,test_XX_i,test_XX1_i,test_XX2_i])
  #############################################
    confnorm_i,cor,ncor = calculate_confusion_matrix(test_Y_i,test_Y_i_hat,classes)
    acc[snr] = 1.0 * cor / (cor + ncor)

    plot_confusion_matrix(confnorm_i, labels=classes, title="ConvNet Confusion
Matrix (SNR=%d)(ACC=%2f)" % (snr,100.0*acc[snr]),save_filename='image[%d]'%
snr)
```

```
    acc_mod_snr[:,i] =
obj_numpy.round(obj_numpy.diag(confnorm_i)/obj_numpy.sum(confnorm_i,axis=1),
3)
    i = i +1
  print(acc)

  #plot acc of each mod in one picture
  dis_num=4
  for g in range(int(obj_numpy.ceil(acc_mod_snr.shape[0]/dis_num))):

    assert (0 <= dis_num <= acc_mod_snr.shape[0])
    beg_index = g*dis_num
    end_index = obj_numpy.min([(g+1)*dis_num,acc_mod_snr.shape[0]])

    plt.figure(figsize=(12, 10))
    plt.xlabel("Signal to Noise Ratio")
    plt.ylabel("Classification Accuracy")
    plt.title("Classification Accuracy for Each Mod")

    for i in range(beg_index,end_index):
      plt.plot(snrs, acc_mod_snr[i], label=classes[i])
      print(snrs, acc_mod_snr[i])
      for x, y in zip(snrs, acc_mod_snr[i]):
        plt.text(x, y, y, ha='center', va='bottom', fontsize=8)

    plt.legend()
    plt.grid()
```

1.6 Conclusion

For AMCs, multiple methodologies have been proposed in the literature based on LB, FB, and DL approaches. Still, as per the literature, DL outperforms FB techniques. This chapter focused on automatic feature extraction using DL frameworks. Four state-of-the-art models (CNNs, CLDNNs, MCCLDNNs and BMCCLDNNs) have been considered. Results have shown that BMCCLDNNs outperform other models and achieve significant classification accuracy. This chapter also discussed the implementation of all four models, data preparation for training and, finally, prediction.

Acknowledgments

We wish to acknowledge the researchers who made their open implementation available on the web for extension.

Bibliography

[1] Zhu Z and Nandi A 2015 *Automatic Modulation Classification: Principles, Algorithms and Applications* (Hoboken, NJ: Wiley)

[2] Teng C-F, Chou C-Y, Chen C-H and Wu A-Y 2020 Accumulated polar feature-based deep learning for efficient and lightweight automatic modulation classification with channel compensation mechanism arXiv:2001.01395

[3] Meng F, Chen P, Wu L and Wang X 2018 Automatic modulation classification: a deep learning enabled approach *IEEE Trans. Veh. Tech-nol.* **67** 10760–72

[4] Chen S, Zhang Y, He Z, Nie J and Zhang W 2020 A novel attention cooperative framework for automatic modulation recognition *IEEE Access* **8** 15673–86

[5] Xu J L, Su W and Zhou M 2010 Likelihood-ratio approaches to automatic modulation classification *IEEE Trans. Syst., Man, and Cybernet., Part C: Appl. Rev.* **41** 455–69

[6] Chavali V G and Da Silva C R 2011 Maximum-likelihood classification of digital amplitude-phase modulated signals in flat fading non-gaussian channels *IEEE Trans. Commun.* **59** 2051–6

[7] Headley W C and da Silva C R 2010 Asynchronous classification of digital amplitude-phase modulated signals in flat-fading channels *IEEE Trans. Commun.* **59** 7–12

[8] Dobre O A, Abdi A, Bar-Ness Y and Su W 2006 Cyclostationarity-based blind classification of analog and digital modulations *MILCOM 2006 IEEE Military Communications Conf.* (Piscataway, NJ: IEEE) pp 1–7

[9] Camara T V, Lima A D, Lima B M, Fontes A I, Martins A D M and Silveira L F 2019 Automatic modulation classification architectures based on cyclostationary features in impulsive environments *IEEE Access* **7** 138512–27

[10] Swami A and Sadler B M 2000 Hierarchical digital modulation classification using cumulants *IEEE Trans. Commun.* **48** 416–29

[11] Wu H-C, Saquib M and Yun Z 2008 Novel automatic modulation classification using cumulant features for communications via multipath channels *IEEE Trans. Wireless Commun.* **7** 3098–105

[12] Li W, Dou Z, Qi L and Shi C 2019 Wavelet transform based modulation classification for 5g and UAV communication in multipath fading channel *Phys. Commun.* **34** 272–82

[13] Azzouz E E and Nandi A K 1995 Automatic identification of digital modulation types *Signal Process.* **47** 55–69

[14] Popoola J J and Van Olst R 2011 A novel modulation-sensing method *IEEE Veh. Technol. Mag.* **6** 60–9

[15] Zhang Z, Li Y, Zhu X and Lin Y 2017 A method for modulation recognition based on entropy features and random forest *2017 IEEE Int. Conf. on Software Quality, Reliability and Security Companion (QRS-C) (Piscataway, NJ)* (IEEE) pp 243–6

[16] Zhang X, Chen J and Sun Z 2017 Modulation recognition of communication signals based on SCHKS-SSVM *J. Syst. Eng. Electron.* **28** 627–33

[17] Wang D, Zhang M, Cai Z, Cui Y, Li Z, Han H, Fu M and Luo B 2016 Combatting nonlinear phase noise in coherent optical systems with an optimized decision processor based on machine learning *Opt. Commun.* **369** 199–208

[18] Wong M D and Nandi A K 2004 Automatic digital modulation recognition using artificial neural network and genetic algorithm *Signal Process.* **84** 351–65

[19] O'Shea T J, Corgan J and Clancy T C 2016 Convolutional radio modulation recognition networks *Int. Conf. on Engineering Applications of Neural Networks (Berlin)* (Springer) pp 213–26
[20] Zhang Z, Wang C, Gan C, Sun S and Wang M 2019 Automatic modulation classification using convolutional neural network with features fusion of spwvd and bjd *IEEE Trans. Signal Inform. Process. Networks* **5** 469–78
[21] Zhang Z, Luo H, Wang C, Gan C and Xiang Y 2020 Automatic modulation classification using cnn-lstm based dual-stream structure *IEEE Trans. Veh. Technol.* **69** 13521–31
[22] Li Y, Shao G and Wang B 2019 Automatic modulation classification based on bispectrum and cnn *2019 IEEE 8th Joint Int. Information Technology and Artificial Intelligence Conf. (ITAIC)* (Piscataway, NJ: IEEE) pp 311–6
[23] Wang J, Gui G and Sari H 2021 Generalized automatic modulation recognition method based on distributed learning in the presence of data mismatch problem *Phys. Commun.* **48** 101428
[24] Hou C, Li Y, Chen X and Zhang J 2021 Automatic modulation classification using kelm with joint features of cnn and lbp *Phys. Commun.* **45** 101259
[25] Lyu Z, Wang Y, Li W, Guo L, Yang J, Sun J, Liu M and Gui G 2020 Robust automatic modulation classification based on convolutional and recurrent fusion network *Phys. Commun.* **43** 101213
[26] Wang Y, Wang J, Zhang W, Yang J and Gui G 2020 Deep learning-based cooperative automatic modulation classification method for mimo systems *IEEE Trans. Veh. Technol.* **69** 4575–9
[27] Moulay H, Djebbar A, Dehri B and Dayoub I 2022 Dendrogram-based artificial neural network modulation classification for dual-hop cooperative relaying communications *Phys. Commun.* **55** 101929
[28] Bouchenak S, Merzougui R, Harrou F, Dairi A and Sun Y 2022 A semi-supervised modulation identification in mimo systems: a deep learning strategy *IEEE Access* **10** 76622–35
[29] Wang Y, Gui J, Yin Y, Wang J, Sun J, Gui G, Gacanin H, Sari H and Adachi F 2020 Automatic modulation classification for mimo systems via deep learning and zero-forcing equalization *IEEE Trans. Veh. Technol.* **69** 5688–92
[30] Gao Z, Zhu Z and Nandi A K 2020 Modulation classification in mimo systems with distribution test ensemble *IEEE Access* **8** 128819–29
[31] Chang S, Huang S, Zhang R, Feng Z and Liu L 2021 Multitask-learning-based deep neural network for automatic modulation classification *IEEE Internet Things J.* **9** 2192–206
[32] Qi P, Zhou X, Zheng S and Li Z 2020 Automatic modulation classification based on deep residual networks with multimodal information *IEEE Trans. Cogn. Commun. Netw.* **7** 21–33
[33] Che J, Wang L, Bai X, Liu C and Zhou F 2022 Spatial-temporal hybrid feature extraction network for few-shot automatic modulation classification *IEE Trans. Veh. Technol.* **71** 13387–92
[34] Xu J, Luo C, Parr G and Luo Y 2020 A spatiotemporal multi-channel learning framework for automatic modulation recognition *IEEE Wireless Commun. Letters* **9** 1629–32
[35] Pathak N K and Bajaj V 2023 Automatic modulation classification using bimodal parallel multichannel deep learning framework for spatial multiplexing MIMO system *Phys. Commun.* **59** 102071
[36] West N E and O'Shea T 2017 Deep architectures for modulation recognition *2017 IEEE Int. Symp. on Dynamic Spectrum Access Networks (DySPAN)* (Piscataway, NJ: IEEE) pp 1–6

[37] Emam A, Shalaby M, Aboelazm M A, Bakr H E A and Mansour H A 2020 A comparative study between cnn, lstm, and cldnn models in the context of radio modulation classification *2020 12th Int. Conf. on Electrical Engineering (ICEENG) (Piscataway, NJ)* (IEEE) pp 190–5

[38] Sainath T N, Vinyals O, Senior A and Sak H 2015 Convolutional, long short-term memory, fully connected deep neural networks *2015 IEEE Int. Conf. on Acoustics, Speech and Signal Processing (ICASSP)* (Piscataway, NJ: IEEE) pp 4580–4

[39] https://github.com/wzjialang/MCLDNN 2020 *A Spatiotemporal Multi-Channel Learning Framework for Automatic Modulation Recognition*

Chapter 2

Applying B-value and empirical equivalence hypothesis testing to intellectual and developmental disabilities electroencephalogram data

Mozhdeh Saghalaini, Zahra Ghanbari and Mohammad Hassan Moradi

In this chapter, we employ a two-step process for hypothesis testing to compare individuals with intellectual and developmental disabilities (IDD) and typically developing controls (TDC) using electroencephalogram (EEG) signals. EEG data were recorded from participants during resting-state and music stimuli and pre-processed using a fully automated and robust pipeline. Features were extracted from the pre-processed data and fed into the two-step hypothesis testing framework. The two-step process incorporates both conventional hypotheses testing and an equivalence testing approach, employing an empirical equivalence bound (EEB), estimated from the input data. According to our results, this method enables a precise comparison of the features associated with IDD and TDC, facilitating the identification of distinctive characteristics related to the disorder.

2.1 Introduction

IDD are a group of neurodevelopmental disorders that are characterized by impairments in cognitive, social, and adaptive functioning. These disorders affect millions of people worldwide, and individuals with IDD often require significant support and interventions tailored to their specific needs. Persons with IDD face significant health disparities, which can be attributed to various factors such as limited access to healthcare, unfavorable determinants of health, and exclusion from public health and preventive care initiatives [1–3]. Global reports have consistently highlighted the existing gaps in healthcare services for individuals with IDD. Despite the increasing global population of individuals with IDD, this population remains vulnerable [4]. Therefore, there should be solutions to accurately assign these

doi:10.1088/978-0-7503-5929-0ch2

disorders and their extent. One approach to understand these disorders is through the analysis of brain function, which has been shown to vary between individuals with IDD and TDC [5, 6]. In recent years, neuroimaging studies have provided insights into the neural mechanisms underlying IDD and the differences between IDD and TDC groups [7, 8].

In this book chapter, we utilized an EEG dataset comprising individuals with IDD and controls without IDD. The dataset is the only available EEG dataset of IDD participants observing music stimuli effect [9].

We aimed to compare brain function between IDD and TDC groups using a two-step hypothesis testing procedure introduced by Zhao *et al* [10] in order to differentiate between IDD and TDC subjects, based on EEG-related features. The mentioned two-step procedure involves a conventional hypothesis testing stage and an equivalence testing stage, where an EEB is estimated from the input data. After proper pre-processing and de-noising (which is provided in the dataset), features are extracted. The unique combination of wavelet decomposition and statistical analysis allows for a comprehensive exploration of brain dynamics, facilitating the identification of EEG features that are indicative of IDD [11, 12].

Power spectral density (PSD) is a commonly used feature extraction method for EEG signals, a representation of the distribution of power or energy of a signal as a function of frequency. In EEG analysis, PSD is used to identify frequency bands associated with specific brain activities, such as alpha (8–12 Hz), beta (12–30 Hz), theta (4–8 Hz), and gamma (30–45 Hz) waves. PSD can be estimated using various methods such as Fourier transform, Welch method, and autoregressive modeling. Several studies have employed PSD as a feature extraction method in EEG analysis. For instance, in a study by Behnampour *et al*, PSD was used to extract features from EEG signals of patients with major depressive disorder [13]. The study found significant differences in the PSD of theta and alpha frequency bands between patients and controls without major depressive disorder. In another study by Andrzejak *et al*, the use of PSD as a frequency domain feature extraction approach was demonstrated by the significant differences observed in the PSDs of various EEG signals, including health, interictal, and ictal signals [14]. It was found that the PSD of theta and alpha bands was significantly different between the interictal and ictal states. Based on the previous related work, wavelet transform and PSD features are used for the present study, and fed as the input to our two-step hypothesis testing framework.

Our study builds on previous research that has investigated brain function in individuals with IDD using various neuroimaging techniques [15, 16]. Furthermore, it aligns with recent advances in statistical methods for neuroimaging data analysis, which have emphasized the importance of robust and automated pre-processing pipelines and two-stage hypothesis testing approach [17, 18]. The aim of the present study is to enhance comprehension of this disorder and provide individuals affected by IDD with appropriate intervention or supports.

This book chapter is organized as follows: First, the data and pre-processing will be discussed. In the feature extraction section, extracted features and their arrangement in the feature matrices are expressed. The fourth part is dedicated to the description of the statistical method and its implementation in the Python environment. In the next section, the results and output of the algorithm are presented for different states, and the final section contains a conclusion to this chapter.

2.2 Materials

In this section, the dataset and pre-processing method is illustrated. Additionally, features and a feature extraction approach are explained. Then, we will examine the two-stage procedure for hypothesis testing.

2.2.1 Dataset

The EEG dataset utilized in this study represents a unique resource as it is currently the only available dataset focusing on individuals with IDD exposed to music stimuli. It is worth noting that the IDD population often experiences discomfort when using a traditional dense electrode and wired EEG acquisition systems, potentially compromising data quality. To overcome this, a wireless, portable, and lesser-electrode EEG acquisition system such as EMOTIV's EPOC was used in this study [19]. The device used in this study was designed with a toy-like appearance and incorporated the aforementioned features. This design choice aimed to enhance the acceptability of the device among participants with IDD, ultimately resulting in the acquisition of high-quality data.

The dataset utilized in this study offers a valuable collection of high-quality EEG data from both subjects with IDD and TDC groups. As a result, it serves as a valuable resource for investigating and distinguishing the cortical brain dynamics exhibited by these two groups in response to music stimuli.

EEG signals were collected from a total of fourteen participants, comprising seven TDC and seven individuals with IDD. The EMOTIV EPOC+ data acquisition system was utilized. Data used in this study comprises of 14-channel EEG data that was recorded at a sampling rate of 128 Hz. The electrode placement adhered to the international 10–20 system, including channels AF3, P7, O1, O2, P8, T7, T8, F7, F3, FC5, FC6, F4, F8, and AF4.

The participants were instructed to rest for two minutes, followed by listening to music for another two minutes. The raw and pre-processed data is organized in table 2.1, which contains the participants' ID, age, gender, and file name for the IDD and TDC groups, respectively. Additional information about IDD participants is provided in table 2.2.

IDD participants were aged between 26 and 31 years, with an average age of 28.28 ± 2.05. Their Intelligence Quotient (IQ) ranged from 52 to 68, with a mean IQ of 59.143 ± 5.33, and their Social Quotient (SQ) ranged from 57 to 62, with a mean SQ of 60 ± 2.08. Meanwhile, TDC participants were aged between 20 and 24 years, with an average age of 21.28 ± 1.60.

The experiment encompassed of two distinct stages. During the initial stage, participants were directed to maintain a resting state for two minutes. Subsequently, participants were exposed to a two-minute music stimulus, comprising a tranquil

Table 2.1. Participant's data [9].

Participant ID	Group	Age	Gender	Condition
NDS001	IDD	28	Male	Rest/Music
NDS002	IDD	26	Male	Rest/Music
NDS003	IDD	26	Male	Rest/Music
NDS004	IDD	28	Male	Rest/Music
NDS005	IDD	31	Male	Rest/Music
NDS006	IDD	31	Male	Rest/Music
NDS007	IDD	28	Male	Rest/Music
CGS01	TDC	20	Male	Rest/Music
CGS02	TDC	21	Male	Rest/Music
CGS03	TDC	20	Male	Rest/Music
CGS04	TDC	23	Male	Rest/Music
CGS05	TDC	24	Male	Rest/Music
CGS06	TDC	20	Male	Rest/Music
CGS07	TDC	21	Male	Rest/Music

Table 2.2. Additional behavioral information, such as facial expressions (emotion), music apprehension, IQ, and SQ for IDD participants [9].

Participant ID	IQ	SQ	Music perception	Facial expression
NDS001	68	60	Enjoyed	Calm
NDS002	59	62	Enjoyed	Aggressive
NDS003	59	62	Enjoyed	Aggressive
NDS004	54	57	Enjoyed	Calm
NDS005	52	59	Enjoyed	Relaxed
NDS006	59	58	Enjoyed	Concentrating
NDS007	63	62	Enjoyed	Relaxed

piano melody delivered through earphones. To minimize potential visual distractions, participants were specifically instructed to keep their eyes closed throughout the entirety of the experiment. Continuous EEG data were captured throughout the entire duration and subsequently segregated into separate datasets corresponding to the resting state and the music stimulus period [9].

2.2.2 Pre-processing

The raw EEG data was subjected to a fully automated in-house pre-processing pipeline [20]. The following steps were performed (figure 2.1) [9]:

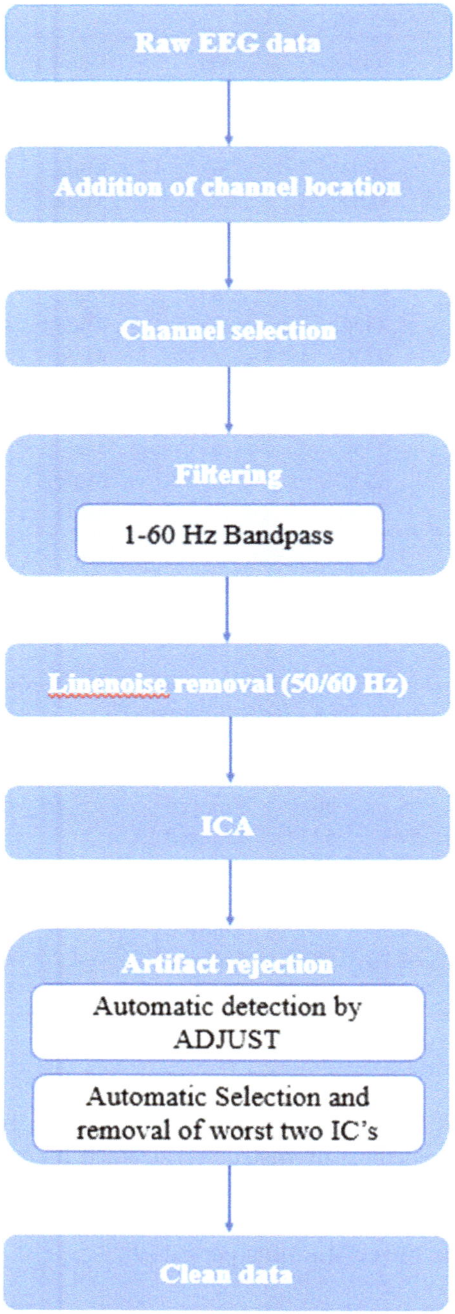

Figure 2.1. Pre-processing pipeline [20].

1. The dataset was loaded into EEGLAB.
2. The 14-channel EEG data were subjected to band-pass filtering, with a lower cut-off frequency set at 1 Hz and a higher cut-off frequency at 30 Hz. To eliminate line noise at 50/60 Hz and reduce artefacts at epoch boundaries, the CleanLine plugin was applied.
3. A custom channel location file was generated, which included the spherical coordinates of the electrodes mapped onto the 10–20 system specifically for the EMOTIV EPOC device.
4. Independent component analysis (ICA) was employed to decompose the data into its constituent independent components, allowing for the identification and subsequent removal of artefacts. The ICA decomposition was performed using the logistic infomax algorithm with a natural gradient feature.
5. Automatic artefact detection was carried out using the ADJUST plugin, which detects eye blinks, vertical and horizontal eye movements, and general discontinuities based on the spatial and temporal characteristics of each independent component. To ensure effective artefact removal while preserving valuable data, the two components deemed most problematic by ADJUST were selected for rejection.
6. Automatic artefact rejection was performed by selecting two independent components with the highest cumulative threshold exceedance across five distinct features.
7. The filtered and artefact-free EEG data were reconstructed after removing the identified components [20].

2.3 Feature extraction

In this section, we describe the feature extraction methods chosen for comparing EEG signals of IDD and TDC groups. These methods are discrete wavelet transform (DWT) and PSD. DWT provides us with both fine-grained and coarse-grained information. Statistical features such as energy, mean, and standard deviation are computed from the wavelet coefficients. According to Guler *et al* [21], these features provide quantitative measures to differentiate between IDD and TDC groups [21]. Furthermore, PSD analysis offers insights into the power distribution across different frequency bands, which can reveal distinct patterns in the IDD group compared to the TDC group. Together, these feature extraction methods contribute to a comprehensive analysis of EEG data, facilitating the identification of meaningful markers associated with IDD.

2.3.1 DWT

The feature extraction method utilized in this study involves the application of DWT to pre-processed EEG data. The primary objective of this method is to capture essential information within the EEG signals that is specifically related to IDD [11]. This technique allows for the identification of important signal characteristics, such as sudden changes, complexities, and similarities [12].

The DWT is a well-established signal processing technique widely used in various fields, including EEG analysis [22]. It decomposes EEG signal into different frequency sub-bands, providing a comprehensive representation of the underlying neural activity. In this study, the fourth-order Daubechies wavelet (db4) was selected as the mother wavelet for the decomposition process. The db4 wavelet has been widely employed in EEG signal analysis due to its effectiveness in capturing both high- and low-frequency components of the signal [23].

Signal decomposition is performed by applying the DWT at multiple levels, resulting in the generation of detail coefficients and approximation coefficients. In our study, signal is decomposed into six levels [24]. The detail coefficients capture the high-frequency components of the signal, whereas the approximation coefficients represent the low-frequency components. This multi-level decomposition allows for a detailed examination of the EEG signal across different frequency bands [11, 23].

To quantify and analyze the extracted wavelet coefficients, statistical features, including energy, mean, and standard deviation, are computed [12]. The energy is calculated by summing the squared values of the wavelet coefficients. It represents a measure of the signal's overall strength. The mean represents the average value of the coefficients, indicating the signal's central tendency. The standard deviation quantifies the dispersion or variability of the coefficients around the mean. The arrangement of DWT feature matrices for IDD and TDC groups in two states of rest and music stimulation illustrated in figures 2.2 and 2.3.

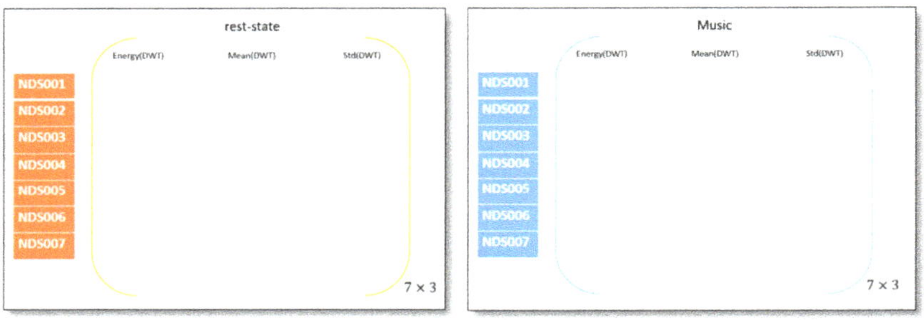

(a) rest-state (b) Music stimuli

Figure 2.2. DWT feature matrix structure in IDD group in (a) rest-state and (b) music stimuli.

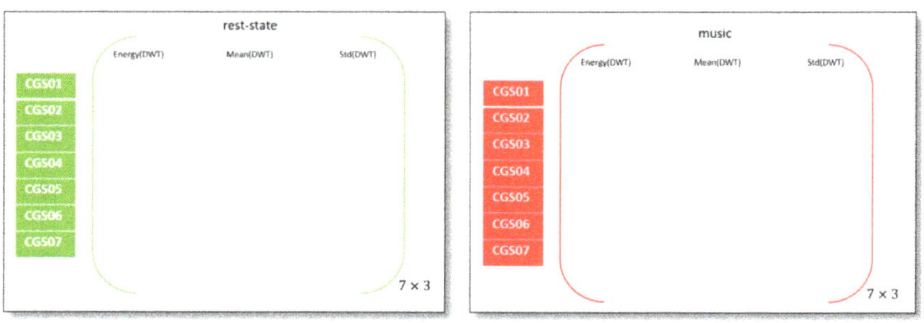

(a) rest-state (b) Music stimuli

Figure 2.3. DWT feature matrix structure in TDC group in (a) rest-state and (b) music stimuli.

2.3.2 PSD

In the context of studying individuals with IDD and TDC, PSD analysis can provide insights into the differences in neural activity between the two groups. Studies have shown that individuals with IDD exhibit alterations in their PSD compared to TDC individuals. For instance, Hatzigeorgiadis *et al* [25] found that individuals with Down syndrome showed lower power in the alpha frequency band compared to TDC individuals. Similarly, Zambrano *et al* [26] found that individuals with autism spectrum disorder had significantly different PSD patterns compared to TDC individuals in several frequency bands, including alpha, beta, and gamma.

In addition to study differences in PSD between IDD and TDC individuals, PSD analysis can also be used to identify specific frequency bands that are associated with cognitive and behavioral processes. For example, the alpha band has been linked to attention and relaxation, while the beta band is associated with motor activity and cognitive processing [27]. Studies have also shown that gamma band activity is related to visual processing and memory [28], while theta band activity is associated with working memory and attention [27].

Overall, PSD analysis is a valuable technique for studying neural activity in individuals with IDD and TDC individuals. By examining the differences in PSD between the two groups and identifying specific frequency bands associated with cognitive and behavioral processes, researchers can gain a better understanding of the neural mechanisms underlying IDD and develop more targeted interventions.

According to the points mentioned before, to analyze the commonalities and differences between the IDD and TDC groups, power spectrum features in the alpha, beta, theta, and gamma frequency bands were extracted from each pre-processed signal.

The PSD feature matrix was generated for the IDD and TDC groups during both states illustrated in figures 2.4 and 2.5.

Extracted feature values (DWT and PSD) for the IDD and TDC group in the music stimulation mode are represented in figure 2.6.

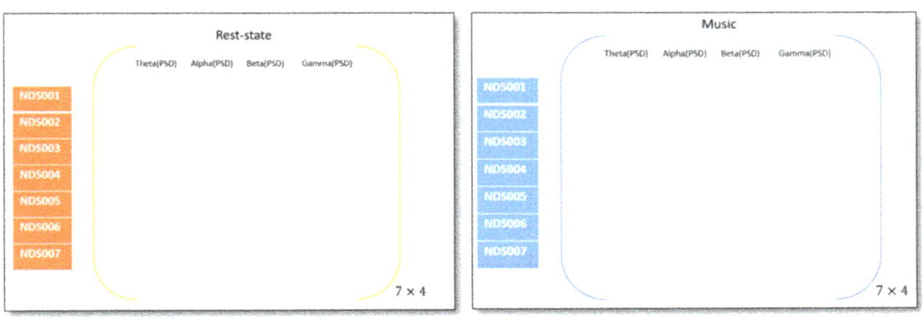

(a) rest-state (b) Music stimuli

Figure 2.4. PSD feature matrix structure in IDD group in (a) rest-state and (b) music stimuli.

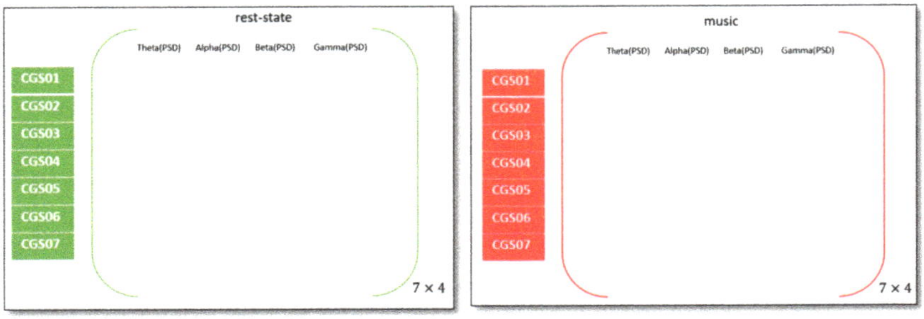

(a) rest-state (b) music stimuli

Figure 2.5. PSD feature matrix structure in TDC group in (a) rest-state and (b) music stimuli.

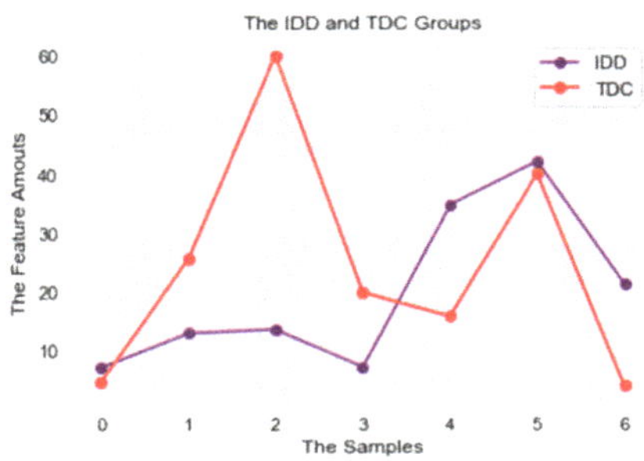

(a) PSD in alpha band

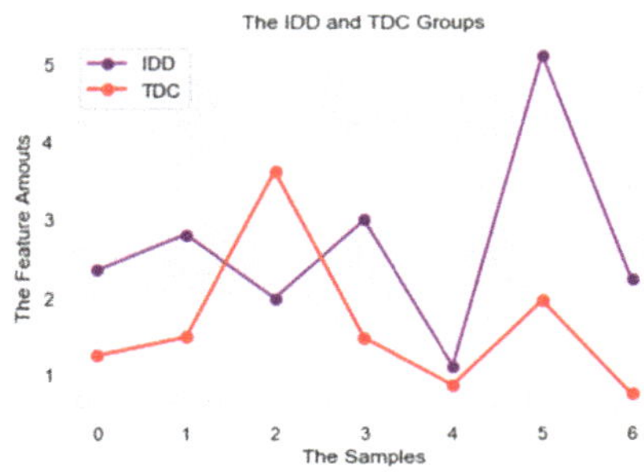

(b) PSD in beta band

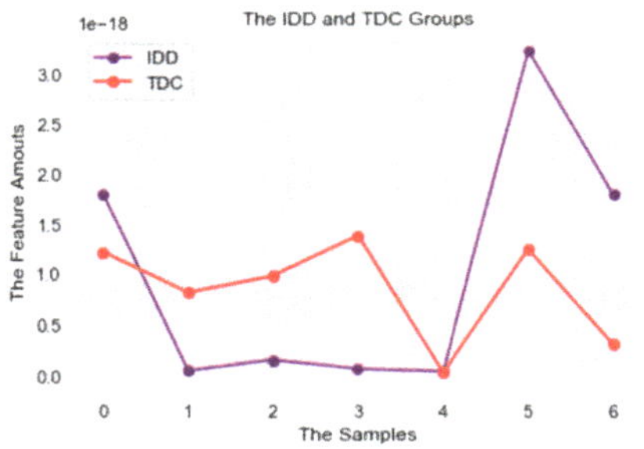

(c) Energy of the DWT coefficients

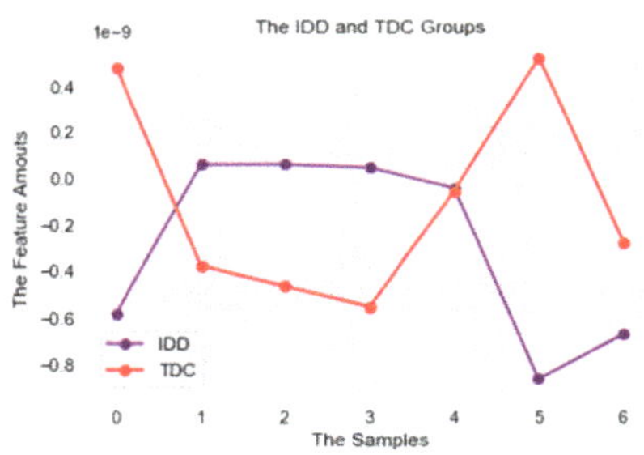

(d) Mean of the DWT coefficients

Figure 2.6. Some of the extracted feature values for music stimuli, (a) PSD in alpha band, (b) PSD in beta band, (c) energy of the DWT coefficients, and (d) mean of the DWT coefficients.

2.4 Statistical method

In this section, we propose a novel two-stage procedure for hypothesis testing that integrates conventional hypothesis testing with an equivalence testing procedure using EEB. The aim is to address the issue of reproducibility and transparency in statistical hypothesis testing by incorporating practical or scientific significance into the interpretation of P-values or confidence intervals. In the present section, the concepts of the B-value and EEB, which are estimated from the data and provide a data-driven approach to improve the reproducibility of findings across studies, are brought up. Moreover, the formulation of the procedure, the choice of equivalence bounds, and the explorations of the statistical properties of the B-value and EEB are explained in this section. Subsequently, a detailed description of the Python code is provided.

2.4.1 The two-stage hypothesis testing based on EEB

The American Statistical Association issued a crucial policy statement in 2016 concerning P-values, emphasizing their proper use and interpretation due to growing apprehensions surrounding the reproducibility and replicability of scientific findings. To enhance the reliability and transparency of statistical hypothesis testing, experts have proposed incorporating P-values or confidence intervals with practical or scientific significance. Equivalence testing is an extension of this idea, aiming to establish non-inferiority or similarity between parameters under the assumption of their apparent inequality [29].

Equivalence testing comes with its set of challenges, especially in defining scientific significance or equivalence, which can be subjective and lacking moral considerations. Determining appropriate equivalence bounds is crucial, as they specify the range within which two groups or parameters are considered equivalent. However, selecting these bounds can be subjective, and predefined bounds may not always be available [30].

To overcome these challenges, two fundamental concepts were introduced: the B-value and the EEB. Selecting the similarity range is crucial in equality tests, and this is determined based on factors such as the effect size. B-value represents the maximum magnitude of the confidence interval bounds. It serves as the smallest symmetric equivalence bound that would lead to rejection of the null hypothesis in an equivalence test. Importantly, reporting B-value allows readers to apply their preferred bounds, similar to choose an error rate when interpreting P-values.

The EEB, on the other hand, offers a data-driven approach to determine equivalence bounds. It represents the minimum equivalence bound that would result in a conclusion of equivalence when the true equivalence holds. The EEB improves the reproducibility of findings across studies, especially in situations where predefined equivalence bounds are not available, or when readers are interested in applying different bounds [10].

The proposed two-stage testing procedure combines traditional hypothesis testing with equivalence testing, utilizing the EEB. The process commences with a standard two-sample t-test to assess differences between groups. If the null hypothesis cannot be rejected, indicating insufficient evidence of a difference, we proceed to the second stage. In this stage, we conduct an equivalence test using the EEB (figure 2.7). This sequential process ensures a comprehensive assessment of both differences and equivalences between the groups under study [10].

In order to examine this method in more detail, consider a two-sample t-test with the following assumptions:

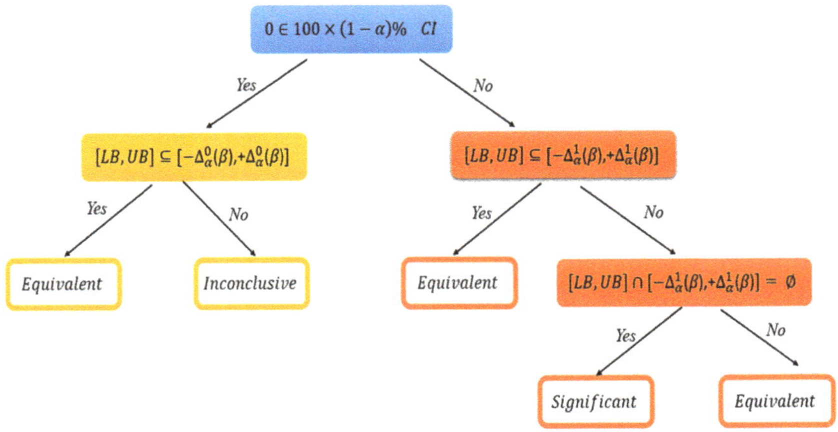

Figure 2.7. Algorithm process of two-stage hypothesis testing method [10]. Adapted from [10] John Wiley & Sons. [© 2022 John Wiley & Sons, Ltd.].

$$\mathrm{H}_0\colon \delta = 0 \quad \text{vs} \quad \mathrm{H1}\colon \delta = 1. \tag{2.1}$$

To perform a conventional hypothesis test, it is required to establish a confidence interval. The confidence level of $100 \times (1 - \alpha)\%$ is represented by the interval [LB_0, UB_0]. Here, delta denotes the mean difference between the two sets. Therefore, we can write:

$$\mathrm{LB}_0 = \hat{\delta} - t_{v,\,1-\frac{\alpha}{2}}\, S, \tag{2.2}$$

$$\mathrm{UB}_0 = \hat{\delta} + t_{v,\,1-\frac{\alpha}{2}}\, S, \tag{2.3}$$

where $t_{v,\,1-\frac{\alpha}{2}}$ is the $100 \times (1 - \alpha)\%$ quantile of a t-distribution with degree of freedom v and $\hat{\delta} = \overline{x_1} - \overline{x_2}$, in which $\overline{x_i}$ is the sample mean of ith group. The value of S is computed using the following formula, where S denotes S_{pooled}:

$$S = \sqrt{\frac{1}{n_1} + \frac{1}{n_2}} \times \sqrt{\frac{(n_1 - 1)S_1^2 + (n_2 - 1)S_2^2}{n_1 + n_2 - 2}}. \tag{2.4}$$

In this equation, S_1 and S_2 represent the standard deviations of the two sample sets, while n_1 and n_2 represent the number of patterns in each sample set. The first step in testing the null hypothesis involves examining whether or not zero falls within the confidence interval [LB_0, UB_0]. If zero falls within this interval, it is not possible to directly conclude whether the two investigated groups correspond. The method proposed by Zhao *et al* [10] involves extracting an equivalence interval (EEB) from the data. EEB is defined as the limit of an equivalence test, which rejects the null hypothesis with a probability of β when the true difference between the means of the two groups lies within the equivalence interval.

Assuming $\delta = \mu_1 - \mu_2$ as the average difference between the two sets being compared, and the null hypothesis of equality between the two averages (H_0: $\delta = 0$) with a significance level of α and a probability of $\beta \in (0,1)$, the resulting C value determines the equivalence interval (EEB). The EEB value at the β level can be calculated as follows:

$$\mathrm{EEB}_\alpha(\beta,|, C) = \inf\{b: F_B(b\,)C, H_0) \geqslant \beta\};\ b \in [0, \infty). \tag{2.5}$$

The above equation represents the EEB value, which is determined based on the significance level, probability $\beta \in (0,1)$, and the result C of the hypothesis testing with the null assumption of $\delta = 0$. The value of C can indicate the presence of zero in the interval. Absence of zero, or zero if the result is not known in advance, leading to the limiting distribution of B. The F_B function represents the conditional cumulative distribution function of B, where the cumulative distribution function of B is:

$$F_B(x,|, H_0) = \begin{cases} 0 & \text{if } x < St_{\nu,\,1-\alpha} \\ 2F_t\left(\dfrac{x}{S} - t_{\nu,\,1-\alpha};\, \nu\right) - 1 & \text{if } x \geqslant St_{\nu,\,1-\alpha} \end{cases}. \tag{2.6}$$

To obtain the EEB value for the two states of presence or absence of zero in the interval $[\mathrm{LB}_0, \mathrm{UB}_0]$, the following formulation can be used:

$$\mathrm{EEB}_\alpha(\beta\,,|, 0 \in [\mathrm{LB}_0, \mathrm{UB}_0]) = \mathrm{S}F_t^{-1}\left(\frac{\beta(1-\alpha)+1}{2};\, \nu\right) + t_{\nu,\,1-\alpha}\Big\},\tag{2.7}$$

$$\mathrm{EEB}_\alpha(\beta\,,|, 0 \notin [\mathrm{LB}_0, \mathrm{UB}_0]) = \mathrm{S}F_t^{-1}\left(1 - \frac{\alpha(1-\beta)+1}{2};\, \nu\right) + t_{\nu,\,1-\alpha}\Big\}. \tag{2.8}$$

The inverse of the cumulative distribution function of the t-Student distribution with ν degrees of freedom is denoted as F_t^{-1}.

The presented method for comparing the mean of two sets is a two-stage test using EEB. The general algorithm of this method is indicated in figure 2.7.

$$\Delta_\alpha^0(\beta) = \mathrm{EEB}_\alpha(\beta\,,|, 0 \in [\mathrm{LB}_0, \mathrm{UB}_0]), \tag{2.9}$$

$$\Delta_\alpha^1(\beta) = \mathrm{EEB}_\alpha(\beta\,,|, 0 \notin [\mathrm{LB}_0, \mathrm{UB}_0]). \tag{2.10}$$

The presented two-stage method for comparing the means of two sets involves a two-sample t-test as the first step. Following this, the variable r is assigned a value of either zero or one based on whether zero falls within the interval $[\mathrm{LB}_0, \mathrm{UB}_0]$ at the significance level of alpha or not:

$$r = \begin{cases} 0; \text{if } 0 \in 100 \times (1-\alpha)\%\ \mathrm{CI} \\ 1; \text{if } 0 \notin 100 \times (1-\alpha)\%\ \mathrm{CI} \end{cases}. \tag{2.11}$$

The second step of the algorithm is dependent on the value of r (figure 2.7). In the second stage, the EEB value is calculated, and the overlap between the confidence interval and the obtained EEB interval indicates the presence of equivalence between the two sets under investigation. It is important to emphasize that using EEB alone may not always produce accurate results if the null hypothesis is incorrectly rejected. Therefore, employing a conditional EEB (subject to β) is a more prudent approach.

2.4.2 Implementation (Python code)

In [10], the two-stage hypothesis testing method is applied to two datasets: the Iris dataset containing the characteristics of three types of iris plants and a dataset related to children with Autism Spectrum Disorder. The authors utilized the R software environment to implement the proposed method and developed a package called 'B-value' within this environment [10]. We implemented this algorithm for EEG data of IDD and TDC subjects in the Python environment.

The following sections consist of description of the Python code, along with a brief explanation of their respective functions in it. First of all, the necessary libraries such as pandas, math, seaborn, numpy, scipy.stats, and matplotlib are imported. Afterward, the constant values of alpha, delta, and beta are defined. The following functions are called in the following order:

- **features** ():

  Within this function, the initial step is uploading data. Subsequently, the user is prompted to select their preferred feature to be compared between the IDD and TDC groups (figure 2.8), consisting of DWT features (energy, mean, and standard deviation of DWT coefficients) and PSD calculated corresponding to the frequency bands (theta, alpha, beta, or gamma) within one of two modes: resting or listening to music.

  Once the desired features have been extracted in one of the available modes, the mean and variance of the corresponding IDD and TDC categories are computed. Subsequently, the degrees of freedom are determined using the following relationship:

$$\nu = n_1 + n_2 - 2. \tag{2.12}$$

The function outputs are the values of the selected features, including the mean, variance, degrees of freedom, and the difference between the two categories.

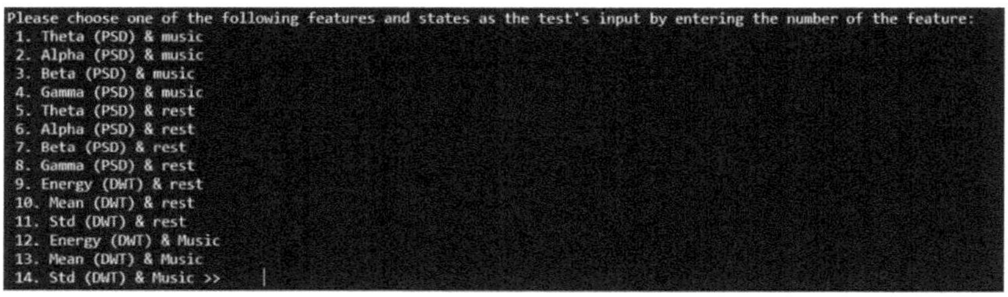

Figure 2.8. Feature and state selection by user.

- **plot_in** (idd, tdc): By accepting the selected feature values for two groups, this function generates a scatter plot illustrating the dispersion of data.
- **stage_1_1** (t, nu, s1_idd, s2_tdc, idd_mean, tdc_mean, idd, tdc): The input parameters for this function include the characteristic values, their mean and variance, the *t*-value, and the degrees of freedom calculated from the *t*-distribution using a significance level of 5% and a degree of freedom of 70 (determined based on the number of data points) [31]. This function serves as the initial step in the two-step method mentioned earlier. It calculates the confidence interval of the test using the *t*-test and verifies whether zero lies within this interval. Based on the result of this investigation, the variable r is assigned a value of zero if zero is present within the interval, or a value of one if it is not.
- **stage_1_2a** (idd_level, s, nu, betha, alpha, t, lower_band, upper_band): This function is executed when zero is included within the obtained confidence interval. In such cases, the function calculates the value of EEB using the provided inputs of beta, alpha, degree of freedom, S_{pooled}, and the range of the confidence interval. It then checks whether the obtained equivalence interval covers the confidence range or not. Based on the result of this check, the function generates the desired output as C.
- **stage_1_2b** (idd_level, s, nu, betha, alpha, t, lower_band, upper_band): The functionality of this function is similar to the previous function when the value of r is 1, indicating that zero is not included in the confidence interval. It is important to note that in the calculation of the EEB value, the inverse of the survival probability function is used instead of the inverse of the cumulative distribution function. These two functions are related to each other as follows:

$$R(t) = 1 - F(t). \tag{2.13}$$

- **plot_out** (lower_band, upper_band, EEB): The objective of this function is to visualize the outputs by drawing the confidence interval and equivalence based on the values obtained from the previous function. Its main purpose is to provide a visual representation of the statistical results.

The following table summarizes the input, output, and performance of the defined functions (table 2.3).

2.5 Results

In this section, we analyze the algorithm's output for PSD of four frequency bands (theta, alpha, beta, and gamma) and three DWT coefficient's features (energy, mean, and standard deviation) in two different states: rest and listening to music. Here are the results for $\beta = 0.8$ and $\alpha = 25\%$. For this purpose, we make the comparison in two ways.

Table 2.3. Summary of the Python code and its functions.

Name	Inputs	Outputs	function
features	—	Algorithm data (selective feature of two groups), mean, variance, degrees of freedom	Data preparation
Plot_in	Attribute values	—	Plot input data
stage_1_1	Algorithm data (selective feature of two groups), mean, variance, degree of freedom, t	The range of the confidence interval, S_{pooled}, r	The first step of the two-step method
stage_1_2a	Beta, alpha, degree of freedom, S_{pooled}, and confidence interval range	EEB, State C	The second step of the two-step method for $r = 0$
stage_1_2b	Beta, alpha, degree of freedom, S_{pooled}, and confidence interval range	EEB, State C	The second step of the two-step method for $r = 1$
plot_out	Confidence interval and equivalence	—	Output plot

2.5.1 Comparison of two groups of IDD and TDC under the same conditions

In this section, each category of features for one of the two states of rest or stimulation of music for two groups of IDD and TDC is examined.

- IDD and TDC groups, theta band, music stimuli (figure 2.9):

$$\text{EEB} = 11.478148532200255.$$

$$C = \text{Equivalent}.$$

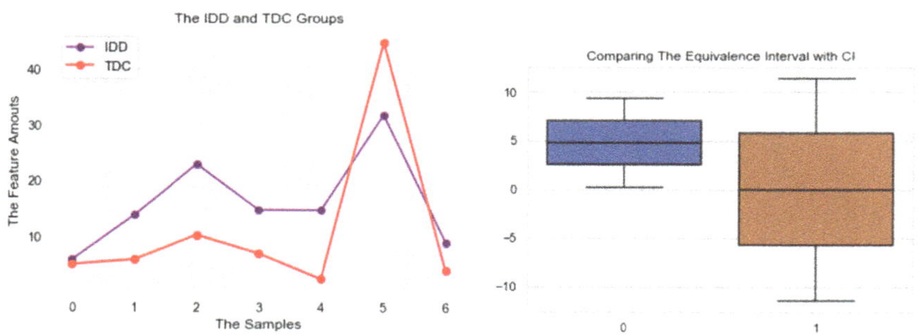

Figure 2.9. Examining the input values of two groups and the confidence interval and equivalence, input characteristic: the PSD of the theta band (music stimuli).

- IDD and TDC groups, alpha band, both states, music stimuli (figure 2.10):

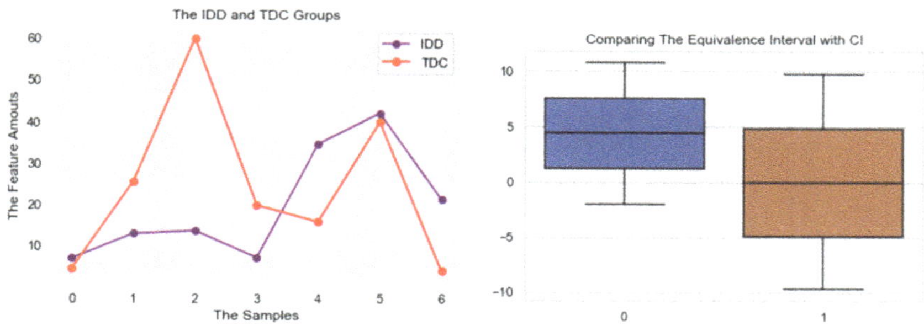

Figure 2.10. Examining the input values of two groups and the confidence interval and equivalence, input characteristic: the PSD of the alpha band (music stimuli).

$$\text{EEB} = 9.744486770349992.$$

$$C = \text{Inconclusive}.$$

- IDD and TDC groups, beta band, music stimuli (figure 2.11):

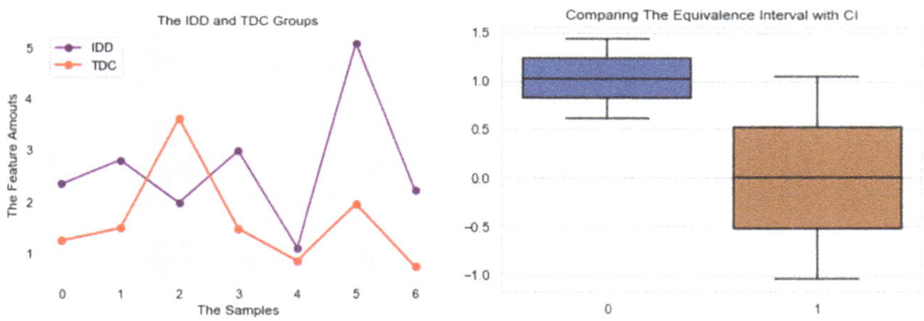

Figure 2.11. Examining the input values of two groups and the confidence interval and equivalence, input characteristic: the PSD of the beta band (music stimuli).

$$\text{EEB} = 1.0422407231880821.$$

$$C = \text{Inconclusive}.$$

- IDD and TDC groups, gamma band, music stimuli (figure 2.12):

$$\text{EEB} = 0.006130132556022247.$$

$$C = \text{Significant}.$$

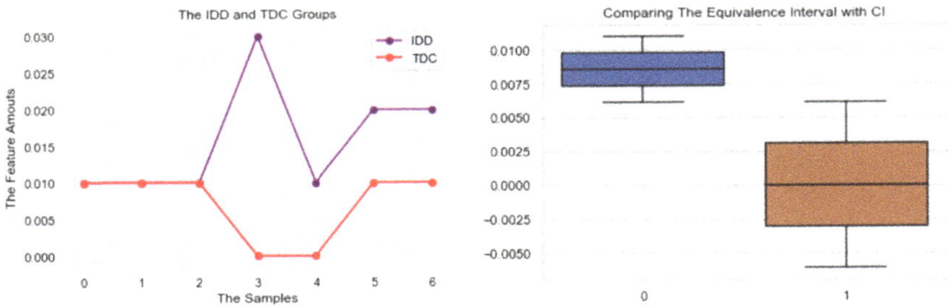

Figure 2.12. Examining the input values of two groups and the confidence interval and equivalence, input characteristic: the PSD of the gamma band (music stimuli).

- IDD and TDC groups, theta band, resting-state (figure 2.13):

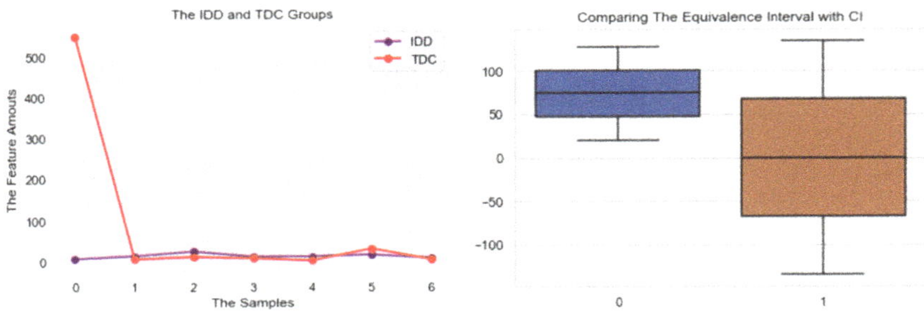

Figure 2.13. Examining the input values of two groups and the confidence interval and equivalence, input characteristic: the PSD of the theta band (resting-state).

$$\text{EEB} = 135.01031573434196.$$

$$C = \text{Equivalent}.$$

- IDD and TDC groups, alpha band, resting-state (figure 2.14):

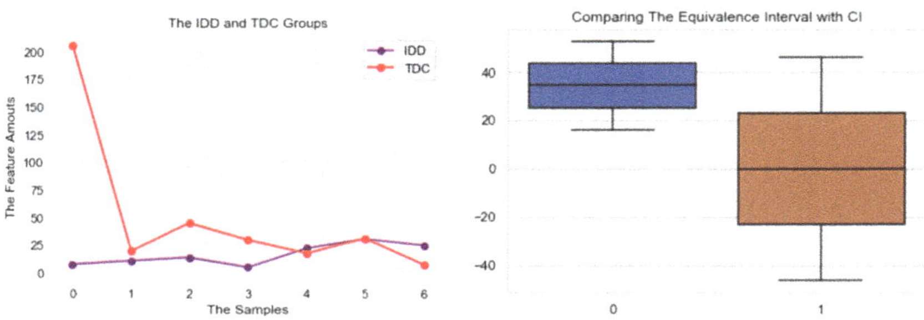

Figure 2.14. Examining the input values of two groups and the confidence interval and equivalence, input characteristic: the PSD of the alpha band (resting-state).

$$\text{EEB} = 46.321735893490086.$$

$$C = \text{Inconclusive}.$$

- IDD and TDC groups, beta band, resting-state (figure 2.15):

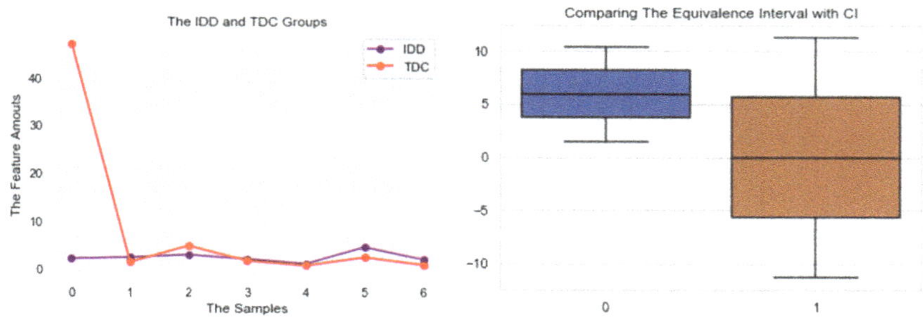

Figure 2.15. Examining the input values of two groups and the confidence interval and equivalence, input characteristic: the PSD of the beta band (resting-state).

$$\text{EEB} = 11.343300543316213.$$

$$C = \text{Equivalent}.$$

- IDD and TDC groups, gamma band, resting-state (figure 2.16):

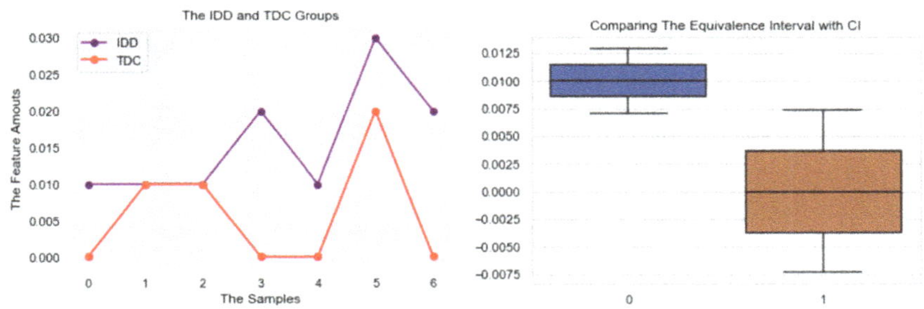

Figure 2.16. Examining the input values of two groups and the confidence interval and equivalence, input characteristic: the PSD of the gamma band (resting-state).

$$\text{EEB} = 0.00736750241870978.$$

$$C = \text{Inconclusive}.$$

- IDD and TDC groups, energy of DWT coefficients, music stimuli (figure 2.17):

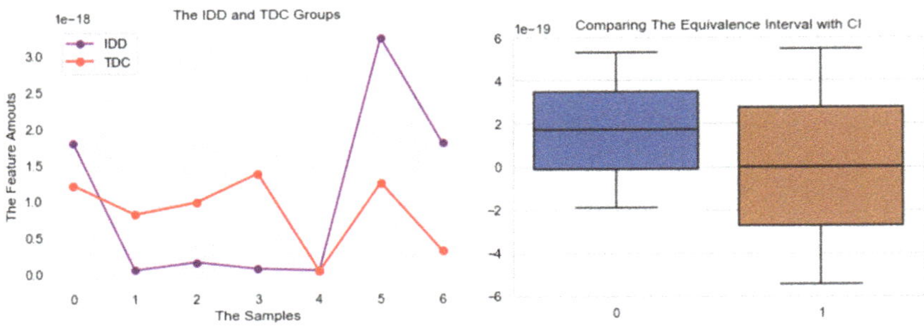

Figure 2.17. Examining the input values of two groups and the confidence interval and equivalence, input characteristic: energy of the DWT coefficients (music stimuli).

$$\text{EEB} = 5.477219777310112\text{e}^{-19}.$$

$$C = \text{Equivalent}.$$

- IDD and TDC groups, mean of DWT coefficients, music stimuli (figure 2.18):

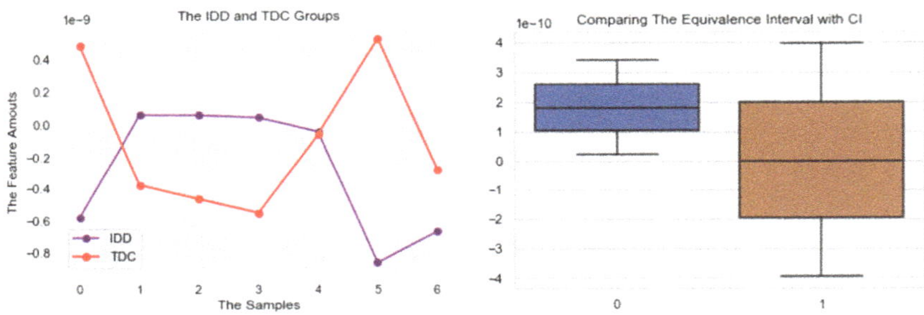

Figure 2.18. Examining the input values of two groups and the confidence interval and equivalence, input characteristic: mean of the DWT coefficients (music stimuli).

$$\text{EEB} = 3.957314643261377\text{e}^{-10}.$$

$$C = \text{Equivalent}.$$

- IDD and TDC groups, standard deviation of DWT coefficients, music stimuli (figure 2.19):

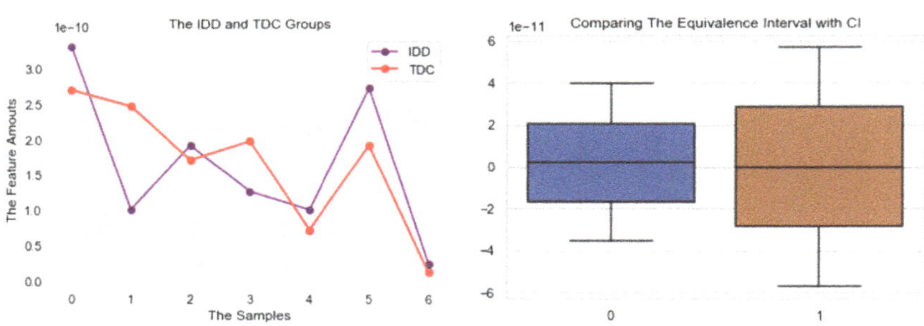

Figure 2.19. Examining the input values of two groups and the confidence interval and equivalence, input characteristic: standard deviation of the DWT coefficients (music stimuli).

$$\text{EEB} = 5.707419626051214e^{-11}.$$

$$C = \text{Equivalent}.$$

- IDD and TDC groups, energy of DWT coefficients, resting-state (figure 2.20):

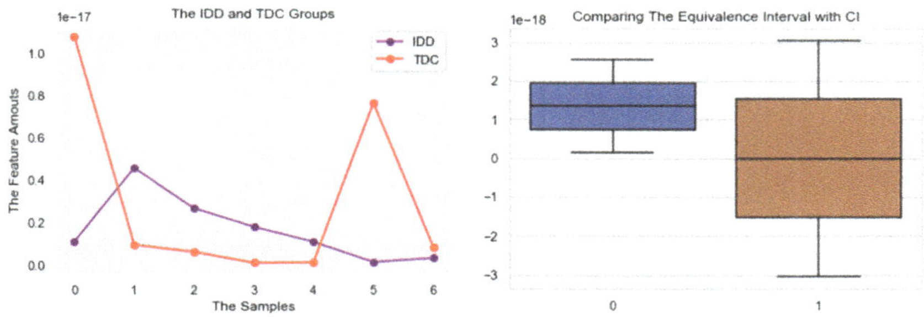

Figure 2.20. Examining the input values of two groups and the confidence interval and equivalence, input characteristic: energy of the DWT coefficients (resting-state).

$$\text{EEB} = 3.057687790518184e^{-18}.$$

$$C = \text{Equivalent}.$$

- IDD and TDC groups, mean of DWT coefficients, resting-state (figure 2.21):

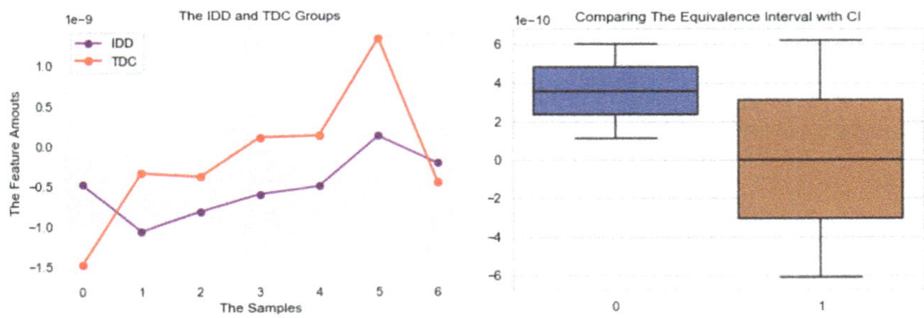

Figure 2.21. Examining the input values of two groups and the confidence interval and equivalence, input characteristic: mean of the DWT coefficients (resting-state).

$$\text{EEB} = 6.175685916585068e^{-10}.$$

$$C = \text{Equivalent}.$$

- IDD and TDC groups, standard deviation of DWT coefficients, resting-state (figure 2.22):

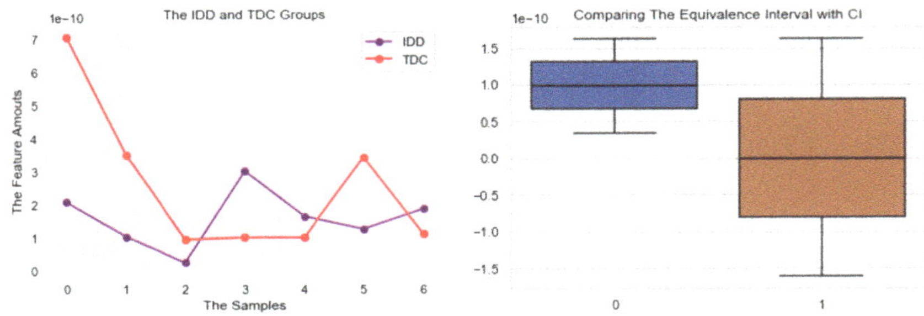

Figure 2.22. Examining the input values of two groups and the confidence interval and equivalence, input characteristic: standard deviation of the DWT coefficients (resting-state).

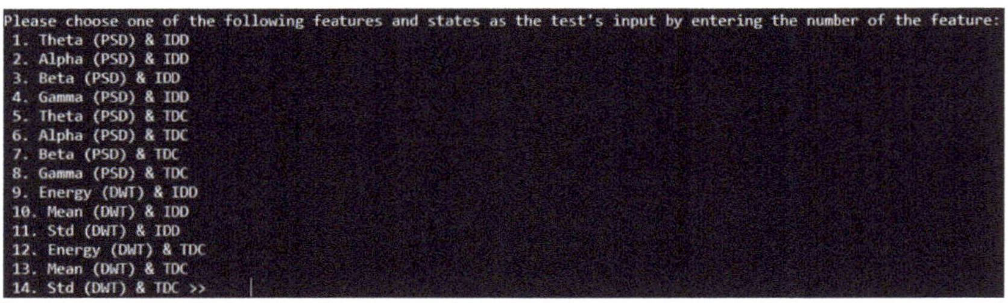

```
Please choose one of the following features and states as the test's input by entering the number of the feature:
1. Theta (PSD) & IDD
2. Alpha (PSD) & IDD
3. Beta (PSD) & IDD
4. Gamma (PSD) & IDD
5. Theta (PSD) & TDC
6. Alpha (PSD) & TDC
7. Beta (PSD) & TDC
8. Gamma (PSD) & TDC
9. Energy (DWT) & IDD
10. Mean (DWT) & IDD
11. Std (DWT) & IDD
12. Energy (DWT) & TDC
13. Mean (DWT) & TDC
14. Std (DWT) & TDC >>
```

Figure 2.23. Situations examined to compare rest-state and music stimuli.

$$\mathrm{EEB} = 1.6164491712825884e^{-10}.$$

$$C = \text{Inconclusive}.$$

2.5.2 Comparison of rest-state and music stimuli for each group of IDD and TDC

In this section, for each of the two IDD and TDC groups, the features extracted in two states of rest and music stimulation are compared. Results indicate the equality of the average in the 3rd, 4th, 5th, 6th, 9th, 10th, 11th, 13th, and 14th states of figure 2.23.

2.6 Conclusion

In this chapter, we used the EEG data of IDD and TDC subjects who were under rest and music stimulation. These data have been subjected to a fully automated in-house pre-processing pipeline by Sareen *et al* [9]. In one of the subsequent studies on this data, which was carried out using a machine learning algorithm to distinguish two groups of IDD and TDC, only the resting state data were examined [32]. Therefore, since the data is the only available EEG data of IDD subjects in the presence of music stimulus, we decided to identify the causes and differences between the EEG signal of IDD and TDC subjects in both resting state and music stimuli. For this purpose, we extracted features using PSD and DWT coefficients. Then, we implemented a two-step hypothesis testing approach in the Python environment.

It should be mentioned that it was previously presented to compare the average features of Iris data with the R programming language [10].

From the analysis of PSD in four frequency bands and in two states of resting-state and music stimuli, it was observed that in the presence of music stimulation in theta band, there is no difference between the two IDD and TDC groups. Meanwhile, in the presence of music stimuli, the gamma band reveals the difference between these two groups. Also, in the comparison of the DWT features of the two IDD and TDC groups, except for the standard deviation of the resting state, no significant difference was observed. Comparing the features extracted from resting-state and music stimuli for each of the IDD and TDC groups, although modes 1, 2, 7, 8, and 12 of figure 2.23 do not show equality, it does not mean they are not equal. It should be noted that all these results were obtained for $\beta = 0.8$ and $\alpha = 25\%$. By setting the first and second type error, the check can be repeated in different situations. According to our results, the method used in this chapter is suitable for distinguishing two groups of IDD and TDC in the presence of music stimuli and resting state, and it can be used in leading research to investigate IDD conditions.

Bibliography

[1] Lin E, Balogh R, Durbin A, Holder L, Gupta N and Volpe T *et al* 2019 *Addressing gaps in the health care services used by adults with developmental disabilities in Ontario* ICES

[2] Australian Government Department of Health and Aged Care 2021 *National roadmap for improving the health of people with intellectual disability* Commonwealth of Australia

[3] South Australia Health 2020 *SA intellectual disability health service: model of care* Government of South Australia

[4] Vi L, Jiwa M I and Lunsky Y *et al* 2023 A systematic review of intellectual and developmental disability curriculum in international pre-graduate health professional education *BMC Med. Educ.* **23** 329

[5] Carper R A, Moses P, Tigue Z D and Courchesne E 2002 Cerebral lobes in autism: early hyperplasia and abnormal age effects *NeuroImage* **16** 1038–51

[6] Just M A, Cherkassky V L, Keller T A, Kana R K and Minshew N J 2007 Functional and anatomical cortical underconnectivity in autism: evidence from an FMRI study of an executive function task and corpus callosum morphometry *Cerebral Cortex* **17** 951–61

[7] Kana R K, Libero L E and Moore M S 2011 Disrupted cortical connectivity theory as an explanatory model for autism spectrum disorders *Phys. Life Rev.* **8** 410–37

[8] Schipul S E, Keller T A and Just M A 2011 Inter-regional brain communication and its disturbance in autism *Front. Syst. Neurosci.* **5** 10

[9] Sareen E, Singh L, Varkey B, Achary K and Gupta A 2020 EEG dataset of individuals with intellectual and developmental disorder and healthy controls under rest and music stimuli *Data Brief* **30** 105488

[10] Zhao Y, Caffo B S and Ewen J B 2022 B-value and empirical equivalence bound: a new procedure of hypothesis testing *Stat. Med.* **41** 964–80

[11] Mallat S 1999 *A Wavelet Tour of Signal Processing: The Sparse Way* (New York: Academic)

[12] Daubechies I 1992 *Ten Lectures on Wavelets* (Philadelphia, PA: SIAM)

[13] Behnampour S, Aarabi A and Pouretemad H R 2021 Extracting features from EEG signals of patients with major depressive disorder using nonlinear dynamics and spectral analysis *Brain Sci.* **11** 222
[14] Andrzejak R G, Lehnertz K, Mormann F, Rieke C, David P and Elger C E 2001 Indications of nonlinear deterministic and finite-dimensional structures in time series of brain electrical activity: dependence on recording region and brain state *Phys. Rev. E* **64** 061907
[15] Kana R K, Libero L E and Moore M S 2011 Disrupted cortical connectivity theory as an explanatory model for autism spectrum disorders *Phys. Life Rev.* **8** 410–37
[16] Schipul S E, Keller T A and Just M A 2011 Inter-regional brain communication and its disturbance in autism *Front. Syst. Neurosci.* **5** 10
[17] Eklund A, Nichols T E and Knutsson H 2016 Cluster failure: Why fMRI inferences for spatial extent have inflated false-positive rates *Proc. Natl. Acad. Sci.* **113** 7900–5
[18] Lindquist M A, Wager T D, Kober H, Bliss-Moreau E and Barrett L F 2019 The brain basis of emotion: a meta-analytic review *Behav. Brain Sci.* **42** e95
[19] EMOTIV EPOC headset (http://emotiv.com/) (accessed 20 January 2020)
[20] Sareen E, Gupta A, Verma R, Achary G K and Varkey B 2019 Studying functional brain networks from dry electrode EEG set during music and resting states in neurodevelopment disorder *bioRxiv 759738* https://doi.org/10.1101/759738
[21] Guler I and Ubeyli E D 2005 Feature extraction techniques for a brain-computer interface *Biomed. Signal Process. Control* **1** 76–86
[22] Percival D B and Walden A T 2000 *Wavelet Methods for Time Series Analysis* (Cambridge: Cambridge University Press)
[23] Vaseghi S 2008 *Advanced Digital Signal Processing and Noise Reduction* (New York: Wiley)
[24] Anwar T 2020 A machine learning approach for recognizing intellectual development disorder using EEG *Int. Conf. on Biomedical Innovations and Applications (BIA)* (Varna, Bulgaria) pp 9–12
[25] Hatzigeorgiadis A, Vlachos F, Paraskevopoulou P and Zacharogiannis E 2018 Power spectral analysis of resting electroencephalogram in Down syndrome *J. Intellect. Disabil.* **22** 1–10
[26] Zambrano D, Hernandez N, Shapson-Coe A, Halder S and Rathore S S 2020 Power spectral density of electroencephalogram signals in autism spectrum disorder: a systematic review and meta-analysis *J. Autism Dev. Disord.* **50** 2858–72
[27] Klimesch W 1999 EEG alpha and theta oscillations reflect cognitive and memory performance: a review and analysis *Brain Res. Rev.* **29** 169–95
[28] Tallon-Baudry C, Bertrand O, Delpuech C and Pernier J 1999 Oscillatory gamma-band (30–70 Hz) activity induced by a visual search task in humans *J. Neurosci.* **19** 424–31
[29] American Statistical Association 2016 Statement on statistical significance and P-values (https://amstat.org/asa/files/pdfs/P-ValueStatement.pdf)
[30] Johnson A and Brown C 2018 Challenges in equivalence testing: defining scientific significance and equivalence bounds *J. Exp. Res.* **52** 321–35
[31] Berman H B *T Distribution Calculator* (https://stattrek.com/online-calculator/t-distribution) (accessed 16 February 2023)
[32] Breitenbach J, Raab D, Fezer E, Sauter D, Baumgartl H and Buettner R 2021 Automatic diagnosis of intellectual and developmental disorder using machine learning based on resting-state EEG recordings *17th Int. Conf. on Wireless and Mobile Computing, Networking and Communications (WiMob)* (Bologna, Italy) pp 7–12

IOP Publishing

Signal Processing with Python
A practical approach
Irshad Ahmad Ansari and Varun Bajaj

Chapter 3

Filter design and denoising technique for ECG signals

Meena Anandan, Manas Kumar Mishra, Akash Kumar Jain, Pandiyarasan Veluswamy and Rohini Palanisamy

An electrocardiography (ECG) is a valuable diagnostic and monitoring tool that records electrical activity of the heart. While acquiring an ECG signal, noises and artifacts such as baseline shift, motion artifacts and electromagnetic interference affect the accuracy and reliability of the diagnostic information in the signal. In this book chapter, filter design and denoising techniques for ECG signal is explored using time domain filters and frequency selective filters. The time domain filters that namely move average filter and derivative filter are explained. In addition, frequency selective filters such as finite impulse response (FIR) filters that use different windowing techniques, infinite impulse response (IIR) filters (Butterworth and Chebyshev) and adaptive filters are discussed. These filtering techniques are described along with their Python implementation and filter characteristics.

3.1 Introduction

An ECG is a diagnostic test that records variations of electrical potential over time. These variations are measured by contraction and relaxation of the atria and ventricles in the heart. To contract and relax the electrical impulse is generated by the sino-articular node present in the right atria to initiate uniform contraction. This electrical impulse from the atria is delayed by the atrio-ventricular node in the right ventricle to avoid uneven contraction in the ventricles. The contraction and relaxation of the cardiac muscles produces an electrical potential that can be recorded using an ECG. This electrical potential results in P, Q, R, S and T waves in which a P wave indicates atria depolarization, a Q R S wave shows the depolarization of ventricles, and a T wave is produced by ventricular repolarization [1]. The abnormalities in these waveforms define the type of cardiac-related disease in a human.

doi:10.1088/978-0-7503-5929-0ch3

Misdiagnosis of cardiac abnormalities can occur if noises and artifacts are added to the system during acquisition of ECG signals. These noises and artifacts are caused due to improper contact of electrodes to the skin, baseline shift, electromagnetic interference, movement artifacts, muscular contraction and instrumentation noise.

Baseline wander is a low-frequency artifact caused by the movement of body parts, breathing, poor electrode contacts and skin–electrode impedance. Movement artifacts result from electrode displacement on the skin that leads to changes in skin–electrode impedance. The impedance introduces baseline variations in the ECG signal [2]. Electromagnetic or powerline interference is generated due to inductive and capacitive coupling of 50/60 Hz powerlines during an ECG signal acquisition. In addition, improper grounding of an ECG acquisition machine or a human and the surrounding electrical equipment could also cause electromagnetic noise [3] [4]. Muscle artifacts are high-frequency noise produced by the electrical activity of muscle contraction, beating of the heart or by a sudden movement of the body [5]. Instrumentation noise is generated due to improper use of electronics in the ECG equipment.

Different types of filters, which are discussed and implemented in this chapter, are used to reduce these noises and artifacts. The low-frequency noise, such as baseline drift and motion artifacts, could be suppressed by using namely band pass or high pass filters [6]. High-frequency noises, such as the electromagnetic interference or powerline interference, are filtered using the band stop filter and low pass filters.

3.2 Filter types

Generally, filters are categorized into analog and digital filters based on the hardware and software implementation. Analog filters can be used to remove the noises, but they introduce nonlinear phase shift in the signal and also increase the size of the instrumentation. Compared with the analog filter, the digital filter is an inexpensive technique that provides more accurate and precise results by removing the noises from the signal. Based on the frequency response of the filters, they are segregated into band pass, band reject or band stop, high pass and low pass.

The low pass filter suppresses a high-frequency signal in stopband and passes low frequencies in passband. Similarly, the high pass filter allows high-frequency signals in passband while suppressing low-frequency signals in the stopband. A band pass filter eliminates the signal in stopband while allowing a specific range of frequencies to pass through the passband. The band reject or band stop filter can assist in removing a certain frequency or band of frequencies from the signal.

In this chapter, the implementation of digital time domain filters and frequency selective filters using Python are discussed to reduce the noises and artifacts from ECG signals.

3.2.1 Time domain filters

Time domain filters are a class of filters that process signals over time. These filters can be implemented using algorithms or techniques in the time domain to get the desired filter characteristics. Through the use of digital filters and signal processing methods, noise can be filtered from the signals in the time domain effectively. The advantage of these filters is that it is not essential to analyze the frequency characteristics of signal and noise. In this section, the basic time domain filters used to remove the noises and artifacts from the ECG signal are discussed.

(a) Moving average filter
(b) Derivative based filter

3.2.2 Frequency selective filters

Frequency selective filters are used for smoothing and filtering the signal by removing or suppressing high- or low-frequency components. These filters are designed in the frequency domain to provide specific, desired filter characteristics. The advantage of the frequency selective filters, when compared with time domain filters, is that they provide more control to the design filter based on the required frequency response [7]. In this section, various frequency selective filters are discussed to eliminate the noises from signal.

(a) Finite impulse response filters
(b) Infinite impulse response filters
    1. Butterworth filters
    2. Chebyshev filters
(c) Adaptive filters

3.3 Python libraries for filter design

The time domain and frequency selective filters can be implemented in Python using the prior available library functions [8]. There are some essential library functions that need to be called during the filter design, and they are described as follows:

(a) SciPy: SciPy is the signal processing library that consists of predefined functions such as *signal.butter*, *signal.iir*, *signal.fir*, *signal.firwin*, *signal.lfilter* and etc. These functions can be called during programming using *import* functions from the library.
(b) NumPy: NumPy is used to define the arrays. This library can be used to perform a mathematical operation on an array, such as defining an array in a multidimensional and random number generation. It can be used to compute large arrays efficiently and is suitable for high performance computing.
(c) Matplotlib: The matplotlib library can be used for data visualization in plots and also to visualize the data in different forms, such as a line plot, scatter plot, bar plot or statistic plot.

These are the essential libraries to be called while writing the Python code to design and implement filtering algorithms for signal analysis.

3.4 Filter specifications

To design a filter, it is essential to define the filter specifications according to the noise features present in the signal. The following are the necessary filter specifications that are needed to design a filter in Python.

(a) Filter order: In a digital filter, the filter order represents the number of coefficients in the filter design and also refers to the complexity or degree of the filter. It is considered an important parameter in the filter design during ECG signal processing. The order also determines the roll-off between passband and stopband frequencies based on the cut-off frequency. An increase of the filter order increases the slope to be steeper, as is an ideal filter frequency response. However, this dynamic introduces an additional phase delay in the output signal.

(b) Cut-off frequency: The cut-off frequency determines the frequencies to be attenuated or that can pass through in the signal. Based on the cut-off, different types of filters could be implemented based on the noise characteristics and applications.

(c) Sampling frequency: The sampling frequency is measured as the number of samples acquired per second. To avoid aliasing effects in the bandlimited signal, the sampling frequency should be greater than twice the maximum frequency of the corresponding bio signal acquisition. As per the Nyquist theorem, the sampling frequency is calculated as,

$$\text{fs} \geqslant 2\ \text{fm} \tag{3.1}$$

3.5 Mapping of the digital frequency

The continuous time frequencies, Ω, can be defined from 0 to ∞, whereas discrete time frequencies, ω, lie from $-\pi$ to π. Both types of frequencies are related through the linear relation,

$$\omega = \Omega T_s \quad \text{where} |\omega| < \pi \tag{3.2}$$

Where T_s is the sampling period used in the conversion of continuous time into discrete time, a unit of Ω is measured in rad/sec, and ω is measured in rad/samples.

$$\begin{gathered} f_s \mapsto 2\pi \\ \frac{f_s}{2} \mapsto \pi \\ \left[-\frac{f_s}{2}, \frac{f_s}{2}\right] \mapsto [-\pi, \pi] \end{gathered} \tag{3.3}$$

This linear mapping connects the continuous frequencies to discrete time frequencies.

3.6 Time domain filters

3.6.1 Moving average window

The moving average window filter is a time domain filter in which the temporal averaging is performed in the signal using a window method. This filter acts as a low pass filter that can be used to reject higher spectral noises in the signal. It is performed by averaging the adjoining values with the defined window. This filter produces a stable output and depends only on the current input and previous input samples. The increase of window size increases the smoothening in the signal but introduces the additional phase delay in ECG samples [9, 10]. The general form of a moving average filter is represented as in equation (3.4),

$$y(n) = \sum_{k=0}^{M} \frac{b_k x(n-k)}{M+1} \tag{3.4}$$

where M is the window size, x and y denote input and output of the filter and b_k represents filter coefficients.

Python Code Implementation

```
# Import necessary library functions from python as mentioned in section
1.3.
df = pd.read_csv('filename.csv')
ECG_signal= df.column2
# Specifty the window size based on your application
WS = 4 # Window size
# Initialize the array for moving average
i = 0
MA = [ ]
for i in range(len(ECG_signal) - WS + 1):
Window = ECG_signal[i:i+WS]
WA = round(sum(Window)/ WS, 2) # Window Average
MA.append(WA)
# Then filtered ECG can be visualized using plt function
plt.plot(MA)
```

The filter characteristics, representative ECG signal and frequency response of a moving average filter is shown in figure 3.1. The magnitude and phase response of a moving average filter are shown in figures 3.1(a) and (b), respectively. From the magnitude response, it can be seen that a designed filter acts as low pass filter that allows low-frequency components by eliminating the high frequency. The transition band between the passband and stopband shows a smoother roll-off. From this observation it is noted that as window size increases, the ripples in the stopband increase and the cut-off frequency is reduced. The phase response of the filter is linear, which indicates a constant time delay across all the frequency components.

Figure 3.1(c) is a plot of a representative raw ECG sample recorded with AD8232 and Arduino with a sampling frequency of 100 Hz. The graph shows the recorded ECG signal with a baseline wandering noise as well as high-frequency noise.

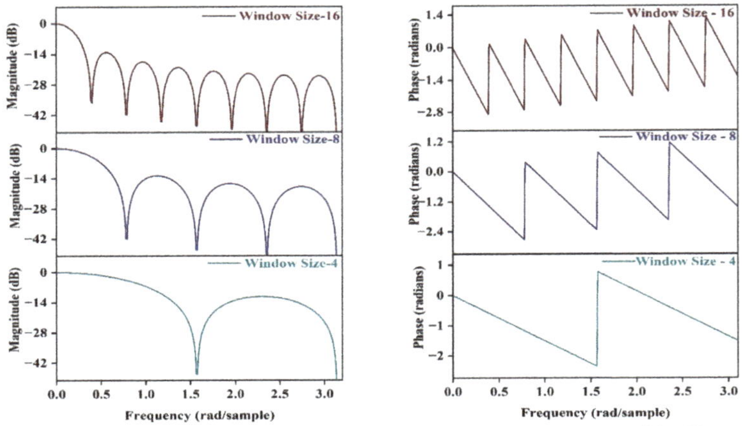

(a) Magnitude Response of the filter

(b) Phase Response of the filter

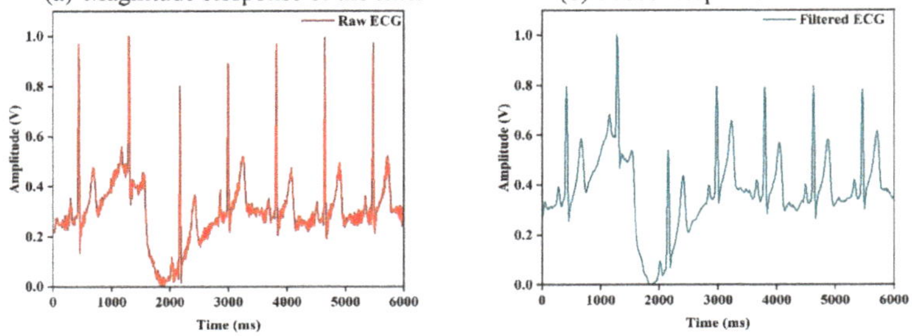

(c) Representative Raw ECG Signal

(d) Representative Filtered ECG Signal

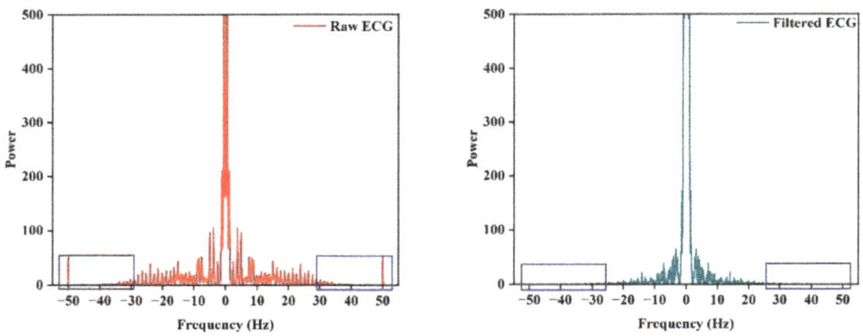

(e) Power Spectral density of raw ECG

(f) Power Spectral density of Filtered ECG

Figure 3.1. Moving average filter (a) Magnitude response of the filter, (b) Phase response of the filter, (c) Representative raw ECG signal, (d) Representative filtered ECG signal, (e) Power spectral density of raw ECG (f) Power spectral density of filtered ECG.

The amplitude of ECG signal ranges from 0.2–0.8 V. The high-frequency noise can be seen clearly between T and P waves.

Figure 3.1(d) shows the representative filtered ECG signal using a moving averaging filter (low pass filter). The high-frequency noise is suppressed and has an amplitude of 0.2–0.8 V. Based on the filter characteristics it can be noted that the ECG signal gets attenuated as the window size increases, which results in reduction of cut-off frequency.

The Fourier transform of a representative ECG and preprocessed ECG can be seen in figures 3.1(e) and (f), respectively. Figure 3.1(e) shows the high-frequency spectral components at 50 Hz, and figure 3.1(f) illustrates the filtered ECG in which the high-frequency components are suppressed and has no frequency components above 20 Hz. Moving average filters are excellent smoothing filters; however, it is clear from the observation that they do not act as ideal low pass filter.

3.6.2 Derivative filter

A derivative filter acts as a high pass filter by suppressing or rejecting the zero frequency component low-frequency components from the signal, such as baseline wandering. The gain of the frequency response increases linearly with the frequency. The derivative filter is analyzed from the signal's rate of change with respect to time [11]. It approximates the derivative operation by calculating the difference between adjacent samples of the signal. The general form derivative filter is given in equation (3.5) as

$$y(n) = \frac{1}{NT}[x(n) - x(n - N)] \tag{3.5}$$

where $y(n)$ is the filtered signal, $x(n)$ is the input signal, T is the time period, N is the order of the filter that varies from 0, 1, 2, ..., M and $x(n - 1)$ represents the previous input data point.

Python Code Implementation:

```
# filter coefficients are calculated based on third order central
difference formula.
b1 = np.array([1,-1]) # first order
b2 = np.array([1/2,0,-1/2]) # second order
b3 = np.array([1/4,0,0,0,-1/4]) # fourth order
# Read ECG signal
df = pd.read_csv('Filename.csv')
# Convolution of ECG signal with filter coeffients.
filtered_signal = np.convolve(ECG_signal,b,mode='same')
# Visualization of filtered ECG signal
plt.plot(filtered_signal)
```

The filter characteristics, representative ECG and frequency response of the derivative filter are shown in figure 3.2. The magnitude response and phase response of the designed filter for first, second and fourth order derivative filter are represented in figures 3.2(a) and (b), respectively. The magnitude response indicates high pass filter characteristics that eliminate low-frequency components. The transition band between the stopband and passband has smoother roll-off. The phase response of the derivative filter is linear in phase and has a constant time delay through all the spectral components. The derivative filter acts as a band pass filter as there is an increase in the filter order.

Figures 3.2(c) and (d) indicate the representative raw ECG signal and filtered ECG signal. From figure 3.2(c), it is noted that both low- and high-frequency

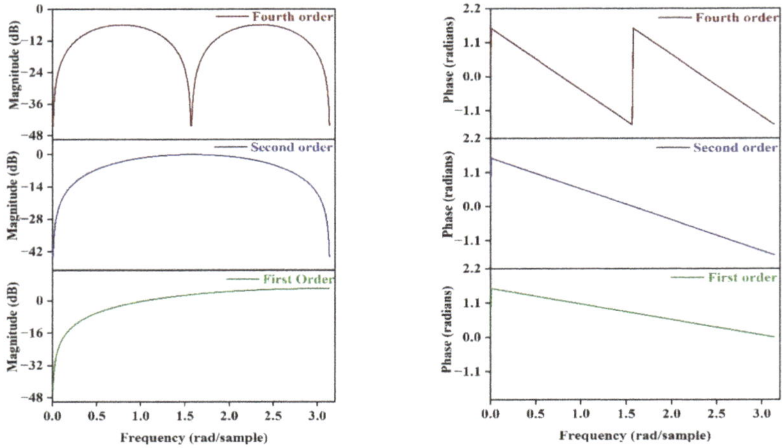

(a) Magnitude response of the filter

(b) Phase response of the filter

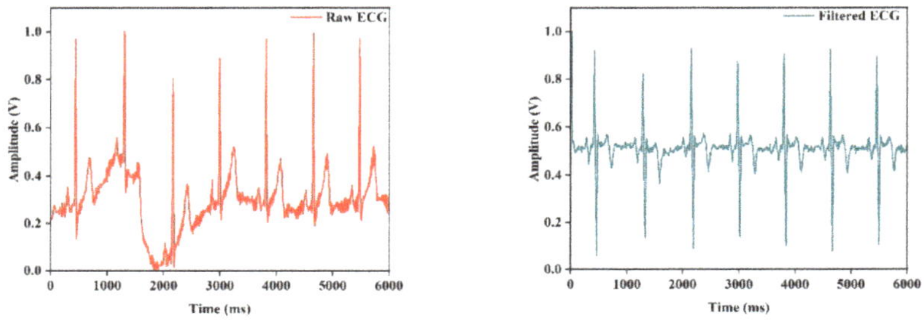

(c) Representative raw ECG Signal

(d) Filtered ECG Signal

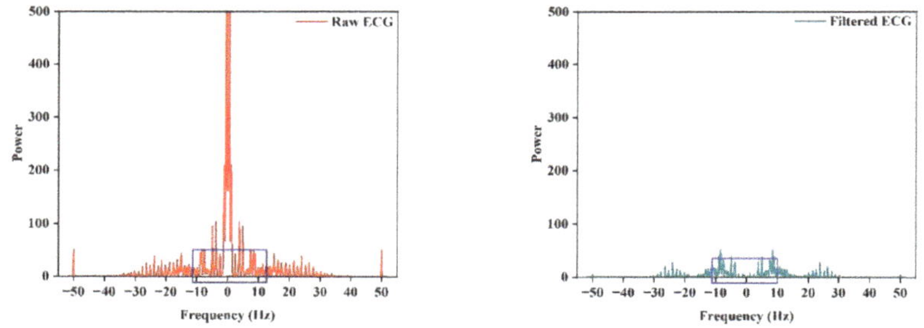

(e) Power spectral density of raw ECG

(f) Power spectral density of filtered ECG

Figure 3.2. Derivative filter (a) Magnitude response of the filter, (b) Phase response of the filter, (c) Representative raw ECG signal, (d) Representative filtered ECG signal, (e) Power spectral density of raw ECG (f) Power spectral density of filtered ECG.

components are present in the signal, and there is an amplitude in the range of 0.2–0.9 V. From figure 3.2(d), it can be observed that the low-frequency components have been suppressed, retaining the high-frequency components in the signal.

In figure 3.2(e), the filter characteristics of the representative raw ECG signal show that the high and low spectral components are present in the entire signal. Figure 3.2(f) represents the power spectral density of the filtered ECG signal. It can be seen that low-frequency components below 5 Hz are removed effectively, whereas high-frequency components are still present after 5 Hz. The baseline wandering has been completely aligned, but the signal gets distorted as there is an increase in the order due to nonlinear phase response. There is also no control over the cut-off frequency to get the desired frequency response. Thus, it is not advisable to use the derivative filter for ECG signal filtering.

3.7 Frequency selective filters

3.7.1 FIR filter

The FIR filter is a frequency selective filter designed to suppress unwanted frequencies from the desired signal. The FIR filter provides a linear phase response and has good stability and flexibility in frequency response design. However, these filters require more coefficient leads to increase in the computational complexity [12]. Generally, the impulse response of any discrete time system in z-transform can be written as,

$$H(z) = \frac{b_0 + b_1 z^{-1} + b_2 z^{-2} + \cdots + b_{M-1} z^{-(M-1)}}{a_0 + a_1 z^{-1} + a_2 z^{-2} + \cdots + a_{N-1} z^{-(N-1)}} \tag{3.6}$$

The denominator of $H(z)$ is replaced by 1 to avoid nonzero poles from the system response. Thus, $H(z)$ can be represented as,

$$H(z) = b_0 + b_1 z^{-1} + b_2 Z^{-2} + \ldots\ldots + b_{M-1} z^{-(M-1)} \tag{3.7}$$

The $H(z)$ has $(M - 1)$ delays and M coefficients and it is in the form of the power series expansion with finite elements. Thus, the finite elements in the impulse response are named as FIR.

The stability of an FIR system can be controlled using z-transform, and $H(z)$ is expressed as

$$H(z) = h[0] + h[1]z^{-1} + h[2]z^{-2} + \ldots . + h[M-1]z^{-(M-1)}$$

$$H(z) = \sum_{n=0}^{M-1} h[n]z^{-n}$$

To obtain the system in complex plane, z is substituted by $e^{j\omega}$

$$H(e^{j\omega}) = \sum_{n=0}^{M-1} h[n]e^{-j\omega n}$$

Normally, the Fourier transform of the impulse response is described in equation format as

$$H_d(e^{j\omega}) = \sum_{n=-\infty}^{\infty} h_d[n]e^{-j\omega n} \tag{3.8}$$

From inverse discrete Fourier transform, $h_d[n]$ can be expressed in terms of $H_d(e^{j\omega})$, where $H_d(e^{j\omega})$ is the desired frequency response of the filter and $h_d[n]$ is infinite length sequence

$$h_d[n] = \frac{1}{2\pi}\int_{-\pi}^{\pi} H_d(e^{j\omega})e^{j\omega n}d\omega \tag{3.9}$$

For converting $h_d[n]$ into $h[n]$, a truncated function, known as window function $w[n]$, is used

$$h[n] = h_d[n]w[n] \quad \forall\ n \tag{3.10}$$

The FIR filter is designed using the sinc function and ideal impulse response. The sinc function represents the ideal low pass filter and has an infinite number of coefficients in the ideal state. To truncate this action, a windowing technique is implemented as a trade-off between the filter ideal impulse response and sinc function [10].

There are different types of the windowing technique that can be implemented, such as the Hamming window, rectangular window, Blackmann window and Hanning window, to suppress the ripples passband and stopband of the filter [13, 14]. From the frequency response of the filter, the filters are categorized into band stop filter or notch filter, band pass filter, low pass filter and high pass filter.

The rectangular window can be defined in equation as,

$$w(n) = \begin{cases} 1 \text{ if } 0 \leqslant n \leqslant m \\ 0 \text{ otherwise} \end{cases} \tag{3.11}$$

The Hanning window can be defined in equation as,

$$w(n) = \begin{cases} 0.5(1 - \cos(2\pi n/m) & \text{if } 0 \leqslant n \leqslant m \\ 0 & \text{otherwise} \end{cases} \tag{3.12}$$

The Hamming window can be realized in equation as,

$$w(n) = \begin{cases} 0.54 - 0.46\cos(2\pi n/m) & \text{if } 0 \leqslant n \leqslant m \\ 0 & \text{otherwise} \end{cases} \tag{3.13}$$

The Blackmann window can be defined in equation as,

$$w(n) = \begin{cases} 0.42 - 0.5\cos(2\pi n/m) + 0.08\cos(4\pi n/m) & \text{if } 0 \leqslant n \leqslant m \\ 0 & \text{otherwise} \end{cases} \tag{3.14}$$

Design of the FIR low pass filter

The cut-off frequency is calculated based on the sampling frequency. The sampling frequency of 100 Hz is mapped to 2π. The cut-off frequency taken here is 40 Hz.

$$H_d(e^{j\omega}) = \begin{cases} 1 \text{ if} |\omega| \leqslant \dfrac{4\pi}{5} \\ 0 \text{ otherwise} \end{cases} \tag{3.15}$$

$$
\begin{aligned}
h_d[n] &= \frac{1}{2\pi}\int_{-\pi}^{\pi} H_d(e^{j\omega})e^{j\omega n}d\omega \\
h_d[n] &= \frac{1}{2\pi}\int_{-4\pi/5}^{4\pi/5} 1.\ e^{j\omega n}d\omega \\
h_d[n] &= \frac{\sin\left(\frac{4\pi}{5}n\right)}{\pi n} \\
h_d[n] &= \begin{cases} \dfrac{\sin\left(\frac{4\pi}{5}n\right)}{\pi n} & \text{if } |n| > 0 \\ \dfrac{4}{5} & \text{otherwise} \end{cases}
\end{aligned}
\tag{3.16}
$$

Python code implementation using the rectangular window

The rectangular window technique is a simple implementation method to filter the noise from the signal. The window has an amplitude of 1 within a specified window length and 0 outside the window. The impulse response is truncated with a rectangular window to get the desired frequency response. The main advantage of this method lies in the simplicity of the implementation. The disadvantage is a poor frequency attenuation in the side lobes, so there will be signal leakage from stopband to passband.

```
# import necessary libraries and read ECG data
df = pd.read_csv('Filename.csv')
ECG_signal = df.column2
Fs = 100
# Specify the tap length to truncate the impulse response
tap_length = np.arange(0, 49)
tap_length1 = np.arange(-24,25)
# Calculate rectangular window using tap length
rectangularwindow = 1
# Calculation of truncated impulse response
filter = np.zeros(len(tap_length1))
for i in range(1,len(tap_length1)):
  if tap_length1[i]==0:
      filter[i] = 4/5
  else:
     filter[i]=np.sin(((4*np.pi)/5)*tap_length1[i])/(np.pi*tap_length1[i]);
# To find filter coefficients based on windowed impulse response
filter_coef = np.multiply(filter, rectangularwindow)
filtered_ecg = np.convolve(filter_coef, x)
```

Python Code Implementation using Inbuilt functions:

```
Filtered_coefficient = signal.firwin(Fo, Fc, Window = 'rectangular', Fs)
# Coefficients and ECG signal are convolved using either filtfilt function
or lfilter function
Filtered_ECG = filtfilt(Filtered_coefficient, 1, ECG_signal)
```

The built-in Python functions for FIR filters are available in the *SciPy* library package where different FIR filter functions are used directly, such as ***firwin, firwin2*** and ***firls*** with corresponding window functions.

The filter characteristics, representative ECG signal and frequency response of the FIR filter with a rectangular window are shown in figure 3.3. Figures 3.3(a) and (b) show the magnitude and phase response of a rectangular window for the FIR low

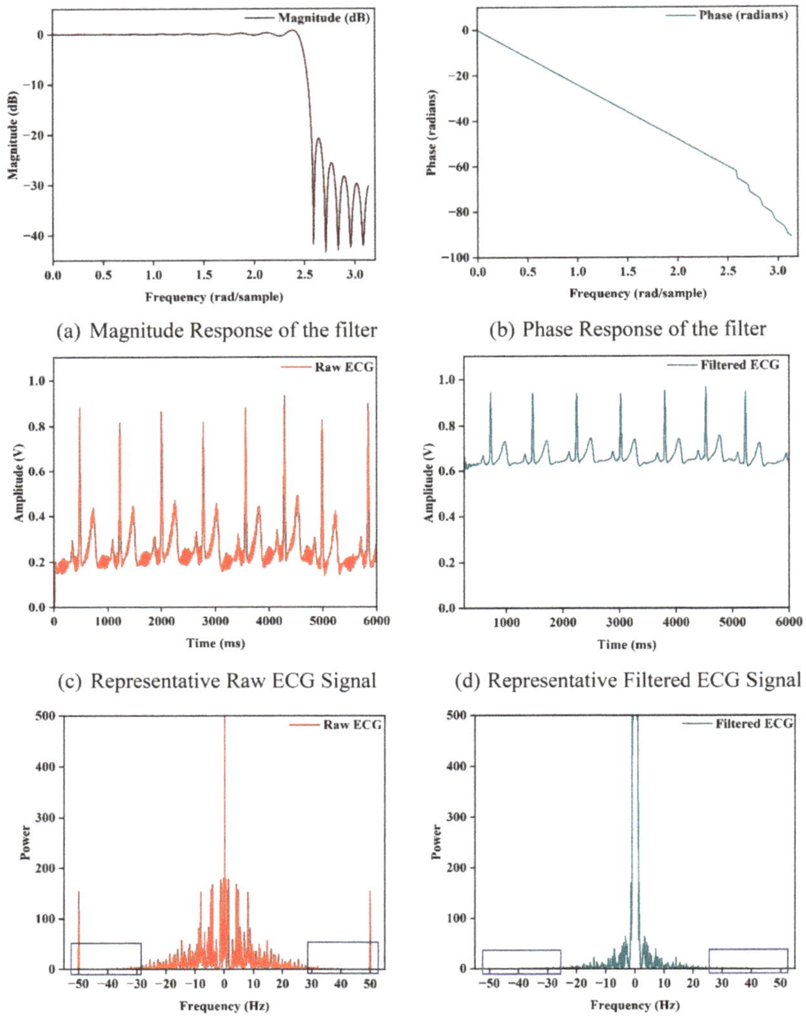

(a) Magnitude Response of the filter

(b) Phase Response of the filter

(c) Representative Raw ECG Signal

(d) Representative Filtered ECG Signal

(e) Power Spectral Density of raw ECG

(f) Power Spectral Density of Filtered ECG

Figure 3.3. FIR low pass filter using a rectangular window (a) Magnitude response of the filter, (b) Phase response of the filter, (c) Representative raw ECG signal, (d) Representative filtered ECG signal, (e) Power spectral density of raw ECG (f) Power spectral density of filtered ECG.

pass filter. The magnitude response is plotted on a logarithmic scale on the y-axis, and the frequency is on the x-axis. The desired frequencies are passed through the passband, and unwanted frequencies are attenuated in the side lobe stopband. The phase response indicates a linear phase that reduces the minimum distortion in the signal's temporal structure.

The raw ECG signal and the filtered ECG signal are shown in figures 3.3(c) and (d), respectively. This ECG signal is plotted in time scale on the x-axis with respect to amplitude on the y-axis. The filtered ECG signal displays a high-frequency suppression in the side lobes of the filter and allows low-frequency components in the desired passband. This method has some signal leakage from the stopband and passband that can be observed between T and P waves, where some distortion is still present.

In figure 3.3(e), the frequency response of the noisy ECG is plotted with frequency on the x-axis and power on the y-axis. The frequency response displays a 50 Hz frequency component present in the ECG signal. In figure 3.3(f), the power spectral density of the filtered ECG signal shows that the high-frequency noises are suppressed effectively.

Python code implementation using the Hanning window

The Hanning window is used for its effective noise removal from the signals. The Hanning window is designed to suppress the sidelobes to 0. It provides balance between the width of main lobe and side lobe suppression.

```
Hanningwindow = 0.5* (1-np.cos(2*np.pi*tap_length/len(tap_length)))
filtercoefhanning = np.multiply(filter, Hanningwindow)
filtered_ecg = np.convolve(filtercoefhanning, x)
```

The filter characteristics, representative ECG signal and frequency response of FIR low pass filter using the Hanning window are shown in figure 3.4.

Figures 3.4(a) and (b) show the magnitude and phase response of the Hanning window, respectively. The magnitude response shows ripples in the side lobes where the unwanted frequencies were suppressed. The phase response shows a linear response in the passband and a nonlinear response in the stopband. The FIR filters have a linear response in passband and ripples in the stopband.

As shown in figure 3.4(c), the raw ECG has both a baseline wander and powerline interference. The high-frequency noise is suppressed in the filtered ECG, as shown in figure 3.4(d). The amplitude of the signal is attenuated, and the voltage ranges from 0.62–0.95 V. It is observed that there is a DC shift produced in the filtered ECG signal due to the filter characteristics.

The frequency response of the raw ECG in figure 3.4(e) shows distribution of power among the frequencies as well as information about multiple frequency components in the signal. The frequency response of the filtered ECG, shown in figure 3.4(f), shows the frequency and power distribution after the filtering process. It can be observed that the high-frequency noise is attenuated effectively.

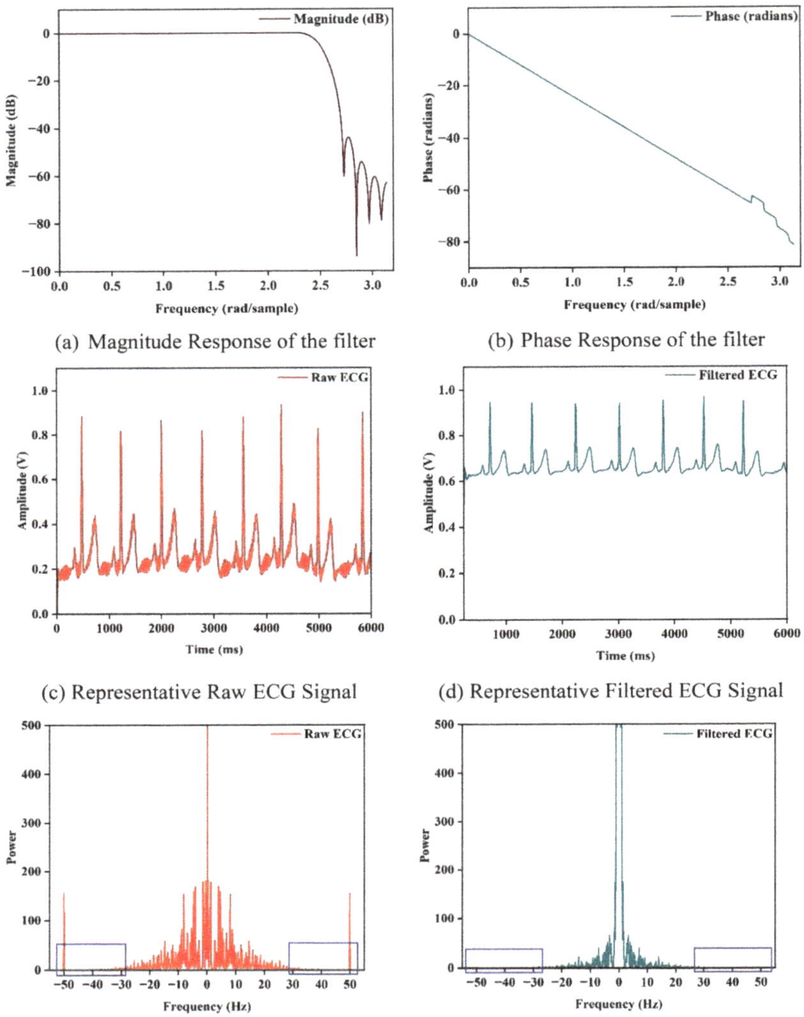

(a) Magnitude Response of the filter (b) Phase Response of the filter

(c) Representative Raw ECG Signal (d) Representative Filtered ECG Signal

(e) Power Spectral Density of raw ECG (f) Power Spectral Density of Filtered ECG

Figure 3.4. FIR low pass filter using the Hanning window (a) Magnitude response of the filter, (b) Phase response of the filter, (c) Representative raw ECG, (d) Representative filtered ECG signal, (e) Power spectral density of raw ECG (f) Power spectral density of filtered ECG.

Python code implementation using the Hamming window

The Hamming window is a symmetrical, bell-shaped window that has value ranges from 0–1. The main lobe of this window has a wider width, compared with other window functions, but has lower side lobes.

```
Hamming_Window = 0.54-0.46*np.cos((2*np.pi*tap_length)/len(tap_length))
filtercoef = np.multiply(filter, Hamming_Window)
filtered_ecg = np.convolve(filtercoef, x)
```

The filter characteristics, representative ECG signal and frequency response of FIR low pass filter using the Hamming window are shown in figure 3.5. Figures 3.5(a) and (b) indicate the magnitude and phase response of a FIR low pass filter with the Hamming window. The magnitude response of the ECG has a cut-off frequency of 40 Hz, and it is observed to have smoother roll-off in the

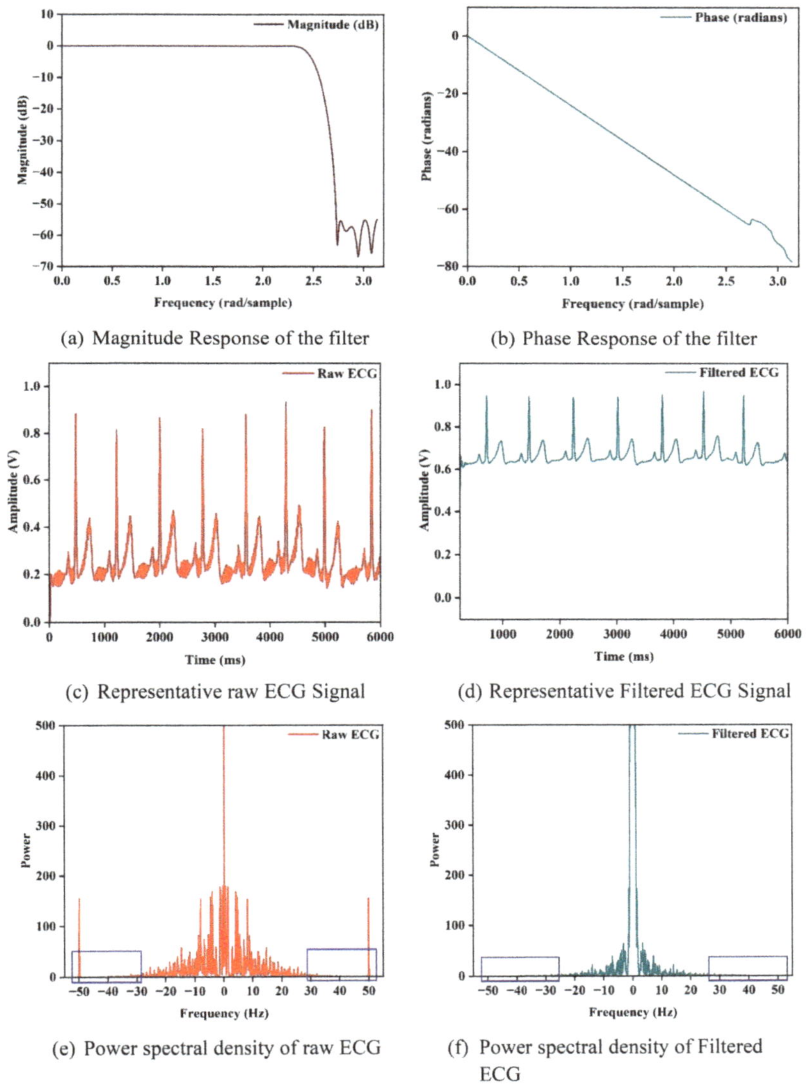

(a) Magnitude Response of the filter

(b) Phase Response of the filter

(c) Representative raw ECG Signal

(d) Representative Filtered ECG Signal

(e) Power spectral density of raw ECG

(f) Power spectral density of Filtered ECG

Figure 3.5. FIR low pass filter using the Hamming window (a) Magnitude response, (b) Phase response, (c) Representative raw ECG signal, (d) Representative filtered ECG signal, (e) Power spectral density of raw ECG (f) Power spectral density of filtered ECG.

transition band between stopband and passband frequencies. The phase response illustrates a linear phase in a lower-frequency range but a nonlinear phase in a higher-frequency range.

The raw ECG signal and preprocessed ECG signal are shown in figures 3.5(c) and (d), respectively. The raw ECG signal has a low-frequency noise and a high-frequency spectral noise and has an amplitude range of 0.2–0.9 V. The filtered ECG signal shown in figure 3.5(d) has an amplitude range of 0.6–0.9 V. The high-frequency noise in the signal is completely eliminated. It is observed that an increase in the filter order smooths and attenuate the signals.

Figures 3.5(e) and (f) represent the frequency response of a raw and filtered ECG, respectively. The raw ECG signal has frequencies throughout the duration of the signal. As can be seen, different frequencies are present in the entire signal range. The filtered ECG signal has a frequency response of up to 40 Hz, and, after that, signals have been eliminated by the designed filter.

Python code implementation using the Blackmann window

The Blackmann window reduces the side lobes in frequency response and eliminates the unwanted frequencies from the signal. This windowing technique has narrower main lobes to increase the frequency resolution and provides a steeper roll-off in the transition band.

```
BlackmannWindow = (0.42 - 0.5*np.cos(2*np.pi*tap_length/len(tap_length))) +
(0.08*np.cos(4*np.pi*tap_length/len(tap_length)))
filtercoefBlackmann = np.multiply(filter, BlackmannWindow)
filtered_ecg = np.convolve(filtercoefBlackmann, x)
```

The filter characteristics, representative ECG signal and frequency response of FIR filter with the Blackmann window are shown in figure 3.6. Figures 3.6(a) and (b) represent the filter characteristics, such as the magnitude response and phase response of the FIR low pass filter using the Blackmann window. The magnitude response has ripples in the stopband and a narrower main lobe to reduce undesirable frequencies in the signal. The phase response is linear up to the passband frequency, and the exhibited ripples in the stopband indicate minimum distortion in the desired signal response.

The representative raw ECG signal shown in figure 3.6(c) represents the raw ECG with high- and low-frequency noises. It is observed that the noise has been attenuated effectively in the Blackmann window, compared with other windows, as in figure 3.6(d). The DC shift is produced in the signal due to the filter characteristics. It is noted that the increase of filter length increases the attenuation in the ECG signal.

The power spectral density of the raw ECG in figure 3.6(e) and filtered ECG in figure 3.6(f) show the power and frequency distribution throughout the signal and the effects of filter implementation in the ECG signal. As can be seen in figure 3.6(f), frequencies above 40 Hz have been suppressed effectively by the window function.

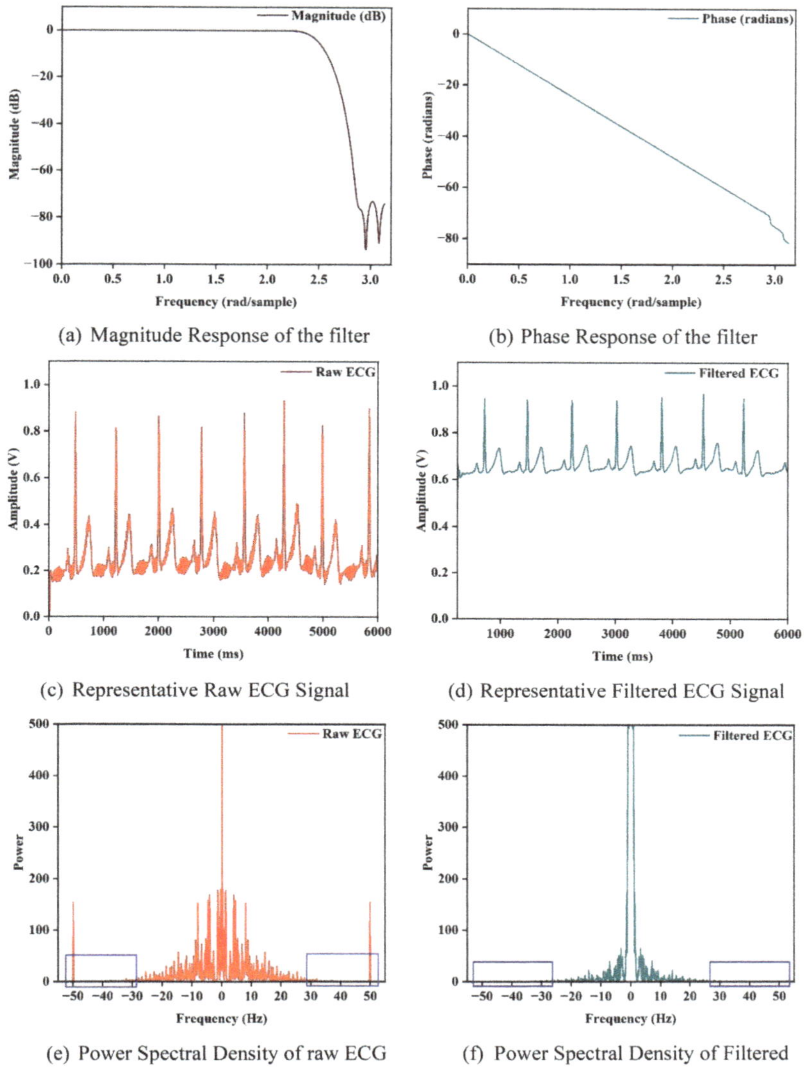

(a) Magnitude Response of the filter

(b) Phase Response of the filter

(c) Representative Raw ECG Signal

(d) Representative Filtered ECG Signal

(e) Power Spectral Density of raw ECG

(f) Power Spectral Density of Filtered ECG

Figure 3.6. FIR low pass filter using the Blackmann window (a) Magnitude response, (b) Phase response, (c) Raw ECG signal, (d) Filtered ECG signal, (e) Power spectral density of raw ECG (f) Power spectral density of filtered ECG.

3.7.2 IIR filters

An IIR filter is a digital filter that suppresses or removes noises from the signal. It works on a feedback mechanism that takes past and present output and past inputs to increase efficiency. IIR filters are computationally efficient and cheaper to implement. To design a stable IIR filter, the poles should lie inside the z-plane and zeros should be less than or equal to poles. The main advantage of IIR filters is that they are efficient to meet filter responses, such as roll-off, and can be implemented with fewer filter orders [15].

Generally, the impulse response of any discrete time system in z-transform can be written as,

$$H(z) = \frac{b_0 + b_1 z^{-1} + b_2 z^{-2} + \cdots + b_{M-1} z^{-(M-1)}}{a_0 + a_1 z^{-1} + a_2 z^{-2} + \cdots + a_{N-1} z^{-(N-1)}} \quad (3.17)$$

When the impulse response of the system has infinite coefficients in z-domain, then it is known as an IIR filter. An IIR filter form in z-transform can be written as,

$$H(z) = \frac{\sum_{i=0}^{M} b_i z^{-i}}{1 + \sum_{i=1}^{N} a_i z^{-i}} \quad (3.18)$$

Since an IIR filter has infinite response, it requires some constrain on the $h(n)$ such that it will be realizable. The constraints are given as follows,

$$h[n] = 0 \quad \forall\, n < 0 \quad (3.19)$$

$$\sum_{n=0}^{\infty} |h[n]| < \infty \quad (3.20)$$

IIR filters are realizable only if $h[n]$ is causal and converging. It is nonlinear phase filters that imply there would be a group distortion present.

There are many types of digital IIR filters based on the filter response. They are the Butterworth filter, Chebyshev filter and elliptic filter.

3.7.2.1 The Butterworth filter

The Butterworth filter has a maximum of the flattest response in the passband and is also known as the maximum flat magnitude filter. It exhibits a smoother roll-off from passband to stopband, which is achieved by distributing poles around a circle in a complex plane. The Butterworth filter has a more linear response compared with other IIR filters [16]. The magnitude response of the Butterworth filter is represented as,

$$|H(e^{j\omega})|^2 = \frac{1}{1 + \left(\frac{\omega}{\omega_c}\right)^{2N}} \quad (3.21)$$

where N is the filter order and ω_c is the cut-off frequency.

In this section, implementation of the Butterworth low pass filter to remove noise is discussed using Python. Due to the complexity of the equations, direct poles have been calculated for a simpler understanding.

Python Code Implementation:

```
Fs = 100 # sampling Frequency
Fc = 40 # cut-off frequency
N = 5 #order of the filter
Nyqrate = Fs/2 # Nyquist rate
Fc_norm = Fc/Nyqrate #Normalized cutoff frequency
Wc = 2*np.pi*Fc_norm #analog cutoff frequency
P = np.zeros(N, dtype=complex)
for k in range(1, N):
  P[k] = np.exp(1j * np.pi * ((2 * k) - 1 + N) / (2 * N)) #poles of
butterworth filter
b = np.poly(Pk) #Polynomial coefficients
b = np.real(b) #Real Polynomial coefficients
b /=np.abs(np.polyval(b,1)) #scaling of coefficients
df = pd.read_csv('Filename.csv') #loading ECG Data
ECG_signal = df.column2
Filtered = signal.lfilter(b, Fs, ECG_signal) #Filter using lfilter function
plt.plot(Filtered)
plt.show()
```

Python Code Implementation using Inbuilt functions:

```
Filtered_coefficient = signal.butter(Fo, Fc, 'lowpass') #calculate filter
coefficients
Filtered_ECG = filtfilt(b, a, ECG_signal) #filter the noise using filtfilt
function
```

The filter characteristics, representative ECG signal and frequency response of the IIR Butterworth low pass filter are shown in figure 3.7.

The magnitude and phase response of the Butterworth low pass filter is shown in figures 3.7(a) and (b), respectively. The magnitude response exhibits low pass filter characteristics, where the frequencies below the cut-off frequencies are suppressed. It has a relatively smooth passband and smoother roll-off in the transition band. The phase response of the filter is linear in phase, indicating that there is a constant delay throughout the frequencies.

Iin figures 3.7(c) and (d), the raw ECG and the filtered ECG of the Butterworth low pass filter can be seen. As mentioned earlier, the raw ECG has both high-frequency noise and low-frequency noise and has an amplitude range of 0.2–0.9 V. The high-frequency noise gets eliminated and has an amplitude in the range of 0.5–0.9 V. It is observed that there is a DC shift produced in the signal due to the filter characteristics.

The frequency response of both the raw and filtered ECG are shown in figures 3.7(e) and (f). The raw ECG has frequencies throughout the signal components. As can be seen from the Fast Fourier transforms of the filtered ECG, the signal below the cut-off frequencies is suppressed and the frequencies above the cut-off frequency are still present in the filtered ECG due to the filter characteristics.

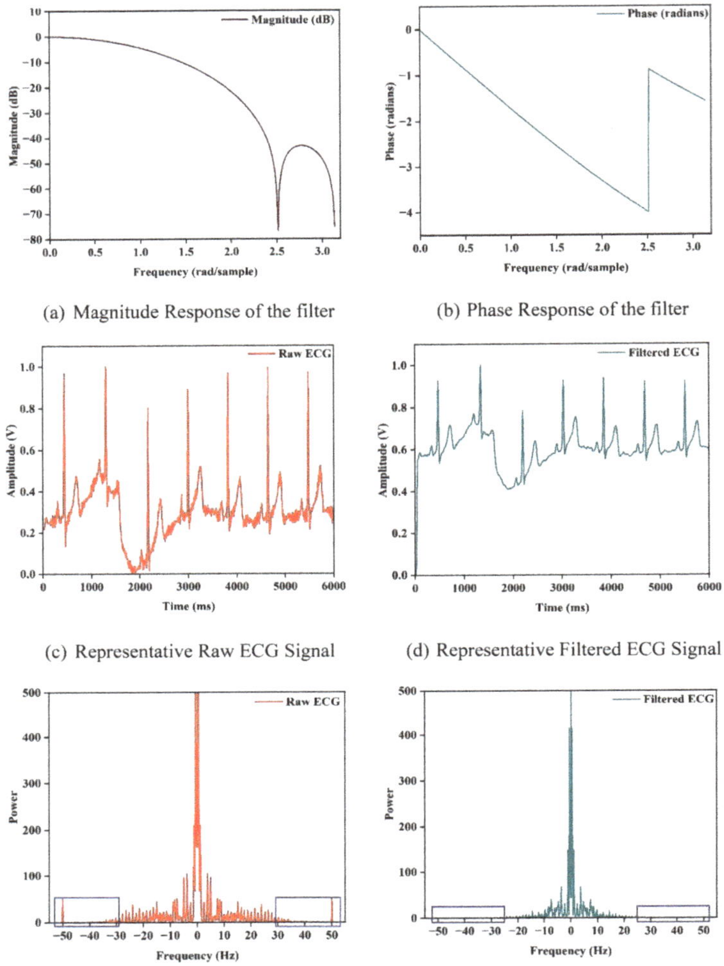

(a) Magnitude Response of the filter

(b) Phase Response of the filter

(c) Representative Raw ECG Signal

(d) Representative Filtered ECG Signal

(e) Power Spectral Density of Raw ECG

(f) Power Spectral Density of Filtered ECG

Figure 3.7. The IIR Butterworth low pass filter (a) Magnitude response, (b) Phase response, (c) Representative raw ECG signal, (d) Representative filtered ECG signal, (e) Power spectral density of raw ECG (f) Power spectral density of filtered ECG.

3.7.2.2 The Chebyshev filter

The Chebyshev filter has a steeper roll-off and is divided into type I or II based on the ripples in the passband and stopband, respectively. The Chebyshev filter can be used only in certain applications due to its irregular response in the stopband. To design and implement the filter, it is crucial to specify the filter order, passband frequencies, stopband frequencies and ripple factor. By allowing the ripple in the passband frequency, it is possible to achieve a faster roll-off by using this filter [17]. The magnitude of the Chebyshev filter is represented as,

$$|H(e^{j\omega})|^2 = \frac{1}{1 + \varepsilon^2 C_n^2(\omega)} \tag{3.22}$$

$$C_n(\omega) = \begin{cases} \cos(n \cos^{-1} \omega) & |\omega| \leqslant 1 \\ \cosh(n\cos^{-1} \omega) & |\omega| > 1 \end{cases} \tag{3.23}$$

where $C_n(\omega)$ is the nth order Chebyshev polynomial, and ε represents the ripple factor.

Python Code Implementation

```
# Filter specifications
Passband_edge = 30
Stopband_edge = 45
ripple = 0.1
Fs = 100
T = 1 / Fs
# Order of the filter (can be calculated with filter specifications or can
be specified directly
N = int(math.acosh(np.sqrt((10 ** (0.1 * Stopband_edge) - 1) / 10 ** (0.1 *
Passband_edge))) / (math.acosh(omega_s / omega_p)))
(or)
N=3
#Prewarping analog frequencies
omega_p = 2*Fs*np.tan(Passband_edge/2)
omega_s = 2*Fs*np.tan(Stopband_edge/2)
#calculate ripple factor
epsilon = np.sqrt(10**(0.1 * ripple) - 1)
# Calculate major and minor semi-axes
a = np.cosh((1 / N) * np.arccosh(1 / epsilon))
b = np.sinh((1 / N) * np.arcsinh(1 / epsilon))
# Calculation of poles
poles = np.zeros(N, dtype=np.complex128)  # Initialization the poles array
for k in range(N):
    pi_k = (np.pi / 2) + (((2 * k + 1) / (2 * N)) * np.pi) #angle of poles
    poles[k] = a * np.cos(pi_k) + 1j * b * np.sin(pi_k)
b = np.poly(poles) #polynomial coefficients
a = np.zeros_like(b)  # Initialization of `a` as zeros
a[0] = 1
df = pd.read_csv('filename.csv') #loading of ECG data
ECG_signal = df.column2
Filtered_ECG = signal.lfilter(b,a,ECG_signal)  #filtering of noise
```

Python Code Implementation using Inbuilt functions:

```
Filtered_coefficient = signal.cheby1(Fo, Fc, 'lowpass') #calculate filter
coefficients
Filtered_ECG = filtfilt(b, a, ECG_signal) #filter the noise using filtfilt
function
```

The built-in ***cheby1*** function is used to find filter coefficients for the Chebyshev type I filter. Likewise, there are other functions available, such as ***cheby1ord, cheby2*** and ***cheby2ord***. The ***cheby2*** and ***cheby2ord*** are used for the Chebyshev type II filter.

The filter characteristics, representative ECG signal and frequency response of the IIR Chebyshev low pass filter can be seen in figure 3.8.

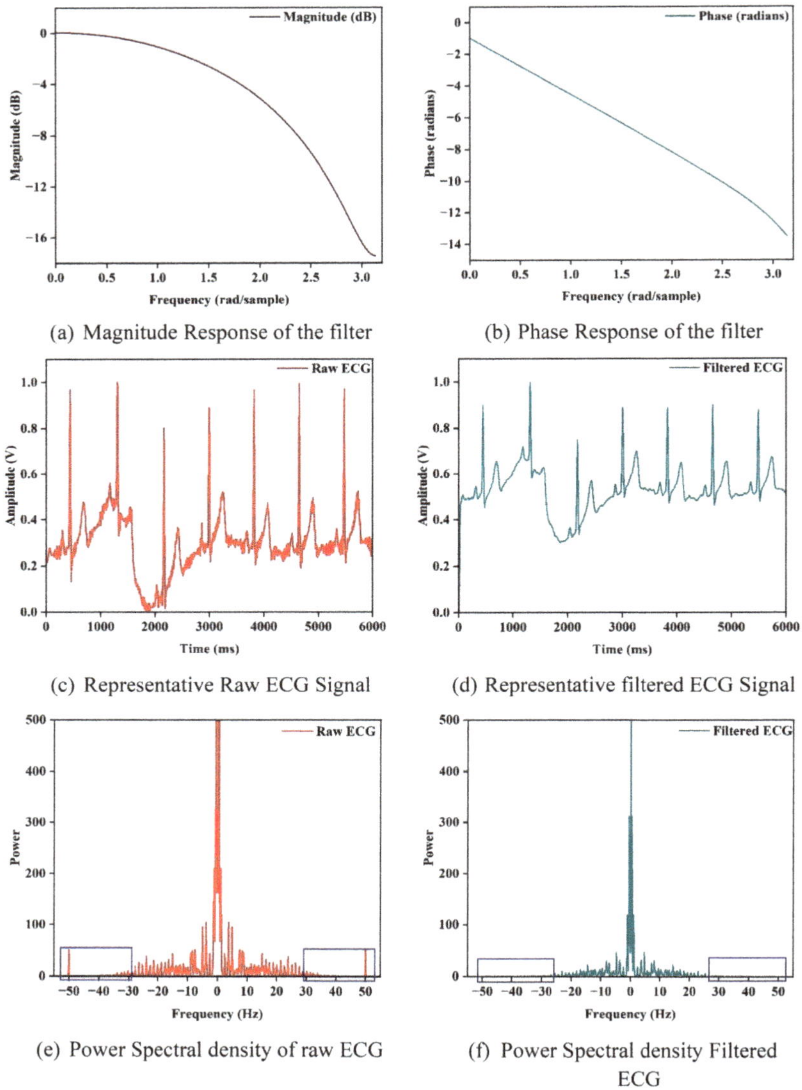

(a) Magnitude Response of the filter
(b) Phase Response of the filter
(c) Representative Raw ECG Signal
(d) Representative filtered ECG Signal
(e) Power Spectral density of raw ECG
(f) Power Spectral density Filtered ECG

Figure 3.8. The IIR Chebyshev low pass filter (a) Magnitude response of the filter, (b) Phase response of the filter, (c) Representative raw ECG signal, (d) Representative filtered ECG, (e) Power spectral density of raw ECG (f) Power spectral density of filtered ECG.

The magnitude and phase response of the Chebyshev low pass filter are shown in figures 3.8(a) and (b), respectively. The magnitude response has a flat passband and ripples in the stopband frequencies. The phase response of the designed filter is linear, as there is constant time delay, and also indicates the stability of the filter [18]. The increase of the order of the filter increases a steeper roll-off in the transition band. This steeper roll-off also increases the time complexity.

The filtered and raw ECG signal can be seen in figures 3.8(c) and (d). The obtained raw ECG signal consists of both a baseline wandering and high-frequency noise. In a filtered ECG signal, it is noted that a high-frequency noise has been effectively eliminated by a low pass filter. The increase of the filter order decreases the cut-off frequency, which results in the loss of signal information.

The power spectral density of the raw and preprocessed ECG signal is shown in figures 3.8(e) and (f). The preprocessed ECG signal has frequencies throughout the bandwidth. It can be seen that a frequency above 30 Hz is suppressed to eliminate the high-frequency noises.

3.7.3 Adaptive filter

Adaptive filters are also a type of digital filter in which the parameters adjust accordingly to the change in the input signal or required output. The filter reduces the difference between the desired output and actual output. Based on features and applications, adaptive filters are categorized into the Kalman filter, recursive least square filter, and least mean square filter (LMS) [19].

The LMS filter adaptive filter can be described in vector form as an equation (3.24),

$$y(k) = x^T(k)w(k) \tag{3.24}$$

where k is the discrete time index, $(.)^T$ represents transposition, x is the input vector and $y(k)$ is the filtered signal.

These adaptive filters are also known as time varying filters that can be used in real time applications to improve the performance based on different conditions. Normally ECG signals are nonlinear and nonstationary signals. Thus, compared with other algorithms, time varying algorithms provide better results. The main disadvantage of adaptive filters, when compared with other filters, is that they require more computational resources.

To design an adaptive LMS filter [20], it is required to know the desired output with which to calculate the error of the noisy ECG signal.

Python Code Implementation:

```
# Load ECG Data
df = pd.read_csv('Filename.csv')
ecg = df.column2
# Load Reference data for noise
reference = pd.read_csv('Filename.csv')
# Filter Specifications
Fl = 4 # filter lengtth
Step_size = 0.001
iterations = 13
def lms(ecg, reference, Fl, Step_size, iterations)
    ecg = np.asarray(ecg, dtype=float) # ECG signal
    reference = np.asarray(reference, dtype=float) #reference signal
    ecg = ecg / np.max(np.abs(ecg)) # Normalization of ECG_signal
    reference = reference / np.max(np.abs(reference)) # Normalization of
Reference signal
    num_frames = len(ecg) // Fl # dividing ECG signal into number of frames
    weights = np.zeros(Fl) # initializing weights

    for _ in range(iterations):
        for i in range(num_frames):
            ecg_frame = ecg[i * Fl: (i + 1) * Fl] #extracting current frame
from ECG signal

noise_frame = reference[i * Fl: (i + 1) * Fl] #extracting current frame
from reference signal
            output = np.dot(weights, noise_frame) #calculating filter
output by combining weights with noise frame
            error = ecg_frame - output # calculate error signal by
substracting filter output from ECG frame
            error_outer = np.outer(error, noise_frame)
            weights += step_size * np.mean(error_outer, axis=0) #updating
filter outputs
            filtered_ecg = np.convolve(ecg, weights[::-1], mode='same')
#Applying filter weights to the ECG signal
      return weights,filtered_ecg
filter_coefficients, filtered_ecg = lms(ecg, reference, Fl, Step_size,
iterations.
```

The filter characteristics, representative ECG signal and frequency response of the adaptive filter using LMS algorithm are shown in figure 3.9. The magnitude and phase response of the adaptive LMS filter are illustrated in figures 3.9(a) and (b). The magnitude response of the adaptive LMS filter indicates that that low pass filter characteristics suppress high-frequency noises. It is apparent that baseline wandering is still present due to low-frequency noise in the ECG signal. The phase response displays a linear phase change in passband region whereas non-linear phase change in the stop-band region. The desired frequency response can be achieved based on the filter length.

The raw ECG signal and preprocessed ECG signal are shown in figures 3.9(c) and (d), subsequently. The amplitude of the filtered ECG has a range of 0.4–0.8 V. It is observed that a high-frequency noise is eliminated, and a DC shift is introduced in

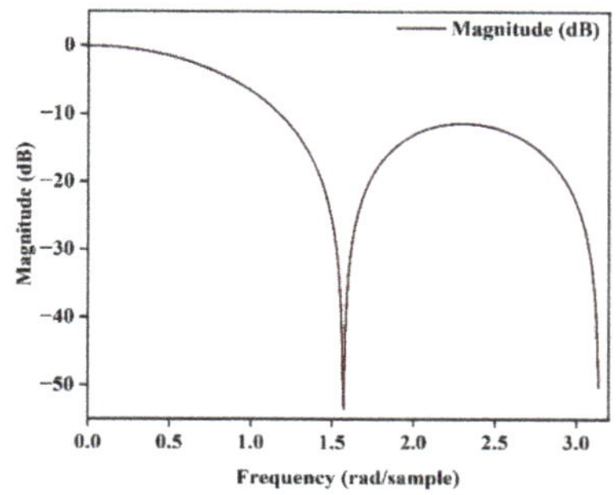

(a) Magnitude response of the filter

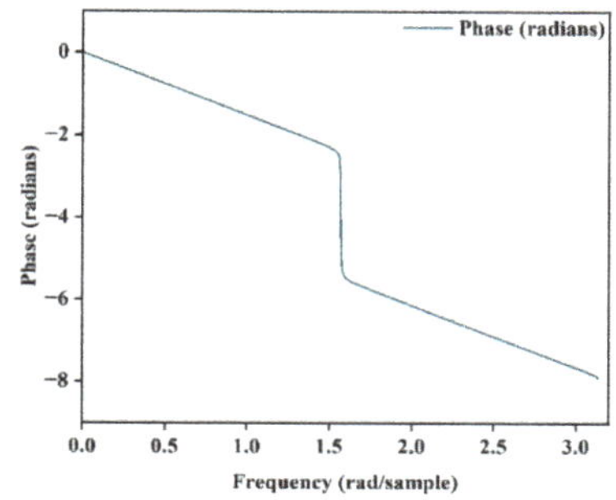

(b) Phase response of the filter

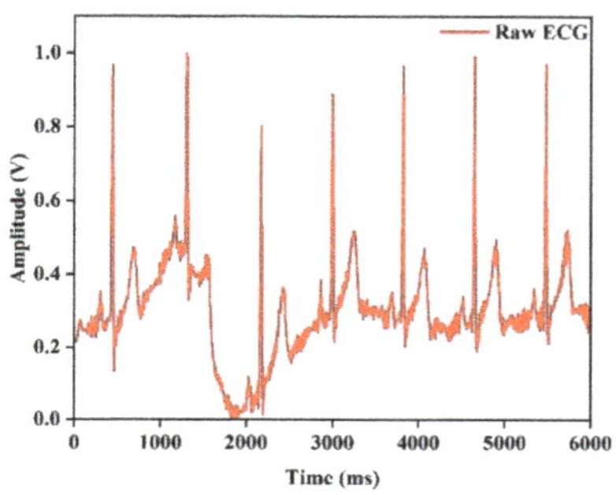

(c) Representative Raw ECG

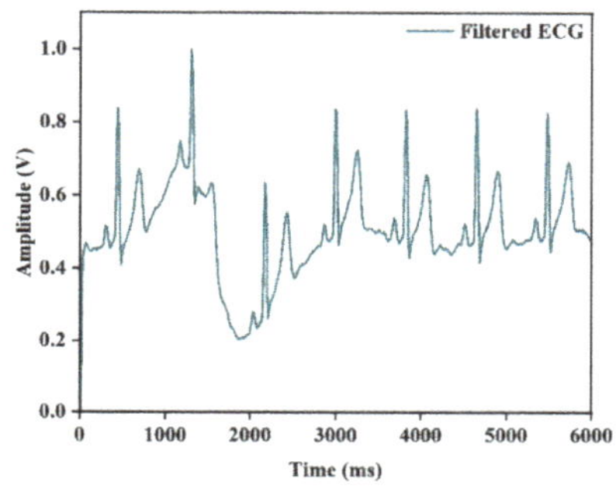

(d) Representative Filtered ECG

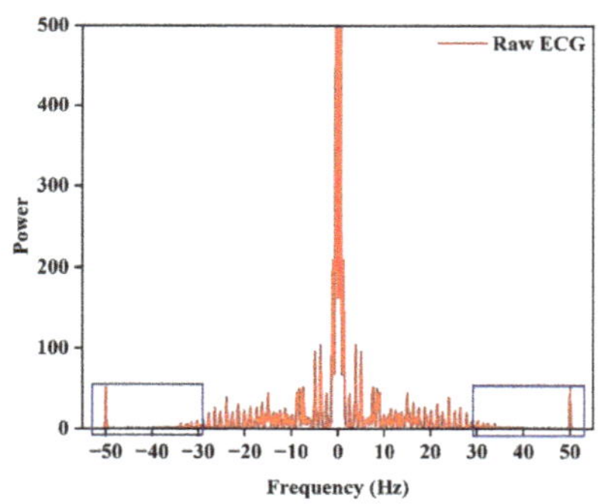

(e) Power spectral density of raw ECG

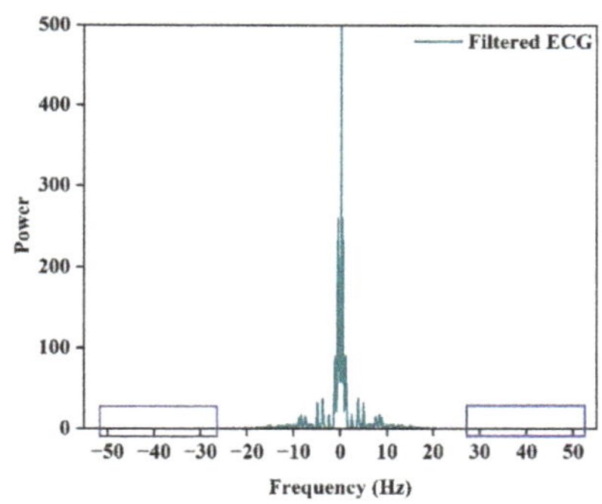

(f) Power spectral density of prepressed ECG

Figure 3.9. The adaptive filter using an LMS algorithm (a) Magnitude response (b) Phase response (c) Raw ECG signal, (d) Preprocessed ECG signal, (e) Power spectral density of raw ECG (f) Power spectral density of filtered ECG.

the signal after the filtering process. The output signal is produced based on the desired data and error data points. The filtered ECG signal has an amplitude range of 0.4–0.8 V.

The power spectral density of the raw ECG signal in figure 3.9(e) and filtered ECG signal in figure 3.9(f) represent the power distribution among frequencies in the entire spectrum. Figure 3.9(e) has a high-frequency noise of 50 Hz as well as low-frequency noises. From figure 3.9(f), it can be seen that a high-frequency noise above 20 Hz is completely suppressed. These filters can be implemented in real time applications due to their adaptive nature.

3.8 Conclusion

In this chapter, the different noises and artifacts present in an ECG signal have been discussed. Depending on the signal and noise characteristics, a filter is designed and implemented to remove or suppress the noises from the ECG signal using Python due to its versatility. The different types of time and frequency domain filters and how they are used to reduce low-frequency and high-frequency noises from a signal efficiently are discussed. Among the time domain filters, the moving average filter is better when compared with the derivative filter, as the former suppresses noises effectively. Among the frequency selective filters, the FIR filter with the Hamming window shows a better frequency response when compared with other windowing techniques. Among the IIR filters, the Butterworth filter has a flat passband when compared with the Chebyshev and adaptive filters that also provide a good frequency response where the filter can change according to the real time environment. These designed filters can be implemented in real time applications.

Bibliography

[1] Berkaya S K, Usyal A K, Gunal E S, Ergin S, Gunal S and Gulmezoglu M B 2018 A survey on ECG analysis *Biomed. Signal Process. Control* **43** 216–35

[2] Awal M A, Mostafa S S, Ahmad A and Rashid M A 2014 An adaptive level dependent wavelet thresholding for ECG denoising *Biocybern Biomed. Eng.* **34** 238–49

[3] Singh B and Kaur M 2009 Powerline interference reduction in ECG using combination of MA method and IIR notch *Int. J. Recent Trends Eng* **2** 125–9

[4] Srinivasa M G and Pandian P S 2019 Elimination of power line interference in ECG signal using adaptive filter, notch filter and discrete wavelet transform techniques *Int. J. Biomed. Clin. Eng.* **8** 32–56

[5] Chatterjee S, Thakur R S, Yadav R N, Gupta L and Raghuvanshi D K 2020 Review of noise removal techniques in ECG signals *Inst. Eng. Tech.* **14** 569–90

[6] Luo S and Johnston P 2010 A review of electrocardiogram filtering *J. Electrocardio* **43** 486–96

[7] Litwin L 2000 FIR and IIR digital filters *IEEE Potentials* **19** 28–31

[8] SciPy 2023 Signal processing (scipy.signal). https://docs.scipy.org/doc/scipy/reference/signal.html

[9] Pandey V and Giri V K 2016 *High frequency noise removal from ECG using moving average filters Int. Conf. on Emerging Trends in Electrical Electronics & Sustainable Energy Systems (ICETEESES): Proc (Sultanpur)* 191–5

[10] Oppenheim A V and Schafer R W 2009 *Discrete Time Signal Processing* (Upper Saddle River, NJ: Pearson) 195–268

[11] Rangayyan R M 2002 *Biomedical Signal Analysis—A Case Study Approach* (Piscataway, NJ: Wiley-IEEE Press) 1 99–114

[12] Parks T W and Burrus C S 1987 *Digital Filter Design* (New York, NY: Wiley-Interscience) 1 17–149

[13] Sulaiman I A, Hassan H M, Danish M, Singh M, Singh P K and Rajoriya M 2022 Design, comparison and analysis of low pass FIR filter using window techniques method *Mater. Today* **49** 3117–21

[14] Kumar V, Bangar S, Kumar S N and Jit S 2014 *Design of effective window function for FIR filters Int. Conf. on Advances in Engineering & Technology Research (ICAETR – 2014): Proc (Unnao)* 1–5

[15] Zhang X and Jiang S 2021 *Application of Fourier transform and Butterworth filter in signal denoising 6th Int. Conf. on Intelligent Computing and Signal Processing (ICSP): Proc (Xi'an)* 1277–81

[16] Rabiner L R and Gold B 1975 *Theory and Application of Digital Signal Processing* (Englewood Cliffs, NY: Prentice Hall) 1 205–92

[17] Podder P, Hasan M M, Islam M R and Sayeed M 2014 Design and implementation of Butterworth, Chebyshev-I and elliptic filter for speech signal analysis *Int. J. Comput. Appl.* **98** 12–8

[18] Nayak C, Saha S K, Kar R and Mandal D 2019 An efficient and robust digital fractional order differentiator based ECG pre-processor design for QRS detection *IEEE Trans. Biomed. Circuits Syst.* **13** 682–96

[19] Aiboud Y, Mhamdi J E, Jilbab A and Sbaa H 2015 *Review of ECG signal de-noising techniques 2015 Third World Conf. on Complex Systems (WCCS) (Marrakech, Morocco)* 1–6

[20] Sharma I, Mehra R and Singh M 2015 *Adaptive filter design for ECG noise reduction using LMS algorithm 2015 4th Int. Conf. on Reliability, Infocom Technologies and Optimization (ICRITO) (Trends and Future Directions) (Noida)* 1–6

Chapter 4

Electroencephalogram signal processing with Python

M Taghaddossi, P Moradi and M H Moradi

Analyzing electroencephalogram (EEG) signals is essential to aid the objective of many studies. As the registered EEG signals are contaminated with various noises, data preprocessing methods are employed to clean the signals.

In this chapter, will discuss some of the methods and tools in Python to clean and process the EEG data. We will use Python libraries MNE, NumPy, Matplotlib, and Pandas to preprocess and make data usable for further machine learning algorithms and models. In detail, we exposed raw EEG signals in the time and frequency domain at the first stage. For the preprocessing stage, we filtered and resampled data, repaired artifacts of the signals by independent component analysis (ICA), handled the bad channels, rejected noisy data spans visually, and reset the EEG reference. We also segmented continuous data into epochs and estimated the evoke response, which is important in evoke-related potential (ERP) studies. Time-frequency analysis and source localization modeling were performed in the processing stage as well.

As a result, we showed the power of the Python language and MNE package in EEG signal analysis and illustrated their proficiency in EEG signal preprocessing and processing stages with informative and fascinating images.

4.1 Introduction

Millions of neurons work together to control human behaviors and emotions. The information goes all over the brain and between the brain and body with the help of neurons. Knowing about the human brain and its cognitive behavior can help us to develop our knowledge about various brain-related topics. For this foundation, we need to analyze images or signals recorded from the brain. EEG is an effective modality that acquires brain signals in a non-destructive, painless, and side effect–less manner [1].

doi:10.1088/978-0-7503-5929-0ch4

These signals are really sensible from the outside environment and subject body. Too many strong noises contaminate the EEG signals. Therefore, it is really vital and important to clean them before doing any work. To do so, we need an interactive and practical software environment for processing and work with them because the first step after EEG signal acquisition is to import the signals to an environment where we can finally start to process them to extract meaningful information. Here we demonstrate how interactively the Python environment and its packages can help us to deal with EEG signals.

4.2 Principal and primary actions in EEG signal processing

In the following section, we make sure that our required packages are installed properly on our system. We indicate how to import EEG signals of any format into the Python integrated development environment (IDE). We also demonstrate how to save the EEG data to other well-known types of variables to expand our ability in any processing on EEG signals in this environment. We modify the imported EEG signals and extract data from it for further analysis.

4.2.1 Importing EEG signals to Python environment

Various setups can record EEG signals, and each setup has its version of recording data, such as .eeg, .edf, .cnt, .bdf .set, .data, and so on. The MNE package has different functions for importing these data. In addition, the EEG data version, saved by the MNE package in Python, is '.fif'.

The MNE package has some sample data. To import the '.fif' data into Python we use the command below. To start coding by the MNE package, we must import this package at the first line of the following code. In the second line, we enter the address of the sample data in a single quotation. These data contain EEG and electroencephalogram (MEG) signals recorded simultaneously when the subject is doing an audiovisual task. If it is your first time working with the MNE package, you may need to download these sample data by running the code '`mne.datasets.sample.data_path`'. The file path is separated by '\' except for the name of the EEG file in this line. The function `read_raw_fif` reads the fif EEG files at the third line. The information of EEG data will go to the `data_raw` object. Some operations of the data, like filtering, need the data to be copied onto the RAM. We should set a `preload` parameter equal to `True` in the function `read_raw_fif` in this manner. To print information about the size, time, and amount of the data in the `data_raw` object, we can print it in the third line below [2].

```
Import MNE
sample_MNE_file='D:\sampledata\MNEsampledata/sample_audvis_filt040_raw.fif'
data_raw = mne.io.read_raw_fif(sample_MNE_file, preload=True)
print(data_raw)
```

Hopefully, the MNE package has specific functions for importing the EEG data that is in other formats. For example, the `read_raw_brainvision()` function is designed to import data recorded by the BrainVision setup. `mne.io.read_raw_cnt()` function is designed to import data recorded by the Neuroscan setup, and so on. In addition, the `read_raw_eeglab()` function is used to import EEG data saved by EEGLAB. This function makes it possible to move them from the EEGLAB MATLAB toolbox to the Python IDE [3].

4.2.2 Saving the details of EEG data on other variables

To know more information about the imported data, like the name of EEG channels, reference channel, sampling rate, and minimum and maximum frequency that exist in the imported signals [4].

```
print(data_raw.info)
```

There are many other practical attributes have been defined for imported raw data. The `ch_names` attribute can acquire the list of channel names. An array of sample times is acquired by the `times` attribute. To save the total number of samples or lengths of the data into an int32 Python variable, you can use `n_times` or `__len__()` attributes [4].

```
data_ChNames = data_raw.ch_names
data_sample_times = data_raw.times
data_length  = data_raw.n_times
data_length = data_raw.__len__()
```

For printing and storing the information of channels in a new Pandas DataFrame variable, we can use the `describe` feature with True input at `data_frame` parameter. The result of this information is shown in figure 4.1 [4].

```
data_raw_ch_info = data_raw.describe(data_frame = True)
```

The `time_as_index` attribute and its argument help to find the sample number of a specific time in seconds during the EEG time series. We can get sample numbers of times in the signal using these attributes [2, 4].

For example, the sampling frequency of the imported data is 150.2 Hz. We can get a sample number of times at the 10th s during the signal by passing the code [3]:

```
data_sample_number = data_raw.time_as_index(10)
```

ch	name	type	unit	min	Q1	median	Q3	max
0	MEG 0113	grad	T/m	-4.96656e-11	-1.33429e-11	-1.02446e-11	-6.9275e-12	3.9895e-11
1	MEG 0112	grad	T/m	-4.10456e-11	2.01406e-12	4.1008e-12	6.32802e-12	5.14872e-11
2	MEG 0111	mag	T	-2.92349e-12	-4.76169e-13	-1.33556e-13	1.45888e-13	4.28169e-12
3	MEG 0122	grad	T/m	-5.50763e-11	-1.88795e-11	-1.49954e-11	-1.13765e-11	4.33039e-11
4	MEG 0123	grad	T/m	-5.39089e-11	-1.84953e-12	1.17418e-12	4.19633e-12	6.07474e-11
5	MEG 0121	mag	T	-3.07159e-12	-2.06786e-13	1.92746e-13	5.30874e-13	4.2601e-12
6	MEG 0132	grad	T/m	-3.50219e-11	-3.00855e-12	3.54313e-13	3.68639e-12	4.4988e-11
7	MEG 0133	grad	T/m	-4.23815e-11	-3.10204e-12	3.09094e-13	3.70464e-12	4.35865e-11
8	MEG 0131	mag	T	-2.3438e-12	-3.08342e-13	6.25706e-15	2.88308e-13	2.93148e-12
9	MEG 0143	grad	T/m	-6.49462e-11	-1.81681e-12	1.1598e-12	4.27692e-12	5.06435e-11
10	MEG 0142	grad	T/m	-3.34385e-11	-1.96731e-12	2.48155e-13	2.61101e-12	3.33852e-11
11	MEG 0141	mag	T	-2.48273e-12	-4.48349e-13	-1.92308e-13	4.85008e-14	3.38508e-12
12	MEG 0213	grad	T/m	-5.27217e-11	-7.76549e-12	-3.42275e-12	8.44294e-13	3.4074e-11
13	MEG 0212	grad	T/m	-4.33097e-11	-1.42479e-11	-1.03726e-11	-6.44812e-12	2.22905e-11
14	MEG 0211	mag	T	-1.80916e-12	-7.4001e-14	2.25597e-13	5.1852e-13	2.28821e-12
15	MEG 0222	grad	T/m	-4.99137e-11	-3.15157e-12	4.79398e-13	4.0696e-12	3.68559e-11
16	MEG 0223	grad	T/m	-2.78695e-11	-6.29119e-12	-2.55663e-12	1.01326e-12	2.45479e-11
17	MEG 0221	mag	T	-1.5675e-12	1.42465e-13	4.15575e-13	6.96674e-13	2.85692e-12
18	MEG 0232	grad	T/m	-5.64236e-11	-9.50884e-12	-6.30291e-12	-2.96365e-12	2.56782e-11

Figure 4.1. Details of recorded channels information stored in a DataFrame variable.

or to get the sample numbers of the 10th and 20th s:

```
data_sample_number_1 = data_raw.time_as_index([10,20])
```

The sample numbers of the 10th to 20th s with 1 s steps can be obtained by code:

```
data_sample_number_2 = data_raw.time_as_index(list(range(1,10)))
```

The sample numbers of the 10th to 20th s with 0.2 s steps can be obtained by code [2]:

```
import numpy as np
data_sample_number_3 = data_raw.time_as_index(list(np.arange(1,10,0.5)))
```

`close` attribute can be used to clean up the raw object from variable explorer. To make a copy from the raw object attribute to a new variable, `copy` attribute can be used [2].

4.2.3 Modifying the imported EEG data

In modifying the data to access the previous steps, we frequently use the attribute `copy` before any modification [2].

In some of the data recording sessions, electrocardiogram (ECG) and electro-oculogram (EOG) signals are recorded by other sensors. These data are recorded beside EEG data and stored in other rows in the dataset. We can use the `pick_types` attribute to exclude and include some channel types if we want to work with EEG data [2].

For example, we can omit the MEG data in the imported data and keep just EEG and EOG data by typing the code below [2].

```
data_raw_EEG_EOG = data_raw.copy()
data_raw_EEG_EOG.pick_types(meg=False, eeg=True, eog=True)
```

Besides restricting data with channel types, we can save new data with specific channels. For this purpose, `pick_channels` is used for reserving some channels by name, and `drop_channels` is used for removing some channels by name [2].

For example, in the code typed below, we first saved two datasets. The second code line results in the dataset with two channels removed, and the fourth code line results in the dataset with just one EEG channel and one EOG channel [2].

```
data_raw_EEG_EOG_2chremoved = data_raw_EEG_EOG.copy()
data_raw_EEG_EOG_2chremoved.drop_channels(["EEG 037", "EEG 059"])
data_raw_EEG_EOG_2channel = data_raw_EEG_EOG.copy()
data_raw_EEG_EOG_2channel.pick_channels(["EEG 060", "EOG 061"])
```

We can also use `pick_channels` and `reorder_channels` attributes to sort channels in our desired order. To do so, we write the name of channels in our desired order at the input argument [2].

To rename the channels of the data, we can use `rename_channels` attributes. For example, we changed the name of the 'EOG' channel to the 'left eye movement' channel in the code below [2].

```
data_raw_EEG_EOG.rename_channels({"EOG 061": "left eye movement"})
```

If some channel types are inaccurate in the dataset, you can correct them using the `set_channel_types` attribute. For example, we noticed the last EEG channel in the dataset is EOG, and its type is labeled wrong. We can type the following code to correct it [2].

```
data_raw_EEG_EOG. set_channel_types ({"EOG 060": "right eye movement"})
```

Sometimes we need to crop part of the data in seconds and save it in another variable. We can use the attribute `crop(timin,tmax)` that crops the EEG data from the time `tmin` in seconds to the time `tmax` [2].

For example, we want to reserve the EEG data in the second minute of the data. The code below can help us with this purpose [2].

```
data_raw_EEG_EOG_time_limited = data_raw_EEG_EOG.copy()

data_raw_EEG_EOG_time_limited.crop(tmin=60, tmax=120)
```

Occasionally we need to crop some parts of the data into various time windows. To append these data altogether, we can use the append attribute [2].

The code below shows us how to reach this purpose. We copied the first time window data and then appended the other time window data to it. The final appended data is also in object type like the primary data [2].

```
data_raw_EEG_EOG_time_limited_1 = data_raw_EEG_EOG.copy()

data_raw_EEG_EOG_time_limited_1.crop(tmin=10, tmax=15)

data_raw_EEG_EOG_time_limited_2 = data_raw_EEG_EOG.copy()

data_raw_EEG_EOG_time_limited_2.crop(tmin=25, tmax=30)

data_raw_EEG_EOG_time_limited_3 = data_raw_EEG_EOG.copy()

data_raw_EEG_EOG_time_limited_3.crop(tmin=65, tmax=74)

data_raw_EEG_EOG_time_limited_1.append([data_raw_EEG_EOG_time_limited_2,
data_raw_EEG_EOG_time_limited_3])
```

4.2.4 Extracting data from a raw object

We should extract data from the raw object for analysis or plot the EEG data by Python functions outside the MNE package. There are several ways to extract EEG data into a NumPy array. One of these ways is indexing the raw object by square brackets [2].

For example, we want to extract the 10th to 15th s of the data and save the result into a new outside MNE variable. The code below picks the 10th and 15th s of the data as starting and ending times and turns them into starting and ending sample numbers. It also gets channels 315–374 indices, of which we want to save their data. Finally, the result data save into two NumPy array elements of a tuple. The first array of the `raw_EEG_selection` tuple is the extracted EEG data, and the second one is the time array corresponding to the first data [2].

```
data_raw_info = data_raw.info

sfreq = data_raw_info["sfreq"]

start_stop_seconds = np.array([10, 15])

start_sample, stop_sample = (start_stop_seconds * sfreq).astype(int)

channel_index = list(range(315,375))

raw_EEG_selection = data_raw[channel_index, start_sample:stop_sample]
```

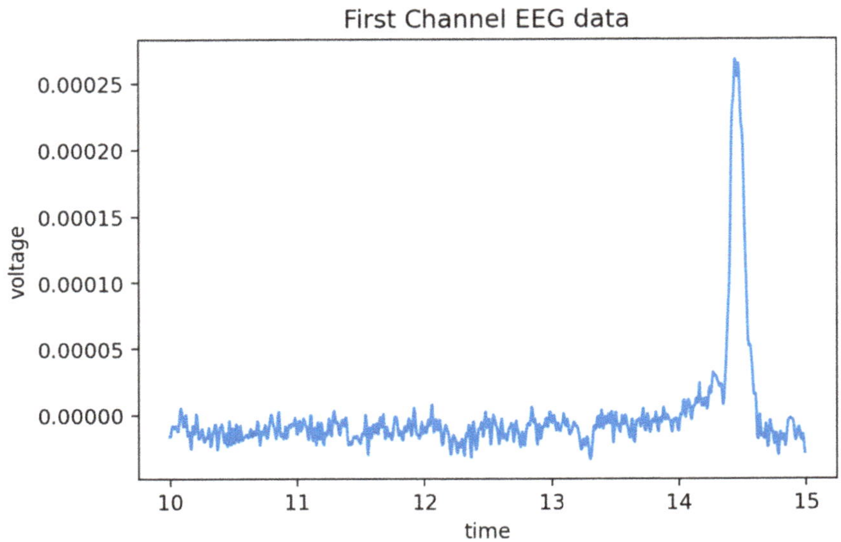

Figure 4.2. Result of EEG data of the first channel from the 10th to 15th s in plot window.

Using the matplotlib package, we can show the extracted data in the plot window. For example, the code and figure 4.2 show data from the first EEG channel in the dataset [2].

```
import matplotlib.pyplot as plt
x = raw_EEG_selection[1]
y = raw_EEG_selection[0][0,:].T
plt.plot(x, y)
plt.title("First Channel EEG data")
plt.xlabel("time")
plt.ylabel("voltage")
plt.show()
```

In the above code, we extract channels by their numbers, but we can also extract channels by their names. It just needs to put the channels' names in a list. For example, in the following code, we extracted EEG data of the two last channels and drew both in a single plot window and at the same time span of the last mentioned code (figure 4.3) [2].

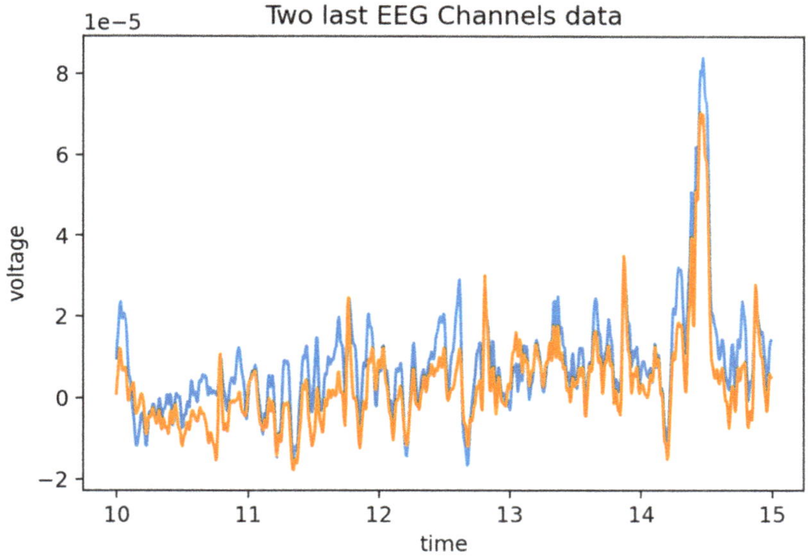

Figure 4.3. Result of EEG data of the two last channels from the 10th to 15th s in the single plot window.

```
channel_names = ["EEG 059", "EEG 060"]
raw_EEG_selection = data_raw[channel_names, start_sample:stop_sample]
x = raw_EEG_selection[1]
y = raw_EEG_selection[0].T
plt.plot(x, y)
plt.title("Two last EEG Channels data")
plt.xlabel("time")
plt.ylabel("voltage")
plt.show()
```

The second helpful method of extracting data from MNE raw data is working with the `get_data()` attribute. This method does not return an array of times as a result. If no parameters are specified as input for this function, it extracts data from all channels in the output. For example, `'data'` variable in the code below has a total raw dataset in a NumPy array (rows are channels, and columns are samples) [2].

```
data = data_raw.get_data()
```

There are several predefined parameters for this attribute. return_times parameters enables returning an array of times as a result. The array of times is saved in the `time` variable in the code below, and an array of the total dataset is saved in the `data` variable in the code below [2].

```
data, times = data_raw.get_data(return_times=True)
```

Specific channels extract using the `picks` parameter. It can accept integer values as row numbers of the dataset or a list of specific channel names of channel types. `Start` and `stop` parameters are sample ranges to extract data of a timespan. The code below is an example of how to specify values as the input of the `get_data` attribute [2].

```
first_channel_data = data_raw.get_data(picks=0)

eeg_and_eog_data = data_raw.get_data(picks=["eeg", "eog"])

two_EEG_channel_data = data_raw.get_data(picks=["EEG 030", "EEG 031"],
start=1000, stop=2000)
```

The `to_data_frame()` attribute can help us to extract data into a Pandas DataFrame. The mentioned parameters, such as `picks`, `start`, and `stop`, are also applicable in this attribute. Columns are channels, and rows are samples in the extracted data frame by this attribute. Its first column carries time in seconds with four decimal numbers [2].

4.2.5 Saving objects

We may need to save the obtained data in each preprocessing or processing step of EEG data analysis. The `save()` attribute can help us with this goal. We need to use this attribute to save raw objects and specify the file name we want to save in quotations ending with the '.fif' suffix. We can specify the path to save our raw object before its name. If we do not specify the file pass, Python saves it in the current directory. The code below helps us save the `data_raw_EEG_EOG` variable, an MNE object, to a file with the same name in the 'D' drive [2].

```
data_raw_EEG_EOG.save('D:\data_raw_EEG_EOG.fif')
```

We can use the `save()` method if we extract data into the NumPy array variable and the `to_csv()` method if we extract data into the Pandas DataFrame variable. For example, to save the `two_EEG_channel_data` variable, which is a NumPy array, we use a method like below. In this code, the variable name must be specified in `arr` predefined parameter, and the name of the file we want to save the variable in must be specified in the predefined parameter `file` [2].

```
np.save(file = 'two_EEG_channel_data' , arr=two_EEG_channel_data)
```

In the code below, we extracted EEG data into a Pandas DataFrame variable and then saved it into a '.csv' file. The pass and name of the file must be specified in quotations after the parameter 'r' [2].

```
start_end_secs = np.array([10, 15])
start_sample, stop_sample = (start_end_secs * sfreq).astype(int)
data_raw_EEG = data_raw.to_data_frame(picks=["eeg"], start=start_sample,
stop=stop_sample)
data_raw_EEG.to_csv(r'D:\data_raw_EEG.csv')
```

4.2.6 Working with events

Events are significant in EEG signal processing. They help us know which parts of the data correspond to which EEG signal recording states. We need the event information to subselect the EEG data. Therefore, as the first step of working with events, we extracted and saved the event array from the imported data in a raw object with the help of the `find_events` function and `stim_channel` parameter. This function gets the name of the specified channel as event data and stores it in a new NumPy array variable. For instance, we imported a sample of raw data named '`sample_audvis_raw`', the same audiovisual data we imported before any filtering. One of its channels, named '`STI 014,`' carries the event information. By running the following code, type and time of events are stored in the NumPy array `events` [5].

```
import numpy as np
import mne
sample_MNE_data = 'D:\sampledata\MNE sample data/sample_audvis_raw.fif'
data_raw = mne.io.read_raw_fif(sample_MNE_data, verbose=False)
events = mne.find_events(data_raw, stim_channel="STI 014")
```

The function `write_events` can store the event information in the `events` variable. The first input of this function gets the file name you want to save the events information in, ending with '.fif'. The second input gets the `events` information in the events variable. In the following code, we save event information in a new fif file and the same directory [5].

```
Events_file_name ='D:\sampledata\MNE sample data/sample_audvis_raw_eve.fif'
mne.write_events(Events_file_name, events)
```

By another function named `read_events`, we can import the event information we previously saved into a NumPy array variable. In the following code, the `events_from_file` variable has the same content as the `events` variable [5].

```
events_from_file = mne.read_events(Events_file_name)
```

If some event tags are not desired for us, and we want to exclude them from the data, we can easily use the function `pick_events` with the exclude parameter. In the following code, we excluded the tag '32' from the data [5].

```
events_without32 = mne.pick_events(events, exclude=32)
```

To combine some tags, the function `merge_events` can help us. For example, in the following code, we combined tags 1 to 3 and labeled them as 3 [5].

```
merged_events = mne.merge_events(events, [1, 2, 3], 1)
```

Each of the tags or event IDs, which are some range of numbers, has a meaning in the experience and shows part of the task in the EEG recording session. To define the real meaning of these event IDs, we can design a dictionary in Python. For example, in the sample imported data, the meaning of each tag is as follows [5]:

```
event_dictionary = {
"left auditory stimuli": 1, " right auditory stimuli ": 2,
"left visual stimuli": 3, "right visual stimuli": 4,
"smiley": 5, "button pressing": 32}
```

To visually check our recorded data, we can plot the events. This work ensures that our tags are added to the data correctly. The function `plot_events` and its predefined parameter can plot each one of the tags versus the sample number or time if we add sampling frequency. The `sfreq` parameter in the following code is the sampling frequency. `first_samp` is the offset between the sample number and sample index in Neuromag systems. `event_id` is also the dictionary that contains the information of tags [5].

```
fig = mne.viz.plot_events(events, sfreq = data_raw.info["sfreq"],
first_samp = data_raw.first_samp, event_id = event_dictionaty)
```

In more than 200 s of recording data, each one of the tags or event IDs with their description are shown in figure 4.4 [5].

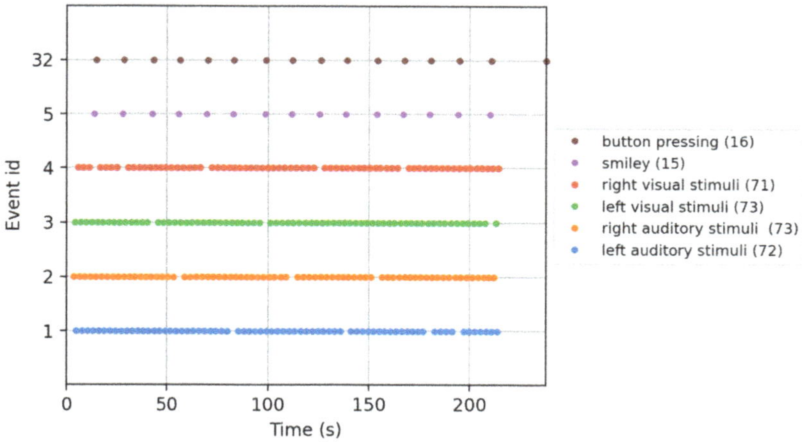

Figure 4.4. The plot of the occurring time of event IDs with their description in the right-side dictionary.

4.3 Exposure signals in time and frequency

When analyzing EEG signals, we will need to visualize signals to look deeper and focus on the EEG signal behaviors. For instance, after loading the data into the Python environment, we want to show them in the time and frequency domain. We showed in figures 4.2 and 4.3 a small part of the data with the help of the matplotlib package. By adding the data using the MNE package in the Python environment, the raw object has several built-in functions for plotting [6].

4.3.1 Displaying data information in time and frequency

The `plot` attribute is a versatile method for displaying continuous data. We can call this attribute to check data quality in an interactive window without any additional input parameters [6].

```
sample_MNE_data = 'D:\sampledata\MNE sample data/sample_audvis_raw.fif'
data_raw = mne.io.read_raw_fif(sample_MNE_data, verbose=False)
data_raw.plot()
```

This attribute opens a plot window with valuable biosignal display tools (figure 4.5). Each row of the figure displays signals from each channel. You can increase or decrease number of the channels using the '+ channels' and '– channels' tool keys at the top of the plot window. Shortening or lengthening the time window, which displays the time-lapse, can be done by pressing the '+ channels' and '– channels' tool keys. We can also manipulate the amplitude of signals for better visualization by pressing the magnifier icons at the top of the plot window. The blue frame at the bottom of the window with a small green outlined frame shows us

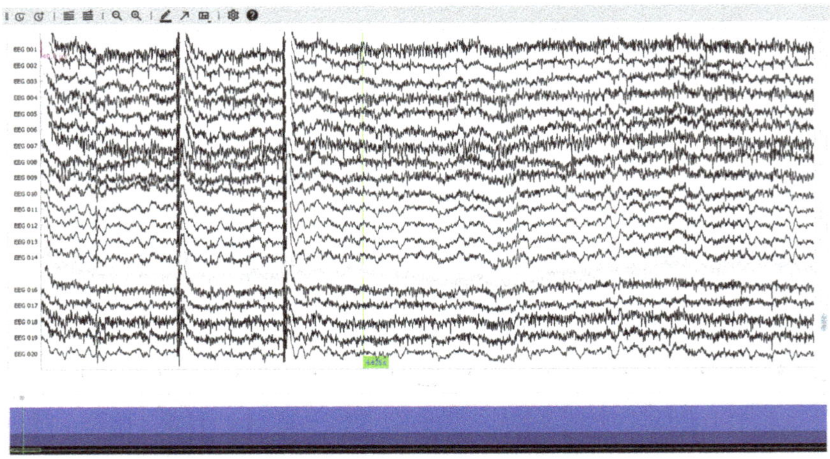

Figure 4.5. Displaying signals in the time domain in the plot window.

which part of the data is being displayed in the window. We can move it along times and channels easily by arrow keys too [6].

If you click anywhere in the plot window, the window will show the exact time of the point we clicked at the bottom of the window. For example, this time is shown in the green frame in figure 4.5. To mark channels as bad, we can click on their signal. Their color turns gray by clicking on them, as shown in figure 4.5, which marks these channels as bad channels. MNE puts these channels into the 'bads' field and updates the raw object information after closing the plot window. For example, the channel 'EEG 015' turned gray because we clicked on it. This channel name is added to the 'bads' field list if we type the following code after closing the plot window [6].

```
data_raw.info['bads']
```

We can also see the signals of the identical channel types in butterfly mode. This work could be done by tapping the 'b' key on the keyboard, which toggles the display mode between the all-channel viewing mode and the butterfly mode. In butterfly mode, signals of all channels of the same type are plotted on top of each other. Figure 4.6 shows data from a raw object in butterfly mode. The first signal (in light blue) is the butterfly plot of the signals recorded from the gradiometer sensors of the MEG system. The second signal (in dark blue) is the butterfly plot of the signals recorded from the magnetometer sensors of the MEG system. The third, fourth, and fifth signals are plots of EEG sensors, EOG sensors, and trigger channels in butterfly mode, respectively [6].

To reveal the frequency content of the data, we should compute the spectral density using the 'compute_psd' attribute and plot its result. The following code saves the frequency content of the raw data in the spectrum_raw object and plots this frequency content in the second line. The parameter pick gets channels we want to

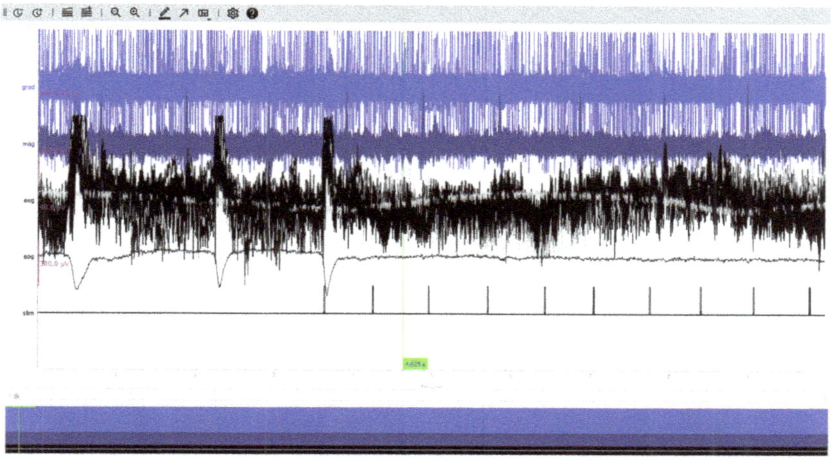

Figure 4.6. Butterfly mode of displaying signals in the time domain in the plot window.

display their spectrum in a Python list variable. Here we assigned 'data' to this parameter to plot the spectrum of all channels [6].

We can also exclude displaying the spectrum of the channels marked as bad channels by the exclude parameter. To plot the average spectrum of the channels, we can equate the average parameter to the True Boolean value (average = True) [6].

```
spectrum_raw = data_raw.compute_psd()
spectrum_raw.plot( picks="data", exclude="bads")
```

As the imported data for plotting contains different types, each channel types' spectrums are displayed in three different windows (figure 4.7). Each spectrum color in this figure shows the position of its channel in the right-side head model. The dashed lines indicate the upper and lower boundaries of the applied filter [6].

4.3.2 Topomap displaying

We can plot spectral power across sensors on the scalp topography, known as a topo map. The plot_topomap attribute helps us to reach this goal. Using this attribute without applying any values in the bands parameter, draws the topo map in the five most common frequency bands (figure 4.8). The unity of power in this figure is in decibels, like figure 4.7 [6].

```
spectrum_raw.plot_topomap(ch_type="eeg")
```

Drawing the power spectrum of all sensors in their position on the head is possible using the plot_topo attribute, but by default, this attribute only shows MEG

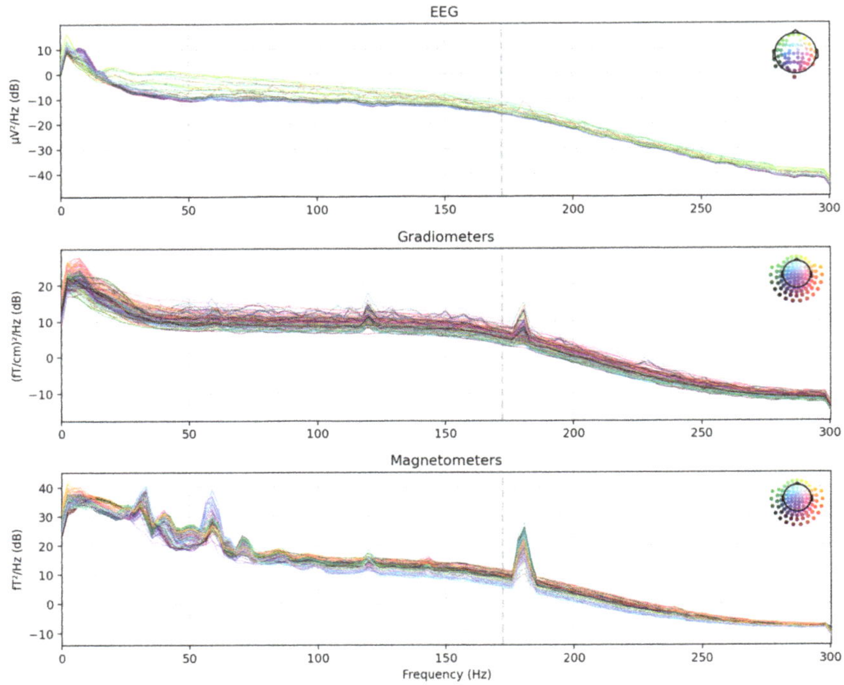

Figure 4.7. The power spectrum of all channels in a raw dataset separated by their type in three different figures.

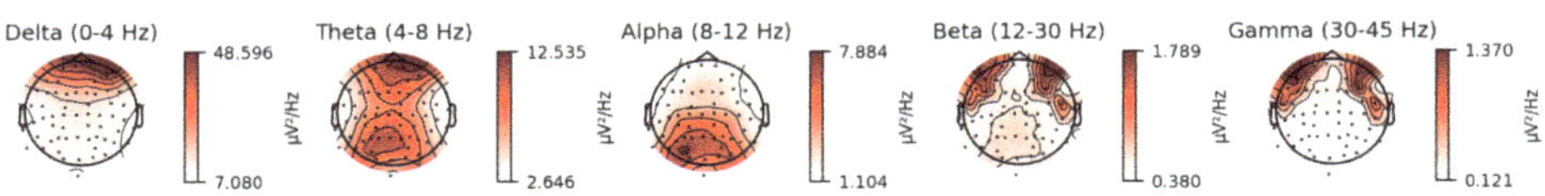

Figure 4.8. Topomap spectral power in five different frequency bands.

channels. To show this function for EEG channels, we should restrict it by the following code. In this code, we chose the color of the head model and the spectrum background to be white, and the color of the spectrum as black. The power spectrum of EEG channels in their position can be seen in figure 4.9 [6].

```
spectrum_raw.pick("eeg").plot_topo(color="k", fig_facecolor="w",
axis_facecolor="w")
```

4.3.3 Saving MNE-produced plots and images

Sometimes we will need to save the image of plots resulting from the MNE processing. Saving images is essential, especially in preparing images for our papers. We can use the matplotlib `savefig` function to save images that MNE produces

Figure 4.9. Power spectrum of EEG signals in their position on the head model.

with our desired name and resolution. To do so, we are required to equate the code lines with any `plot` attribute to a variable. This variable has a figure type. Consequently, we can save this variable as an image in our operating system. For example, in the following code we saved the output of the `plot_topomap` attribute to a variable named `plot_topomap`. In the second line, we save the figure, which is the result of the `plot_topomap` attribute by the name and directory assigned in the first input of the savefig function. The resolution of saving the image is specified in the dpi parameter, and its format is specified in the format parameter.

```
plot_topomap = spectrum_raw.plot_topomap(ch_type="eeg")
plot_topomap.savefig('plot_topomap', dpi=200, format='tif')
```

4.4 EEG signals preprocessing

Many artifacts contaminate the EEG signals with their undesired effect. Before extracting any information from EEG signals, we need to clean these signals from any noises. Some artifacts have environmental backgrounds, such as AC power line frequency, unwelcome auditory noise in the building, and electromagnetic noise from nearby elevators and cell phones. Some of these artifacts are instrumentation artifacts like EEG sensor malfunctions and movements of wires. Some of these signals are in biological categories, such as eye movements and blinks, ECG rhythmic signals, electromyogram (EMG) artifacts resulting from body movements, etc. There are different methods for cleaning data out of these artifacts. In the following section, we described some of the most important methods for dealing with EEG noises, known as preprocessing methods.

4.4.1 Setting an EEG reference

EEG signals are the voltage between the EEG electrodes and a reference electrode. The signal recorded by the reference electrode is subtracted from all other signals recorded by other EEG sensors. The reference electrode's location is mostly somewhere in the subject's body near the head. Therefore, the reference electrode signal has environmental impacts that involve other EEG sensor signals but is not electric potentials caused by the brain. The reference electrode position could be on the earlobes, nose, mastoid, and collarbone [7].

Sometimes in the preprocessing steps, we may need to change the EEG reference to another channel signal or the average of all channel signals. This work is called re-referencing [7].

Here we work with the restricted raw data that has only EEG and EOG channels. We can use the `set_eeg_reference` attribute to change the reference by an existing channel and recompute the data with its signal as a new reference. This attribute channel name of the new reference is the `ref_channels` attribute. In the following code, we re-referenced the data with the existing channel named 'EEG 060,' for example [7].

```
data_raw_EEG_EOG = data_raw.copy()
data_raw_EEG_EOG.pick_types(meg=False, eeg=True, eog=True)
data_raw_EEG_EOG_reref = data_raw_EEG_EOG.copy()
data_raw_EEG_EOG_reref.set_eeg_reference(ref_channels=['EEG 060'])
```

We can re-reference data to the average of two channels, for example, referencing data to the average of two earlobes. To do so, we can use the same function as follows [7].

```
data_raw_EEG_EOG_rerefby2 = data_raw_EEG_EOG.copy()
data_raw_EEG_EOG_rerefby2.set_eeg_reference(ref_channels=['EEG 059','EEG 060'])
```

One of the essential steps of preprocessing is re-referencing the data to a virtual reference named average reference. All EEG signals subtract from the average signal of all channels in average re-referencing. This action is mainly done at the last stages of EEG preprocessing [7].

```
data_raw_EEG_EOG_averef = data_raw_EEG_EOG.copy()
data_raw_EEG_EOG_averef.set_eeg_reference(ref_channels=['average'])
```

We can use the `set_bipolar_reference` function for bipolar re-referencing, which operates in the contralateral approaches. Imagine channels named 'EEG 002' and 'EEG 004' are placed on two sides of the head symmetrically, and we want to reference them in a bipolar way. The following code shows how to do this task [7].

```
data_raw_2ch_bipolarref = data_raw_EEG_EOG.copy()

data_raw_2ch_bipolarref.pick_channels(["EEG 002", "EEG 004"])

mne.set_bipolar_reference(data_raw_2ch_bipolarref, anode=['EEG 002'],
cathode=['EEG 004'])
```

Sometimes the signal of the reference channel on the scalp does not save alongside the EEG raw data. We may need to re-reference all data plus that previous reference data to a new scalp electrode. In this way, we use the `add_reference_channels` signal. We assign the name of our previous reference in the ref_channels parameter. The `add_reference_channels` function adds the reference channel we assigned consisting of all zeros (for now) to the raw data. We again use the `set_eeg_reference` attribute to re-reference all channel data plus previous references [7].

For example, imagine the sample EEG raw data are referenced to the 'EEG 061' channel during the recording signal. The signal of this channel does not exist in the dataset. In the following code, we added the 'EEG 061' channel as a previous reference channel to the dataset by the `add_reference_channels` function. Afterward, we used the `set_eeg_reference` attribute to re-reference all channels to channel name 'EEG 060' [7].

```
data_raw_EEG_EOG_addRefChannel =
mne.add_reference_channels(data_raw_EEG_EOG, ref_channels=["EEG 061"])
data_raw_EEG_EOG_addRefChannel_reref =
data_raw_EEG_EOG_addRefChannel.set_eeg_reference(ref_channels=['EEG 060'])
```

4.4.2 Removing bad channels and data spans

The following section focuses on visually cleaning the data. First we show how to detect and clean bad channels, and then we show how to detect and reject bad time spans [8].

4.4.2.1 Bad channel detection

Bad channels in EEG data are the channels that malfunction and provide noisy data. We should look alongside EEG channels to see if some do not have a valid signal. We can put channels in a list in the bads field if we distinguish any channel as noisy by plotting it in the plot window, as shown in section 4.3.1. For example, we want to check which one of the EEG channels, from channel name 'EEG 050' to channel name 'EEG 059' is noisy (figure 4.10). We can use the `pick_channels_regexp` function to reduce the number of EEG channels in the plot window. In the following command, we are working with the `data_raw_EOG` data, which has only EEG signals. The '.' in `regexp` parameter is a wildcard character of extract data in our desired boundary [8].

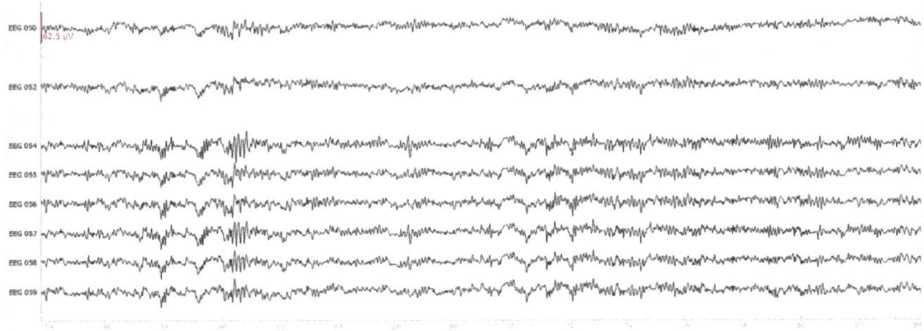

Figure 4.10. Displaying the signals of the EEG channels named 'EEG 050' to 'EEG 059'.

```
data_raw_EEG = data_raw.pick_types(meg=False, eeg=True, exclude=[])

data_raw_EEG_picks = mne.pick_channels_regexp(data_raw_EEG.ch_names,
regexp="EEG 05.")

data_raw_EEG.plot(order=data_raw_EEG_picks, n_channels=len(data_raw_EEG
_picks))
```

We can see the behavior of channels 'EEG 051' and 'EEG 053' does not have the pattern of brain source signals. These channels are added to the 'bads' list after clicking the plot window as we click the signals of these channels to mark them. The following command is used to see, add, and remove channels from the 'bads' list. Adding a single channel and lists of channels is possible by the `append` and `extend` attributes, respectively. Removing the last n channels is possible by `pop(-n)` [8].

```
print(data_raw_EEG.info["bads"])
data_raw_EEG.info["bads"].append("EEG 050")
data_raw_EEG.info["bads"].extend(["EEG 051", "EEG 052”])
data_raw_EEG.info["bads"].pop(-n)
```

The can also change the whole 'bads' list with other channel names by the following command [8].

```
data_raw_EEG.info["bads"] = new_bads_list
```

4.4.2.2 Dealing with bad channels

We can easily remove the noisy channels after detecting them by the `drop_channels` attribute, as noted in section 4.2.3. But in most studies, channel removal spoils sensor numbers and positions similarities between the sessions' and subjects' data. Therefore, the signals of bad channels could be recovered by signals of other non-noisy channels without removing the noisy channels. In the interpolation approach,

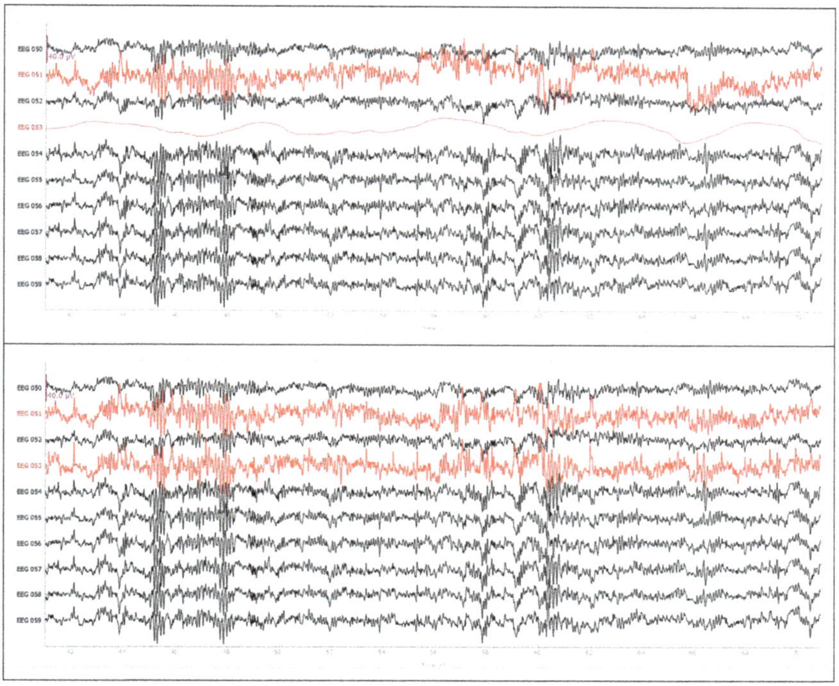

Figure 4.11. Displaying the effect of channel interpolation (upper image: before interpolation, lower image: after interpolation).

we can interpolate the signal of the noisy channels according to the signals of other valid sensors. For interpolation, the MNE package uses the spherical spline method. The `interpolate_bads` attribute helps us to interpolate noisy channels. After interpolating those channels, this attribute can format the 'bads' list by assigning '`True`' to the `reset_bads` parameter or not format it by assigning 'False' to this parameter [8].

In the following command we did not format the '`bads`' list for better displaying variation of noisy channels before and after interpolation [8].

```
data_raw_EEG_interp =
data_raw_EEG.copy().interpolate_bads(reset_bads=False)

data_raw_EEG.plot(n_channels=10, duration=30, start=41.0, title="Before
interpolation", bad_color='r')

data_raw_EEG_interp.plot(n_channels=10, duration=30, start=41.0,
title="After interpolation", bad_color='r')
```

Figure 4.11 contains 10 raw signals recorded from 10 EEG channels; its two signals are marked as 'bad' channels. The top window shows the behavior of these two signals before interpolation and the bottom window shows their behavior after interpolation [8].

4.4.2.3 Annotating data spans

There are several models for data annotation. The first model is manual data annotation in the plot window. We can detect the noisy data spans after visualizing the data in the plot window and moving alongside the time in this window. The annotating frame opens in the plot window by pressing 'a' on the keyboard. As shown in figure 4.12, the annotating frame has some options for adding annotation with our desired name and starting and ending time of the annotated time span. For example, in this figure we define a new annotation named 'low-frequency noise,' as it is evident in the highlighted area that some of the channel signals are contaminated with the low-frequency noise [9].

After the plot window is closed, our defined annotation is saved in the data object. We can check the annotation information of the data by using the `annotations` attribute and typing the command below [9].

```
print(data_raw_EEG.annotations)
```

Another way of manually annotating the data is by using the code. This method is particularly applicable when we have information about the time of event occurrence in the signals. The `annotations` function can help us with this purpose [9].

For example, assume two rest parts exist between the 10th to 20th s and between the 50th to 60th s of the data recording. We have the following commands to annotate this data part as 'rest'. We got the starting time of EEG signals from the object that contains only EEG signals at the first line. We defined starting time, duration, and description of two resting parts of the data at the first to third lines of the command below, respectively. After saving the annotation information in the annotation object at the fourth line, we add this annotation to the EEG data `set_annotations` attribute [9].

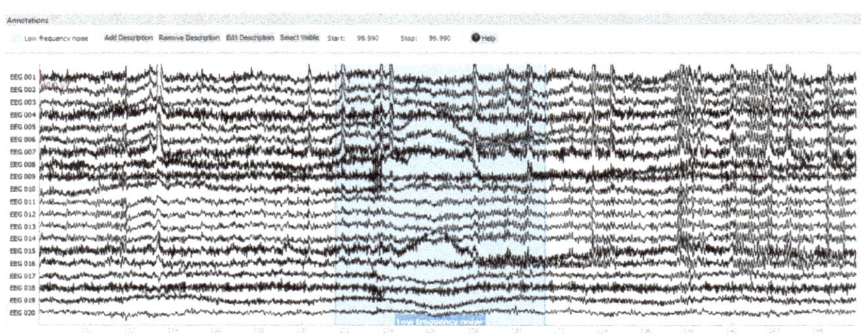

Figure 4.12. Manual annotation of the EEG signals in the plot window. In this figure, we annotated a time span as 'low-frequency noise'.

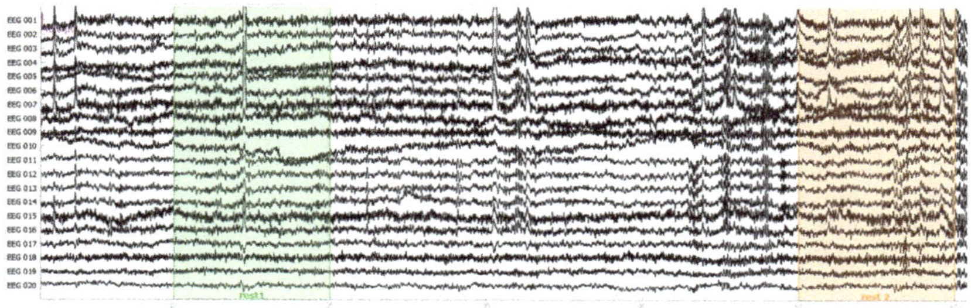

Figure 4.13. Annotating two rest parts of the data by manual annotation in codes.

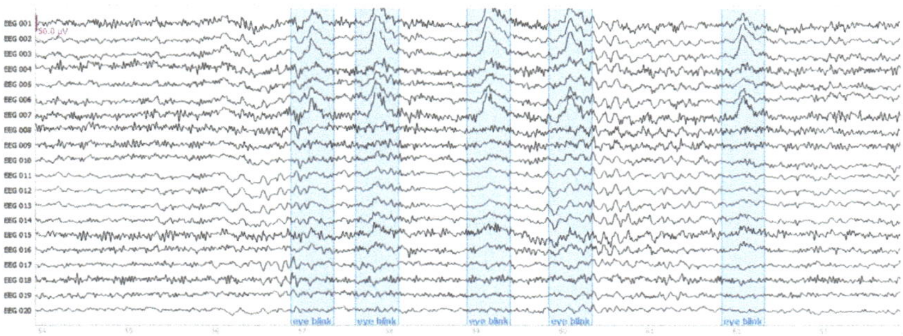

Figure 4.14. Automatic annotation of EEG data parts that are contaminated with EOG artifacts.

```
onsets = [data_raw_EEG.first_time + 10, data_raw_EEG.first_time + 50]
durations = [10, 10]
descriptions = ["rest1", "rest 2"]
block_annots = mne.Annotations(
    onset=onsets,
    duration=durations,
    description=descriptions,
    orig_time=data_raw_EEG.info["meas_date"])
data_raw_EEG.set_annotations(data_raw_EEG.annotations + block_annots)
data_raw_EEG.plot()
```

The result of manual annotation after running the above command is shown in figure 4.13.

There are some automatic ways to annotate the noisy data too. Two functions, `find_eog_events` and `find_ecg_events` are automatic functions to detect parts of the data contaminated with EOG and ECG artifacts. The pattern of employing these automatic ways for detecting EOG artifacts is shown in the following command and figure 4.14. We used total raw data for the automatic detection of eye blinking, but we have annotated the raw data that only contains EEG data. We also set the eye blink duration to 0.5 s, which varies among subjects. We set the names of the annotated time spans to 'eye blink' in the fourth line of the command below [9].

```
eog_events = mne.preprocessing.find_eog_events(data_raw)
onsets = eog_events[:, 0] / data_raw_EEG.info["sfreq"] - 0.25
durations = [0.5] * len(eog_events)
descriptions = ["eye blink"] * len(eog_events)
blink_annot = mne.Annotations(
    onsets, durations, descriptions,
orig_time=data_raw_EEG.info["meas_date"])
data_raw_EEG.set_annotations(blink_annot)
data_raw_EEG.plot()
```

4.4.2.4 Rejecting bad data spans

There is a parameter named `reject_by_annotation`, and it could be used in many preprocessing functions such as `epochs`, `ICA`, find_`eog_events`, find_`ecg_events`, `compute_raw_covariance`, and `compute_psd`. If we use the epoch function and `reject_by_annotation = True`, the function will epoch the dataset and remove any epochs that overlap with the annotation; their description begins with 'bad' [9].

The second method for removing the annotated time spans uses the following designed function. This function crops the data spans not annotated as '`eye blink`' and appends these data spans altogether [10].

```
def reject_bad_segs(raw):
    """ This function rejects all time spans annotated as 'eye blink' and
concatenates the rest"""
    raw_new = []
    for n in range(len(data_raw.annotations)):
        if raw.annotations.description[n] != 'eye blink'
           raw_new.append(raw.copy().crop(tmin=raw.annotations.onset[n],
           tmax=raw.annotations.onset[n+1],include_tmax=False))
    return mne.concatenate_raws(raw_new)
```

4.4.3 Filtering the data and resampling

Filtering is usually the first step of preprocessing the EEG signals. These signals usually exist in a limited frequency band. Therefore we need to suppress the frequency components above and below the cut-off values. The `filter` attribute helps for filtering the data. This attribute gets the lower passband edge at the parameter `l_freq`, upper passband edge at the parameter `h_freq`, channels you want to apply this filter at the `picks` parameter, and method of filtering that could be 'fir' or '`iir`' at the `method` parameter [10].

EEG data have some drifts from baseline in the EEG signals after recording (figure 4.15). We can use highpass filtering of the signals to correct the drifts and clean the signals from this artifact. In the following code, we used a fir filter by '1 Hz' upper passband edge and applied this filter to the total raw data. By default, the method of filtering is set to fir if we do not point to the `method` parameter [11].

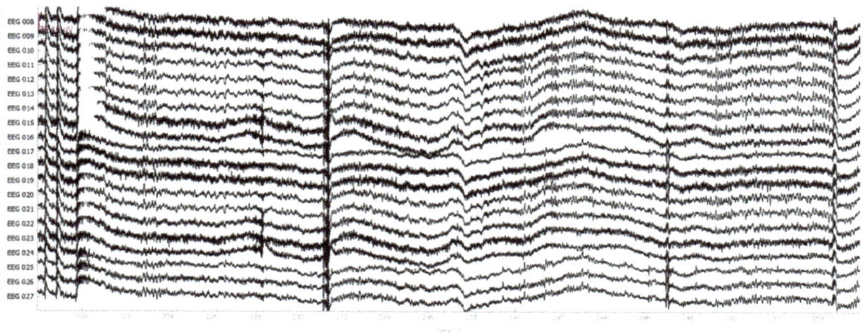

Figure 4.15. A time window of EEG signals that have drifted from the baseline.

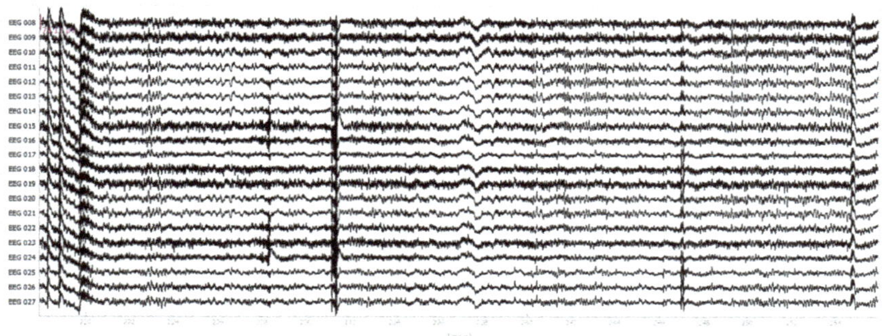

Figure 4.16. Removing drifts in the same time window of EEG signals using 1 Hz highpass filtering.

```
data_raw_filtered = data_raw.copy()
data_raw_filtered.filter(l_freq=1, h_freq=None)
data_raw_filtered.plot()
```

As you can see, there is a drift from baseline in EEG channels 008 to 027 in figure 4.15, but after applying a highpass fir filter, this drift is wiped from the signals, as you can see in figure 4.16 [11].

We can employ `plot_filter` and `create_filter` functions in our code if we need to visualize the filter and check its information. As the `create_filter` function needs data in the NumPy array in its first input, we use the `get_data` attribute to extract data from the raw object, as explained in section 4.2.4. By the following command, we saved the filter information in the `filter_params` variable at the first line and visualized it at the second line in `flim` frequency rage (figure 4.17) [11].

```
filter_params = mne.filter.create_filter(data_raw.get_data(),
data_raw.info["sfreq"], l_freq=1, h_freq=None)

mne.viz.plot_filter(filter_params, raw.info["sfreq"], flim=(0.01, 5))
```

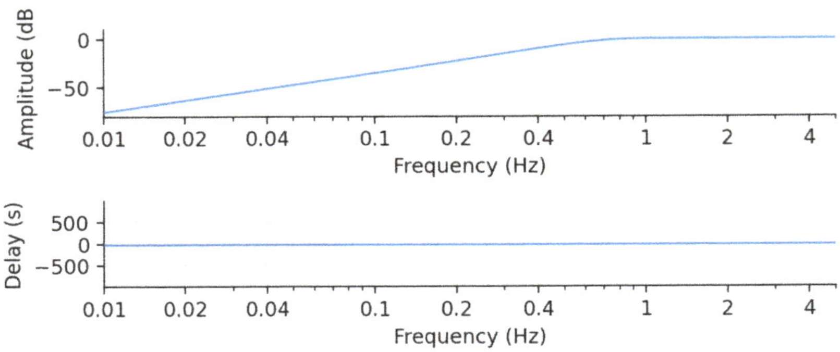

Figure 4.17. Visualizing the designed filter for highpass filtering.

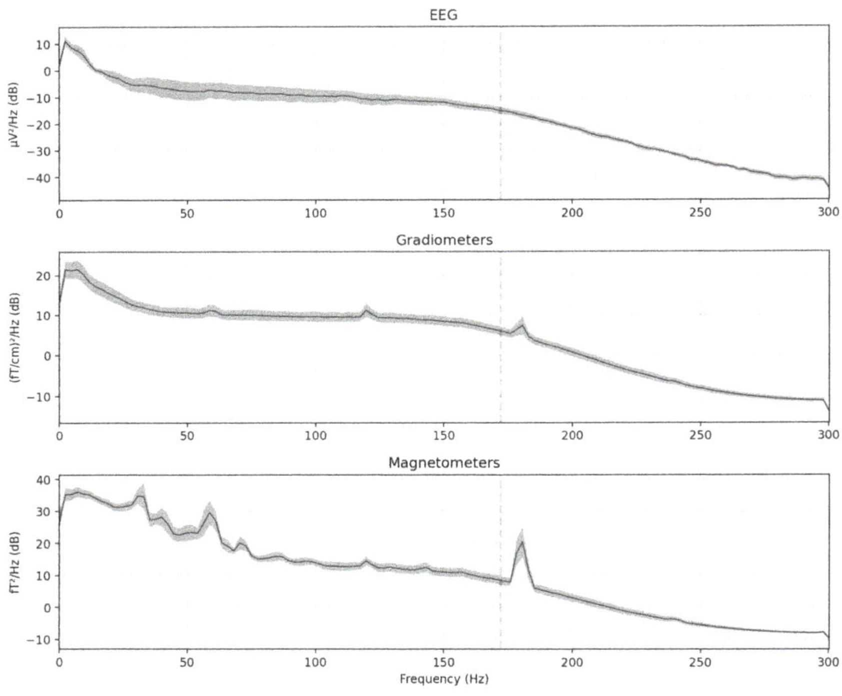

Figure 4.18. PSD of imported raw data, which has picked at line noise frequencies.

If we need to remove frequency components below a specific value, we must assign the number to the `h_freq` parameter. In this way, our filter act as a highpass or band pass [11].

One of the essential noises in EEG signals is 'line noise,' which manifests strong power at AC power line frequency and its harmonics. For instance, our imported raw data is contaminated with 60 Hz line noise frequency. Hence we can see peaks at 60 Hz and their harmonics in the power spectrum density (PSD) of the signals (figure 4.18) [11].

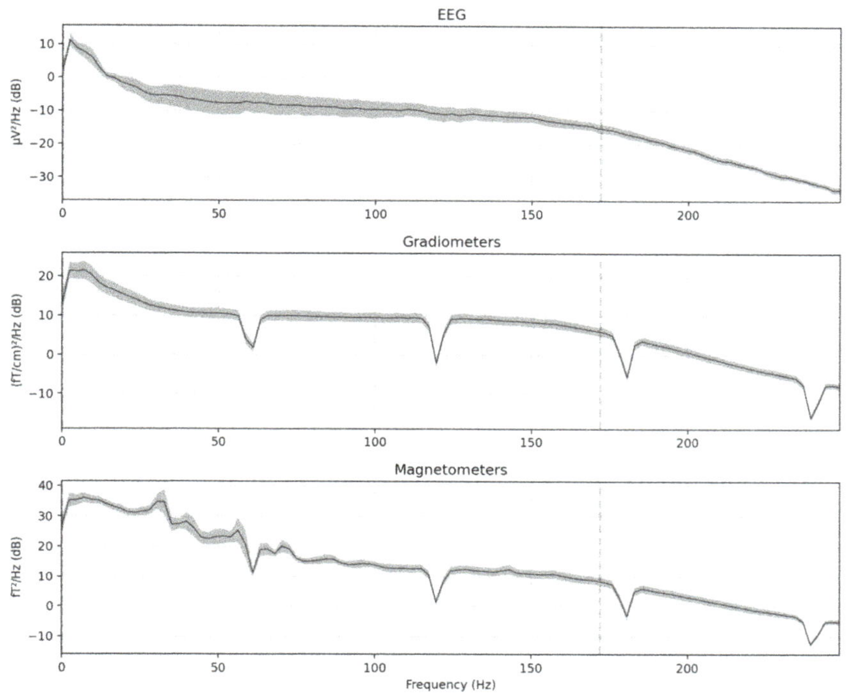

Figure 4.19. PSD of imported raw data after removing line noise.

Notch filtering is used to clean signals from the line noise. In Python, we can use the `notch_filter` attribute, but as evident in figure 4.18, this noise effects the EEG-type signals less than MEG-type signals. Accordingly, using the `picks` parameter, we only apply a notch filter on MEG channels. In the following command, we increased the width of the notch filtering to remove peaks at line noise frequencies. The peaks are wiped in PSD after applying the notch filtering (figure 4.19) [12].

```
freqs = (60, 120, 180, 240)
data_raw_notch_filtered = data_raw.copy()
data_raw_notch_filtered.notch_filter(freqs=freqs, notch_widths=4,
picks=data_raw.ch_names[0:314])
fig = data_raw_notch_filtered.compute_psd(fmax=250).plot(average=True,
picks="data", exclude="bads")
```

We can increase or decrease the amount of data by resampling. To do so, we can use the `resample` attribute. This attribute gets a new sampling rate at the `sfreq` parameter. It should be noted that to reduce the sampling rate (downsampling) we need to import the sfreq rate lower that current sampling rate of the signals. The highest frequency component that exists in the signals is the sampling rate value divided by two from the Nyquist rate. At the following command, we reduce the sampling rate to 200 Hz.

```
data_raw_downsampled = data_raw.copy()
data_raw_downsampled.resample(sfreq=200)
```

4.4.4 Artifact removal by ICA

ICA has been introduced as a suitable method for EEG signal preprocessing because it allows extracting the artifacts from the brain-recorded signals. After applying ICA, the mixed EEG signals are mapped to the independent components. Some of these components reflect the brain source data and some reflect non-brain source data, which should be removed from the data [13].

Many algorithms have been invented for ICA calculation. An MNE Python package can implement ICA by three different algorithms: 'fastica' and 'infomax,' which both are comprehensive algorithms, and the faster and more robust 'pickard' algorithm. MNE uses 'fastica' as a default algorithm. An ICA object must be created at the first step of ICA implementation in MNE. Some general parameters are specified in this step. Afterward, the ICA object is fitted to data for independent component extraction by `fit` attribute. We then should inspect the ICA components and reject the noisy components for cleaning EEG signals from noises such as eye blinking and movements, EMG, heartbeats, channel noises, etc. The remaining components can reconstruct the noises-free EEG signals by using the `apply` attribute [14].

It is argued that applying principal component analysis (PCA) before ICA can improve the ICA results by discarding small trailing eigenvalues and minimizing pair-wise dependencies of the input data [15]. The MNE package scales the input data to unit variance and whitens them to deal with different channel types having different units (MEGs and EEGs). Afterward, MNE decomposes the pre-whitened data using the PCA method before ICA implementation. `noise_cov` parameter is the default value in the ICA function. If `noise_cov = None`, then all channel data are scaled by the standard deviation. If this parameter is equal to `Covariance`, which is noise covariance calculated from the covariance of the input data, all channels are pre-whitened using covariance value. After PCA implementation, the first n components can pass through the ICA algorithm using the value we assign to the `n_components` parameter in the `ICA` function. The value of `n_components` can be a float between 0 and 1, which indicates the fraction of the components number we use in the ICA algorithm [14].

Slow drifts can negatively influence the results of ICA method. Thefore, it is better to filter the input data using a highpass 1 Hz cutoff frequency filtering.

```
data_raw_filt = data_raw.copy().filter(l_freq=1.0, h_freq=None)
```

The `ICA` object is constructed to the filtered input data using the first `n_compo-nents` of the PCA method. As shown in the following code, a random seed value is

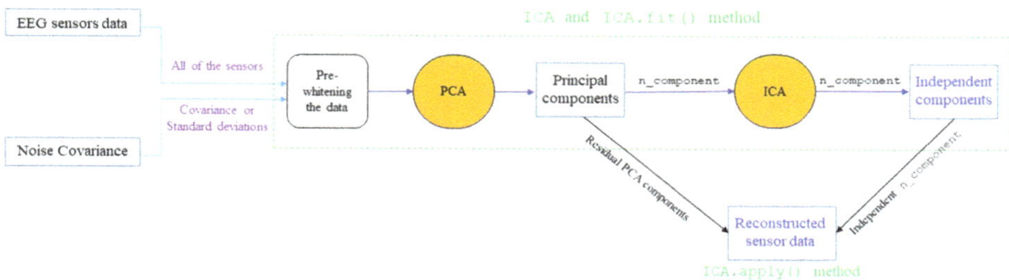

Figure 4.20. ICA decomposition and sensors signal reconstruction in the MNE diagram.

allocated to the `random_state` parameter to get the identical result in ICA implementation with similar input every time. The `'auto'` value, given to the `max_iter` parameter, sets the maximum number of iterations for fitting the ICA object to input data (the second line in the following code). If ICA is implemented using the 'fastica' algorithm, this number is 1000, and if it is implemented using 'infomax' or 'picard' algorithms, it is 500. Changing the algorithm of ICA calculation is possible using the method parameter in the ICA function. Here we just reserved EEG signals and imported them to the ICA method to calculate ICA for EEG signals [14].

```
data_raw_filt_EEG = data_raw_filt.copy()
data_raw_filt_EEG.pick_types(meg=False, eeg=True, eog=False)
ica=mne.preprocessing.ICA(n_components=59, max_iter="auto",random_state=97)
ica.fit(data_raw_filt_EEG)
```

After visualizing the independent components (ICs), we exclude the ICs we recognize as noisy. The name of banned ICs is listed in `ICA.exclude`. The reconstruction of the sensor signal can be executed using all of the ICs (Except ICs exist in `ICA.exclude`) and all of the principle components (PCs) were not contributed in ICA decomposition (the PCA residual components). Figure 4.20 shows the block diagram of the MNE package's ICA decomposition and reconstruction of sensor signals [14].

The `plot_sources` attribute shows the time series of the ICs. Clicking on the ICs signals can omit these ICs and add them to the `ICA.exclude` list. Figure 4.21 is the IC time series. By right-clicking on the name of these ICs, which are in the left column, the details of those ICs pop up in the ICA properties window. We calculated the ICA for all 59 PCs, as shown in the above code, but we displayed ten components in figure 4.21. It is obvious that ICA000 and ICA003 are EOG and ECG artifacts, respectively, because there are eye blinks and ECG rhythm in their time-series. We clicked on their signal to put them in the `ICA.exclude` list [14].

To show more information about these two ICs, we open their properties window. These windows contain a topomap plot, raster plot, power spectrum,

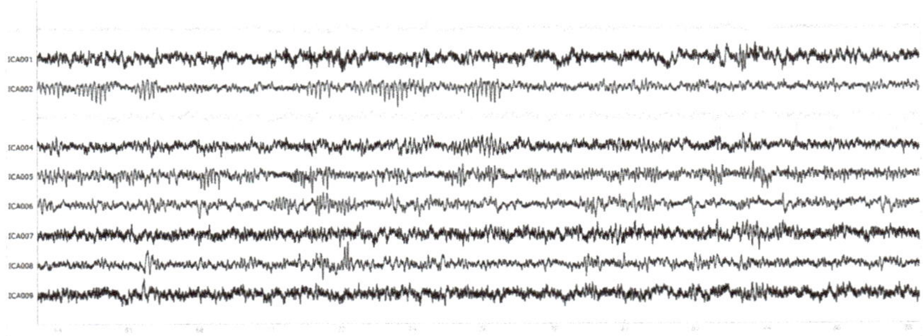

Figure 4.21. ICs in the time window.

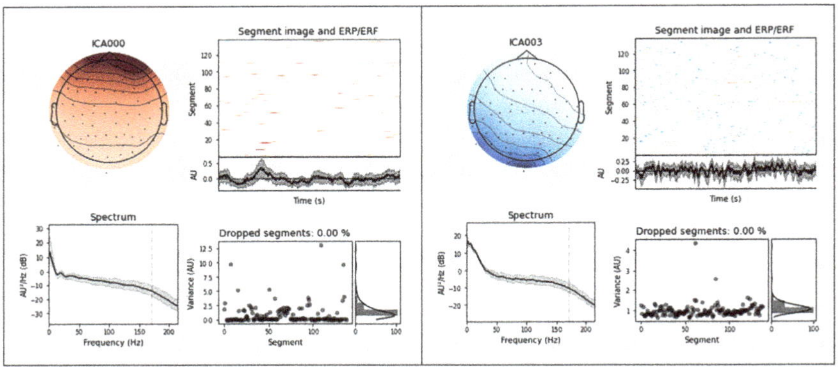

Figure 4.22. More information on the first two ICs in the properties window.

and variance of each EEG segment. We can annotate each one of the ICs in the brain or artifact group by inspecting their behavior in time and within each option of properties window [16]. As we can see in figure 4.22, there is a pattern in the forehead in the topomap of ICA000 which confirms that this IC is the EOG artifact. We can also see from topomap of the component ICA003 that this component is more powerful in near neck channels, which we expect from ECG artifact [14].

As we marked the first two ICs as artifacts, they have joined the bad ICs group. We can see their IDs by typing `print(ICA.exclude)`. We can pass the following code to show all calculated ICs topo map in one figure. The topo maps of the first 20 ICs from all 59 ICs are indicated in figure 4.23. The label of these two ICs are gray as we marked them as bad ICs [14].

```
ica.plot_components()
```

To inspect the properties window of each IC we want, we can pass the following code by putting the ID of ICs in the `picks` attribute [14].

```
ica.plot_properties(data_raw_filt_EEG, picks=[0,1,2,3])
```

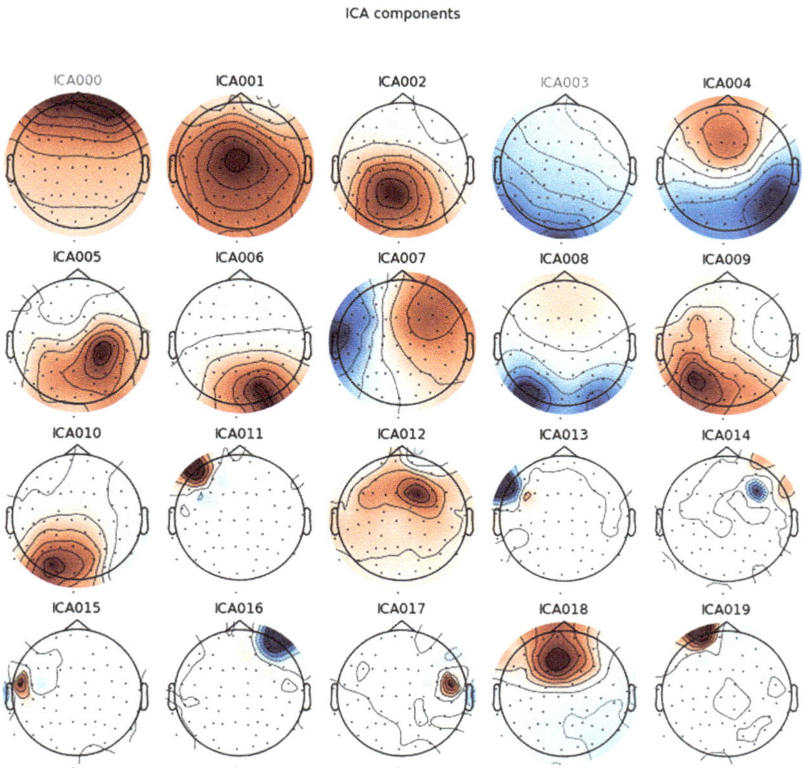

Figure 4.23. Topo map of 20 calculated ICs in one window. Label of noisy ICs are marked gray as we put them in the bad ICs group.

We can plot signals before and after ICA removal to ensure that the ICs we want to remove effectively impact cleaning the signals. The `plot_overlay` attribute shows these signals and helps us in this manner. The `exclude` parameter gets the ICs we want to see their removal effects. We also restricted the type of signals in the plot figure by the `picks` parameter. This attribute only plots the first 3 s of the signals by default, but to show this signal in an arbitrary duration, we can specify the start and ending time in this attribute's `start` and `stop` parameters [14].

Figure 4.24 shows the effect of EOG removal on EEG signals by passing the following code. This figure indicates the effect of EOG removal on the butterfly plot [14].

```
ica.plot_overlay(data_raw_filt_EEG, exclude=[0], picks="eeg")
```

Figure 4.25 indicates the effect of EOG removal on the average of EEG signals [14].

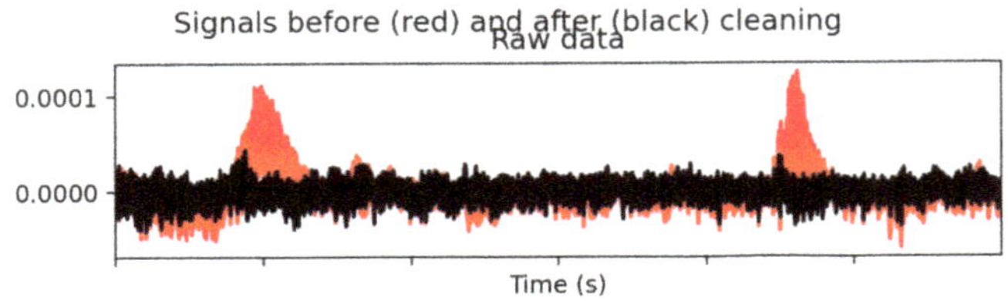

Figure 4.24. Effect of EOG artifact removal on butterfly plot of EEG signals.

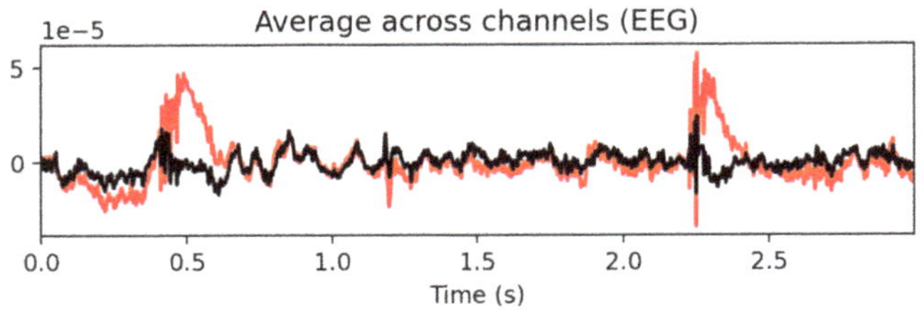

Figure 4.25. Effect of EOG artifact removal on the average of EEG signals.

We need to rebuild the sensor signals after exclusion has been set. The apply attribute helps us in this manner [14].

```
reconst_raw = data_raw_filt.copy()
ica.apply(reconst_raw)
```

4.5 EEG signal processing

We argued the essential methods of brain signal cleaning in section 4.4. Now it is the time to take another step forward to extract information from the signals. In this section, we extract features from the noise-free signals. There are countless feature extraction methods, as there are numerous EEG features. Here we will explain some of the remarkable feature extraction methods.

4.5.1 Evoke responses analysis

Data contain many tags. First, we need to epoch the data so as to keep parts of the data which are important to us.

4.5.1.1 Create epoched data from a raw object and visualize them

After epoching the data, we reach the epoch data structure. Epoch objects are chunks of the EEG/MEG signal with equal duration. They represent data that is time-locked to repeated experimental events. Inside the epochs, data are sorted in the following order [17].

`(n_epoches, n_channels, n_times)`

For example, after importing the raw data and saving it in the `raw_data` object, we can store the event information recorded in the ‘`STI 014`’ channel into a new variable named events. This variable carries the event information, such as the event’s name and time of occurrence, which are essential for epoch extraction [17].

```
data_events = mne.find_events(data_raw, stim_channel="STI 014")
```

Now we can make epoched data from the raw data to data_raw_epoched object by helping epochs function. This function also gets tmin (start time of each epoch) and tmax (ending of each epoch) [17].

```
data_epochs = mne.Epochs(data_raw, data_events, tmin=-0.2, tmax=0.8)
```

The number of each epoch corresponding to each event will show in the console by printing the `epochs` variable [17].

```
print(data_epochs.event_id)
```

We can visualize the epoched signals by plot attribute `epochs.plot()`. This command just shows the epochs in an ordinary window (figure 4.26) [17, 18].

We can also add other parameters in the plot attribute to get more informative plots, which carry the events’ information in each epoch (figure 4.27) [18].

```
data_epochs.plot(events=data_events)
```

As mentioned earlier in section 4.2.6, we defined the description for each one of the events to the dictionary named `event_dict`. We can also epoch the data and

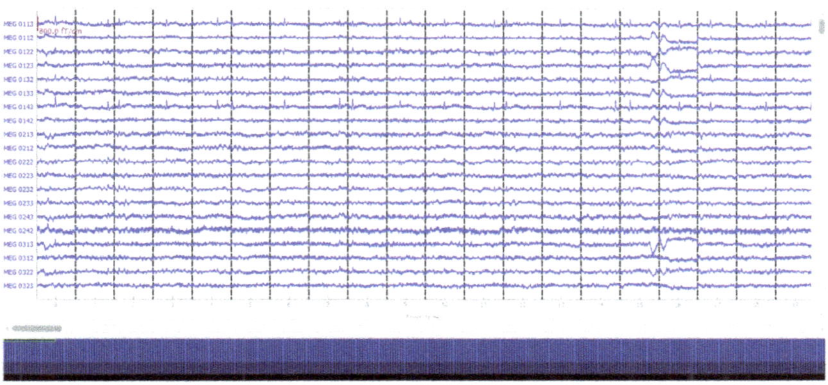

Figure 4.26. The epoched data visualization in time.

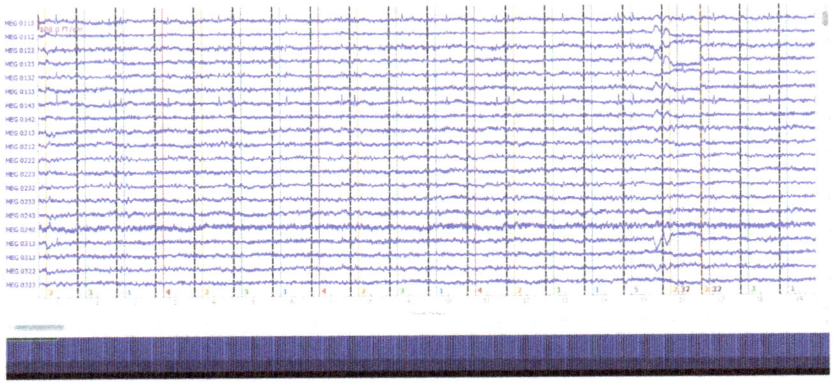

Figure 4.27. The epoched data visualization in time by event IDs.

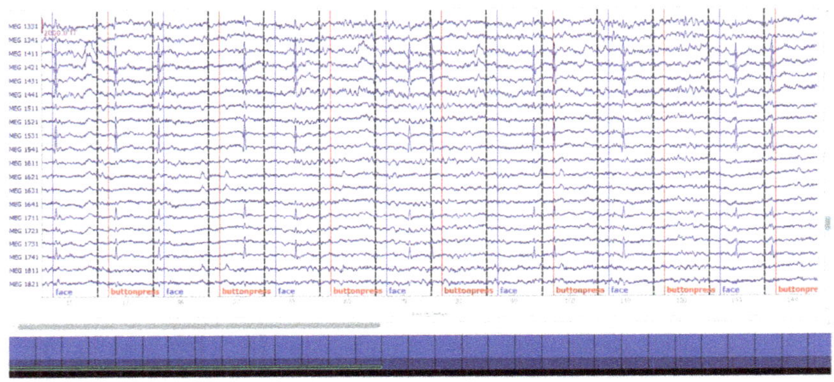

Figure 4.28. Visualization of the epoched data corresponding to 'face' and 'buttonpress' in time.

visualize them using these events' descriptions. The epoched data corresponding to 'face' and 'buttonpress' with our desired colors (red and blue) are shown in figure 4.28 with the help of the following code [18].

```
data_epochs = mne.Epochs(data_raw, data_events,  tmin=-0.2, tmax=0.8,
event_id=event_dict, preload=True)

face_and_buttonpresses = mne.pick_events(data_events, include=[5, 32])

data_epochs["face","buttonpress"].plot(

    events=face_and_buttonpresses,

    event_id=event_dict,

    event_color=dict(buttonpress="red", face="blue"))
```

We can also use compute_psd to plot the power spectrum of epochs. The power spectrum of 'auditory' epochs in EEG signals is shown in figure 4.29 [18].

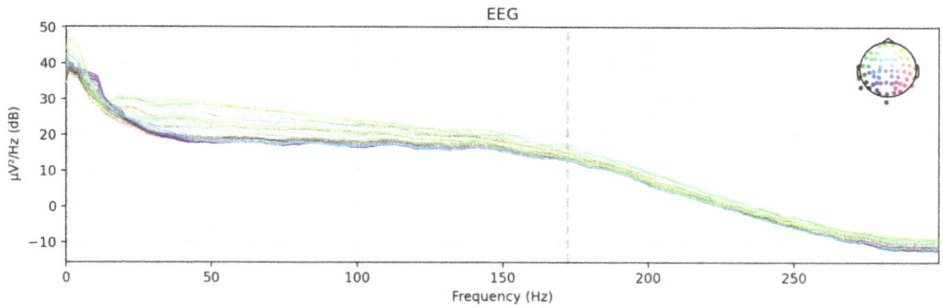

Figure 4.29. Power spectrum of auditory epochs in EEG signals.

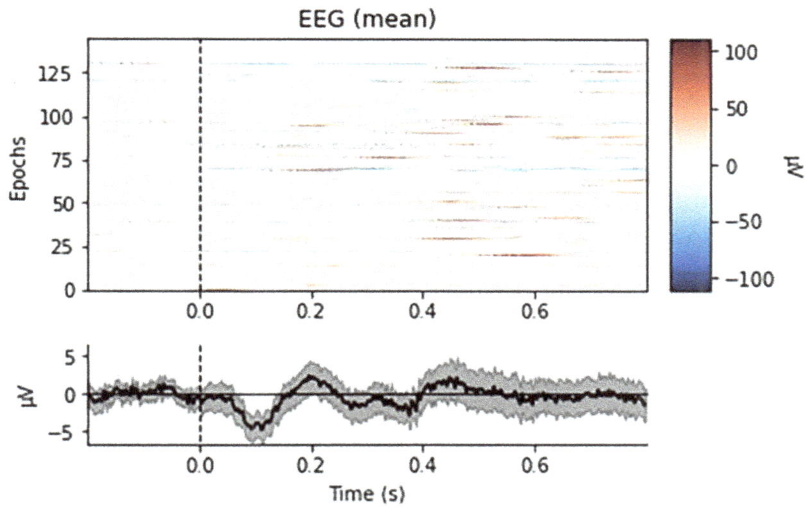

Figure 4.30. Image map of 'auditory' epochs (each row is the mean of EEG signals of that epoch).

```
data_epochs["auditory"].compute_psd().plot(picks="eeg", exclude="bads")
```

Image map is a convenient way to draw many epochs simultaneously. Each row in the image map represents a single epoch, and columns are time. Each value of epochs' signals is shown by a color in the image map. We visualize the signal of each channel on a separate color map. In addition, we can combine epoch signals across channels and visualize them in the image map. In figure 4.30, we drew this image for 'auditory' epochs by the mean value of EEG signals by using the following code [18].

```
data_epochs["auditory"].plot_image(picks="eeg", combine="mean")
```

4.5.1.2 Removing noisy epochs

By inspecting and scrolling signals in the time domain, we may detect that some of the trails are noisy and need to be removed from the dataset. We should click on these trails in the plot window to do so. The color of these trail signals turns red after

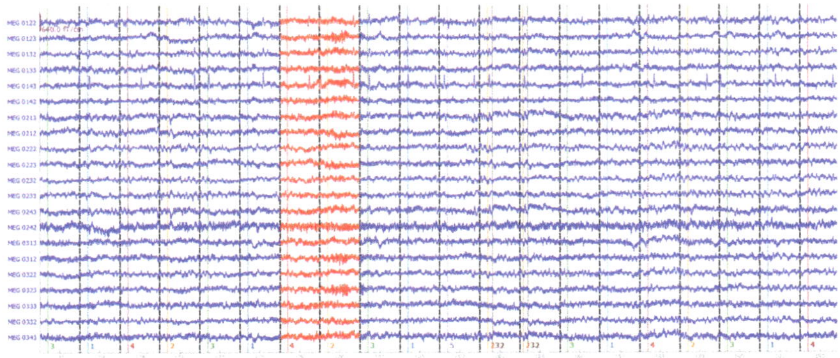

Figure 4.31. Removing noisy trails by visual inspection in a plot window.

clicking on them, and these trails will be removed by closing the plot window. For example, in figure 4.31 we chose trials 179 and 180 as noisy trials, which will be omitted from the dataset after closing this plot window [18].

Sometimes, we see some of the channel's signals in a trail have outlier values. Keeping these trails in our dataset can lead us to the wrong results. Therefore, we need to omit these epochs. There is an automatic method that helps us in these cases. We first need to design rejection criteria in a dictionary. This dictionary rejects epochs based on the peak-to-peak (PTP) signal amplitude value. In other words, we set a value for each channel type in the rejection dictionary. Consequently, we can exclude the epochs that have any signals with a peak higher or lower than that value. In the following code, we omit the epochs in which the EEG signals have a PTP value outside the range of 150 μV and the EOG signals have a PTP value outside the range of 250 μV [19, 20].

```
reject_criteria = dict(eeg=150e-6, eog=250e-6)
data_epochs.drop_bad(reject=reject_criteria)
```

4.5.1.3 ERP analysis

After dropping all bad epochs, it is time to calculate evoke-related responses by creating the `evoked` object. We need to specify the name of `events` and use the `average` attribute to turn out the `evoked` object. For example, in the following code we generated the ERPs of the auditory/left event in the `evoke_visual_right` object. We drew the generated object in the time domain using the `plot` attribute. We used the 1–40 Hz filtered data and reserved just EEG data for displaying a better plot of auditory ERP in the EEG channels (figure 4.32) [20].

```
evoke_visual_right = data_epochs["auditory/left"].average()
evoke_visual_right.plot(spatial_colors=True)
```

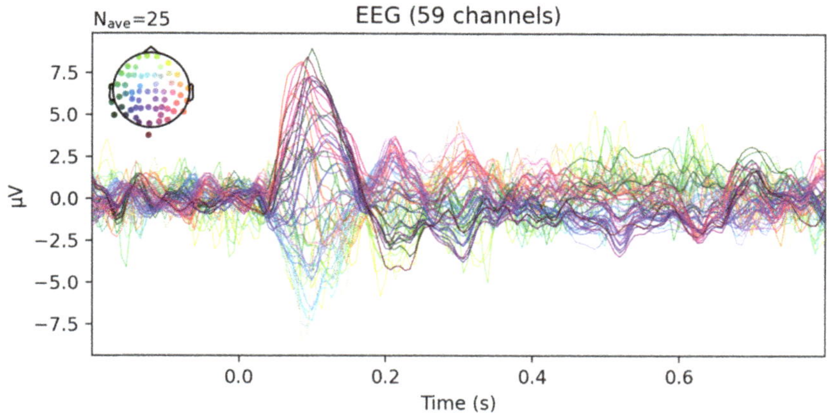

Figure 4.32. ERP time-series of EEG-filtered channels corresponding to left auditory events.

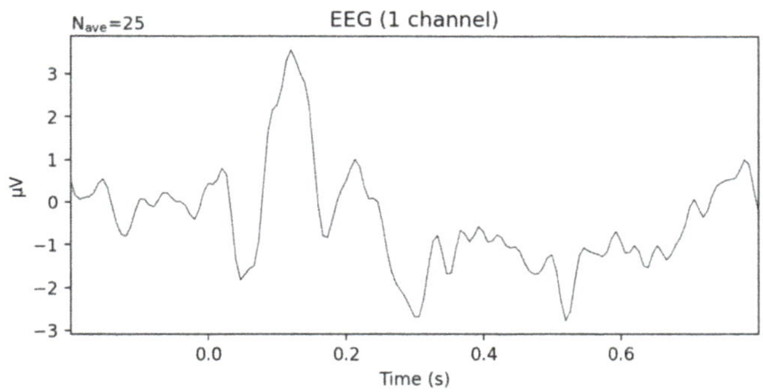

Figure 4.33. Left ear auditory ERP of 'EEG 047' filtered channel.

For inspection of just one channel ERP signal, we can specify the name of that channel in the `picks` parameter like it is shown in figure 4.33 and the following code [20].

```
evoke_visual_right.plot(picks='EEG 047', spatial_colors=True)
```

We can obtain topo map figures of ERPs at our desired specific times or time spans in epochs by the `plot_topomap` attribute. This function gets the time of the topo maps before and after the moment the event occurs. In addition, we can draw the topo maps for the mean of the time duration length we specify at the `average` parameter, in which the center of this time duration is established in the `times` parameter. In the following code, we drew the topo maps for three 0.05 s spans at 0.1 s before, 0.1 s after, and 0.3 s after the event occurs (figure 4.34) [20].

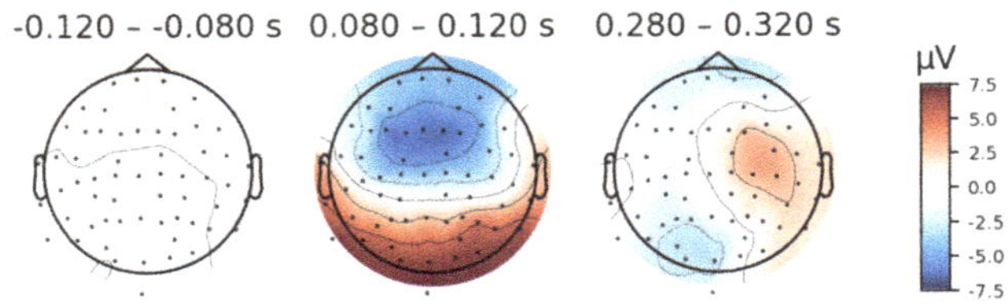

Figure 4.34. Left ear auditory ERP of three 0.05 s spans at 0.1 s before, 0.1 s after, and 0.3 s after the event occurs.

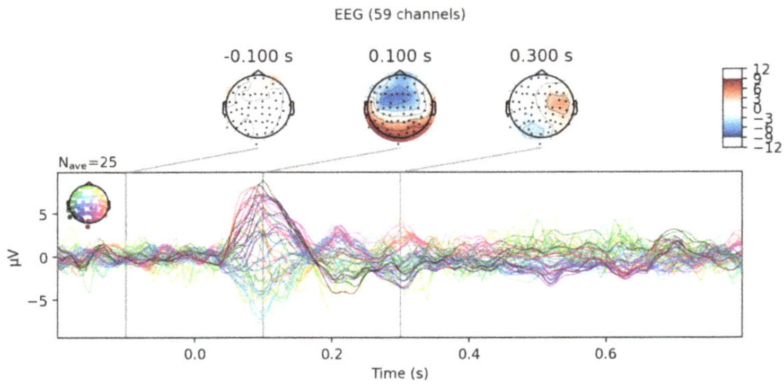

Figure 4.35. Left ear auditory ERP of three specific seconds in butterfly plot and topo map.

```
evoke_visual_right.plot_topomap(times=[-0.1, 0.1, 0.3], average=0.05)
```

We can simultaneously have topo maps and butterfly plots in one figure window by the `plot_joint` attribute (figure 4.35) [20].

```
evoke_visual_right.plot_joint(times=[-0.1, 0.1, 0.3])
```

4.5.2 Time-frequency analysis

There are many methods for time-frequency analysis, like short-time Fourier transform, Wigner–Ville, wavelet, etc. For instance, we compute time-frequency representation (TFR) for epochs with the help of the Morlet wavelet in the `tfr_morlet` function. This function gets epoched data in its first input, with the frequency range in `freqs` and the number of wavelet cycles in `n_cycles`. This function can also return inter-trial coherence, but we turn it off here. In the following code, we calculated the TFR of signal epochs with a left ear auditory event by Morlet wavelets at 1–40 Hz frequency. The result of TFR for the first EEG channel is shown in figure 4.36 [21].

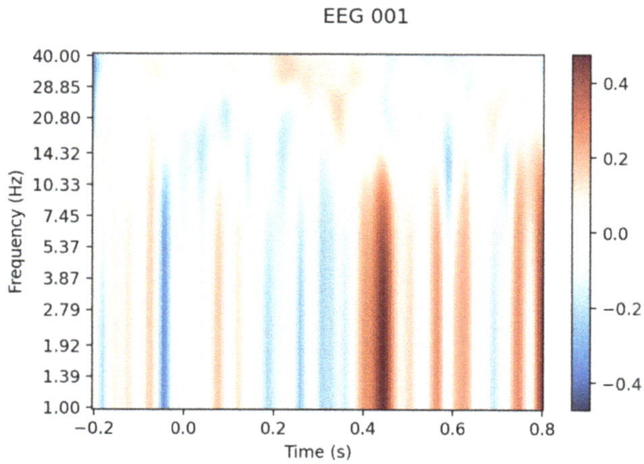

Figure 4.36. TFR of the signal epochs of the first EEG channel related to left ear auditory events.

```
import numpy as np
freqs = np.logspace(*np.log10([1, 40]), num=40)
n_cycles = freqs / 8
power = mne.time_frequency.tfr_morlet(
    data_epochs["auditory/left"],
    freqs=freqs,
    n_cycles=n_cycles,
    return_itc=False)
power.plot([0], baseline=(-0.2, 0), mode="logratio", title=power.ch_names[0])
```

4.5.3 Source localization

In source localization, we project EEG or MEG sensor data into a 3D position inside the head model. In other words, the data are transformed from the recorded time series at the sensor locations map to the time series at the inside head spatial location.

The pre-needed data and functions are [22]:

```
import numpy as np

import matplotlib.pyplot as plt

import mne

from mne.datasets import sample

from mne.minimum_norm import make_inverse_operator, apply_inverse
```

We work with 1–40 Hz filtered sample data with just EEG channels in this section. Then we calculate evoked data that carries just MEG channels and event ‘1’ [22].

```
data_fname = 'D:\MNE sample data/sample_audvis_filt-0-40_raw.fif'
#Importing the raw data
data_raw = mne.io.read_raw_fif(data_fname)
data_raw = data_raw.pick_types(
     meg=False, eeg=True, eog=False, stim = True)
data_events = mne.find_events(data_raw, stim_channel="STI 014")
event_id = dict(aud_l=1)  # event trigger and conditions
tmin = -0.2  # start of each epoch (200ms before the trigger)
tmax = 0.8  # end of each epoch (800ms after the trigger)
baseline = (None, 0)  # means from the first instant to t = 0
data_epochs = mne.Epochs(
    data_raw,
    data_events,
    event_id,
    tmin,
    tmax,
    proj=True,
    picks=("eeg"),
    baseline=baseline)
```

We calculate regularized noise covariance as it's needed in the inverse_operator function.

```
data_noise_cov = mne.compute_covariance(
    data_epochs, tmax=0.0, method=["shrunk", "empirical"], rank=None,
verbose=True)
fig_cov, fig_spectra = mne.viz.plot_cov(data_noise_cov, data_raw.info)

data_evoked = data_epochs.average().pick("eeg")
data_evoked.plot(time_unit="s")
data_evoked.plot_topomap(times=np.linspace(0, 0.8, 5), ch_type="mag")

data_evoked.plot_white(data_noise_cov, time_unit="s") #cleaning evoked data
```

We should import the coordinate system before source localization. Some MNE-Python objects (named forward) carry information about coordinates, and we import them at the beginning of the coding [23]. The following code help us in forward model importing.

```
data_path = sample.data_path()
fmName = data_path / "MEG" / "sample" / "sample_audvis-eeg-oct-6-fwd.fif"
forward_model = mne.read_forward_solution(fmName)
```

Afterwards, we need to make an inverse operator by the following code.

```
inverse_operator = make_inverse_operator(
    data_evoked.info, forward_model, data_noise_cov, loose=0.2, depth=0.8)
```

Sources time series and their locations will be determined by the following code. The MNE package makes it possible to calculate source localization with methods like MNE, dSPM, sLORETA, and eLORETA. All of these methods can be assigned to the `method` parameter [22].

Source time course (stc) output carries information about dipoles activation in time and location, and `residual` output carries source time series [22]. In detail, `stc` is the amplitude of sources over time and carries just the activations amplitude. We need to access `source space (src)` to get the source location. Each one of the time series can be assigned to a vertex (spatial position) in the `FreeSurfer` surfaces, these are made by 3D triangulations on inflated brain representation. Activations amplitude are stored in a matrix in `stc.data`. The values in this matrix represent the strength of our signals in positions (rows) and time points (columns). The surface representation for the left and right hemispheres is stored in the FreeSurfer. We can access the left hemisphere's activations amplitude data by passing `stc.lh_data`. We can access the right hemisphere's activations amplitude data by passing `stc.rh_data`. Both of them are $m \times n$ matrices. m is the number of spatial locations, and n is the number of time points. `stc.vertices` contains the indices of surface locations. `stc.lh_vertno` and `stc.rh_vertno` are location indices on the brain's left and right hemispheres model [22].

```
method = "dSPM"
snr = 3.0
lambda2 = 1.0 / snr**2
stc, residual = apply_inverse(
    data_evoked,
    inverse_operator,
    lambda2,
    method=method,
    pick_ori=None,
    return_residual=True,
    verbose=True)
```

Figure 4.37 illustrates the evoked signals at the sensor and the source level. The upper plot is evoked signals from sensors, and the lower plot is the same signals from sources [22].

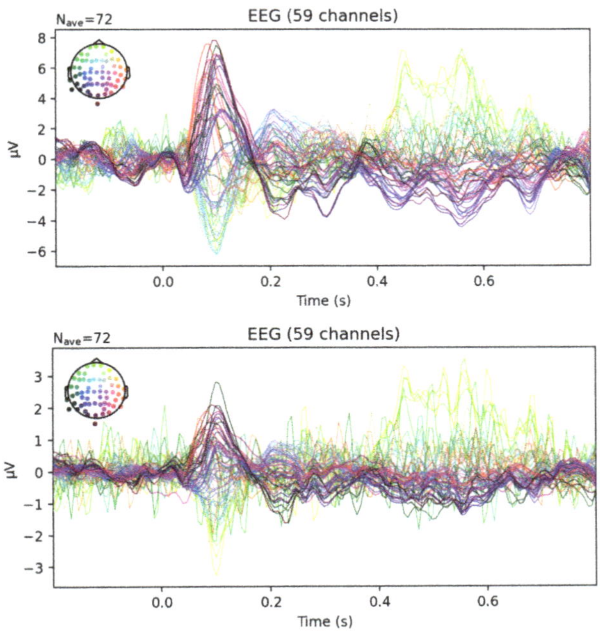

Figure 4.37. Evoked signals at sensors level (upper plot) and at source level (lower plot).

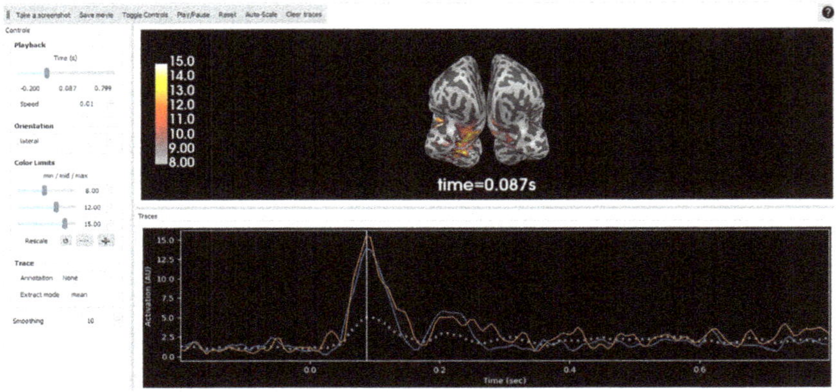

Figure 4.38. Localized source window for the evoked data.

```
upperplot = data_evoked.plot()
lowerplot = residual.plot()
```

By passing the following code, a new window will pop up (figure 4.38). This window carries source-localized images of the brain. In the left column of this window, there are tools for time, color, and moving adjustment. On the right, we can see which brain parts of both hemispheres are more activated for the evoked data over time by changing the time at the lower part [22].

```
subjects_dir = data_path / "subjects"
surfer_kwargs = dict(
    hemi="both",
    subjects_dir=subjects_dir,
    clim=dict(kind="value", lims=[8, 12, 15]),
    views="lateral",
    initial_time=time_max,
    time_unit="s",
    size=(800, 800),
    smoothing_steps=10)
brain = stc.plot(**surfer_kwargs)
```

As we need to access source positions, we go toward the computation of src, which represents the position and orientation of the candidate source locations. We can make the src object by pre-stored information in MNE, as shown in the following code [22].

```
data_path = sample.data_path()
subjects_dir = data_path / "subjects"
subject = "sample"
src = mne.setup_source_space(
    subject, spacing="oct4", add_dist="patch", subjects_dir=subjects_dir)
```

`src[0]['rr']` carries all vertex coordination of the left hemisphere, and `src[1]['rr']` carries all vertex coordination of the right hemisphere. We can access the locations of sources of our imported dataset by passing `src[0]['rr'][stc.lh_vertno]` for the left hemisphere and `src[1]['rr'][stc.rh_vertno]` for the right hemisphere. If we want to access the position of each source signal, we can restrict the imported signals into the apply_inverse function and import data of each sensor time series separately [24].

Bibliography

[1] Kumar J S and Bhuvaneswari P 2012 Analysis of electroencephalography (EEG) signals and its categorization—a study *Proc. Eng.* **38** 2525–36

[2] MNE Developers 2021 *The Raw Data Structure: Continuous Data* (https://mne.tools/0.23/auto_tutorials/raw/10_raw_overview.html#sphx-glr-auto-tutorials-raw-10-raw-overview-py) (accessed 22 June 2023)

[3] MNE Developers 2021 *Importing Data from EEG Devices* (https://mne.tools/0.23/auto_tutorials/io/20_reading_eeg_data.html#sphx-glr-auto-tutorials-io-20-reading-eeg-data-py) (accessed 22 June 2023)

[4] MNE Developers 2023 *mne.io.Raw* (https://mne.tools/0.23/generated/mne.io.Raw.html#mne.io.Raw) (accessed 22 June 2023)

[5] MNE Developers 2023 *Working with Events* (https://mne.tools/stable/auto_tutorials/raw/20_event_arrays.html) (accessed 22 June 2023)

[6] MNE Developers 2023 *Built-in Plotting Methods for Raw Objects* (https://mne.tools/stable/auto_tutorials/raw/40_visualize_raw.html#sphx-glr-auto-tutorials-raw-40-visualize-raw-py) (accessed 22 June 2023)

[7] MNE Developers 2023 *Setting the EEG Reference* (https://mne.tools/stable/auto_tutorials/preprocessing/55_setting_eeg_reference.html) (accessed 22 June 2023)

[8] MNE Developers 2023 *Handling Bad Channels* (https://mne.tools/stable/auto_tutorials/preprocessing/15_handling_bad_channels.html) (accessed 22 June 2023)

[9] MNE Developers 2023 *Rejecting Bad Data Spans and Breaks* (https://mne.tools/stable/auto_tutorials/preprocessing/20_rejecting_bad_data.html) (accessed 22 June 2023)

[10] MNE Developers 2023 *filter* (https://mne.tools/stable/generated/mne.io.Raw.html#mne.io.Raw.filter) (accessed 22 June 2023)

[11] MNE Developers 2023 *Filtering and Resampling Data* (https://mne.tools/stable/auto_tutorials/preprocessing/30_filtering_resampling.html#sphx-glr-auto-tutorials-preprocessing-30-filtering-resampling-py) (accessed 22 June 2023)

[12] MNE Developers 2023 *mne.filter.notch_filter* (https://mne.tools/stable/generated/mne.filter.notch_filter.html) (accessed 22 June 2023)

[13] Rejer I and Gorski P 2013 *Inria HAL* 10.1007/978-3-642-40925-7_11 (posted online 27 March 2017) Independent Component Analysis for EEG Data Preprocessing—Algorithms Comparison (https://inria.hal.science/hal-01496056)

[14] MNE Developers 2023 *Repairing Artifacts with ICA* (https://mne.tools/stable/auto_tutorials/preprocessing/40_artifact_correction_ica.html#sphx-glr-auto-tutorials-preprocessing-40-artifact-correction-ica-py) (accessed Jul. 08, 2023)

[15] Draper B A *et al* 2003 Recognizing faces with PCA and ICA *Comput. Vis. Image Underst.* **91** 115–37

[16] Chaumon M *et al* 2015 A practical guide to the selection of independent components of the electroencephalogram for artifact correction *J. Neurosci. Methods* **250** 47–63

[17] MNE Developers 2023 *The Epochs Data Structure: Discontinuous Data* (https://mne.tools/stable/auto_tutorials/epochs/10_epochs_overview.html#sphx-glr-auto-tutorials-epochs-10-epochs-overview-py) (accessed 09 July 2023)

[18] MNE Developers 2023 *Visualizing Epoched Data* https://mne.tools/stable/auto_tutorials/epochs/20_visualize_epochs.html#sphx-glr-auto-tutorials-epochs-20-visualize-epochs-py (accessed 10 July 2023)

[19] MNE Developers 2023 mne.Epochs (https://mne.tools/stable/generated/mne.Epochs.html#mne.Epochs.drop_bad) (accessed 10 July 2023)

[20] MNE Developers 2023 *EEG Analysis—Event-Related Potentials (ERPs)* (https://mne.tools/stable/auto_tutorials/evoked/30_eeg_erp.html) (accessed 10 July 2023)

[21] MNE Developers 2023 *Time Frequency Sensor Analysis* (https://mne.tools/stable/auto_tutorials/time-freq/20_sensors_time_frequency.html#sphx-glr-auto-tutorials-time-freq-20-sensors-time-frequency-py) (accessed 10 July 2023)

[22] MNE Developers 2023 *Source Localization with MNE, dSPM, sLORETA, and eLORETA* (https://mne.tools/stable/auto_tutorials/inverse/30_mne_dspm_loreta.html) (accessed 11 July 2023)

[23] MNE Developers 2023 *Working with CTF data: the Brainstorm auditory dataset* (https://mne.tools/stable/auto_tutorials/io/60_ctf_bst_auditory.html#sphx-glr-auto-tutorials-io-60-ctf-bst-auditory-py) (accessed 22 December 2023)

[24] MNE Developers 2023 *The SourceEstimate Data Structure* (https://mne.tools/stable/auto_-tutorials/inverse/10_stc_class.html#relationship-to-sourcespaces-src) (accessed 17 July 2023)

IOP Publishing

Signal Processing with Python
A practical approach
Irshad Ahmad Ansari and Varun Bajaj

Chapter 5

AG-PSO: prediction of heart diseases for an unbalanced dataset using feature extraction

Deepika Sainani and Urvashi P Shukla

Healthcare is an essential responsibility in human life. Heart disease is normally associated with an impact on the working of the heart and blood vessels. Early detection techniques for cardiovascular diseases have helped determine appropriate interventions for high-risk individuals, ultimately reducing their risks. This article examines different machine learning models to predict heart disease and improve classification accuracy in the face of an unbalanced dataset and a large sample size. The technique of data augmentation applied is synthetic data augmentation for tabular data (SMOTE). Selecting the appropriate features is a crucial preprocessing step in classification that involves removing attributes that are unnecessary, repeated, and intrusive. By doing so, model performance can be improved, the computational cost reduced, and the issue of the curse of dimensionality can be addressed. To implement the feature selection, binary particle swarm optimization (BPSO) is utilized, as it is a powerful and efficient technique for global search. It is particularly suitable for feature selection problems because of its ability to effectively explore large search spaces, its computational efficiency, ease of implementation, and minimal parameter requirements. As a result, BPSO is a highly effective algorithm for addressing feature selection tasks. The dataset used is the Kaggle archive for heart disease patients. The models have performed well with better accuracy and fewer false positives.

5.1 Introduction

Physical fitness is the primary requirement of happiness, according to Joseph Pilates, who stressed the significance of physical fitness for general well-being. Winston S. Churchill agreed that healthy citizens are a nation's greatest asset and emphasized the importance of maintaining them. Heart failure, a serious health concern, has

doi:10.1088/978-0-7503-5929-0ch5

increased as a result of an unhealthy lifestyle in modern times, along with other ailments [1].

One of the leading causes of death in the globe, cardiovascular disease (CVD), affects both the development and economy of countries. Heart failure results when the heart is unable to pump sufficient blood to fulfill the body's needs. Although the precise causes of CVDs are still unknown, several important factors like age, heredity, body mass index, high blood pressure, smoking, diabetes, and cholesterol levels are thought to have a significant role. Chest pain, discomfort in other body areas, and shortness of breath are typical signs of CVDs. Thus, prompt testing is essential to guaranteeing adequate treatment [2, 3]. The enormous volumes of sequential data generated by these testing methods, however, make accurate analysis in the early stages of the disease a substantial problem. Cardiology operates in a dynamic setting that is always changing due to new developments. As a result, accurate and current CVD statistics are essential for successfully tracking and forecasting the course of this disease.

There has been a lot of study done on the risk factors for heart disease. Finding a trustworthy and practical approach to figuring out these risk factors is still a research problem. To examine the association between different disease-related variables and the data's hidden information, disease prediction techniques are frequently used [4]. These methods seek to unearth insightful information that can help in foreseeing the onset and progression of cardiac disease.

To improve performance in the area of heart disease prediction, a variety of machine learning (ML) and deep learning techniques have been widely applied. By breaking up big datasets into smaller groups, ML can find patterns in the data that can be applied to decision-making [5]. Modern ML techniques have been used to forecast heart failure–related adverse events and evaluate the severity of the condition [6–8]. Researchers have used a variety of existing classifiers to predict heart diseases, such as regression trees, support vector machines (SVM), logistic regression (logistic), K-nearest neighbors (KNN), gated recurrent unit, convolutional neural networks, ensemble methods, deep belief networks, and deep neural networks. These classifiers have demonstrated efficacy in analyzing diverse data aspects and generating precise heart disease predictions. Researchers hope to enhance diagnostic skills and enable early diagnosis of cardiac illness by utilizing the power of these algorithms. The normally used approach for heart disease prediction with the feature selection approach is as shown in figure 5.1.

For accurate prediction through ML, a robust dataset is essential to provide the model with insights into the various traits and characteristics of the problem at hand. However, while training ML models on comparative small datasets, they often suffer from overfitting, resulting in high predictive accuracy on the given input training data but poor performance on new, unseen data. Techniques are used to balance the classes in the dataset and regenerate artificial data using pattern analysis to overcome this problem. By using pattern analysis, scientists can create artificial data that closely replicates the patterns found in the actual dataset. The training set can be expanded by using this synthetic data, which also gives the model a wider variety of examples to draw from. Additionally, class balancing strategies are used

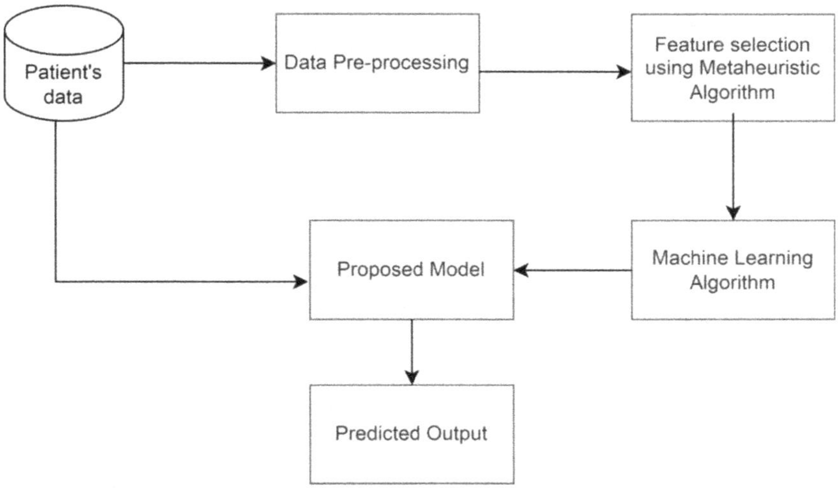

Figure 5.1. Basic flow for heart disease prediction using BPSO.

to correct any notable imbalances in the distribution of various classes within the dataset. In addition to reducing bias towards the dominant class and enhancing the model's capacity to generalize to new data, this makes sure that the model is exposed to a sufficient representation of each class. ML algorithms can get over the constraints provided by tiny datasets, producing more reliable and accurate predictions, by balancing the classes and enhancing the dataset with false data. These methods improve the use of the data that is already available and improve how well the model performs in practical situations. This article significantly contributes in various ways, including:

1. Addressing class imbalance and small dataset size: The availability of small-sized data for the prediction of heart disease presents challenges. This is addressed by applying strategies for creating synthetic data, and the SMOTE algorithm is used to address the issue of class imbalance. This improves the dataset and guarantees a balanced representation of the various classifications.
2. The significance of feature selection: It is a crucial preprocessing step. It makes it easier to comprehend how each feature contributes to the overall accuracy of the forecast. Feature selection raises the accuracy of the models by lowering overfitting and eliminating redundant and noisy data. The study successfully addresses the NP-hard problem of feature subset selection using a metaheuristic method, specifically BPSO.
3. Implementation of basic ML algorithms: The study uses many common ML algorithms to predict cardiac illnesses for preprocessed and chosen feature subsets after the feature selection and preprocessing processes are finished. These algorithms use the knowledge acquired to accurately predict the future from the data.

The study makes a contribution to the field of cardiac disease prediction by addressing issues including class imbalance, minimal dataset size, and feature selection. It highlights how crucial it is to use preprocessing methods and choose the right features to improve the precision and dependability of ML models for predicting heart disease.

The structure of the article is as follows: section 5.1: A synopsis and contribution of the author. This section gives a brief summary of the situation and details the author's particular contribution. The author emphasizes the importance of the research issue and introduces their own method or viewpoint. The next section, section 5.2: Literature survey, offers a thorough literature review in which the author evaluates recent studies and research on the subject of interest. It establishes the state of the art, identifies research gaps, and emphasizes the necessity of the proposed contribution. The significant contribution of the paper is then highlighted in section 5.3: Significant contribution model with detailed steps, which also presents the recommended model or approach in detail. The author describes the step-by-step process, including data preprocessing, feature selection, and the complete model. Finally, section 5.4: Results, discussion, concluding remarks, and future contributions. Here, the results of the experiments or analysis conducted are presented. The author discusses the findings, their implications, and any insights gained from the study. Concluding remarks are provided to summarize the key outcomes and contributions of the research. Additionally, the author has discussed potential future research directions or areas that could benefit from further investigation.

5.2 Literature survey

In this section, the author has carried out considerable research on the various existing ML models designed in the past years. The complete survey is tabulated in table 5.1

5.3 Proposed methodology

In this article, the complete methodology is divided into three sections, as depicted in figure 5.2.

5.3.1 Data augmentation and imbalance (SMOTE)

The SMOTE method is a well-liked oversampling approach used to address data class imbalance as well as the generation of synthetic points. When one class considerably outnumbers the other in terms of examples within a dataset, there is a class imbalance. The dataset is dominated by the majority class, also known as the negative class, while the minority class, also known as the positive class, accounts for a smaller percentage of the dataset. The issue of data imbalance has grown to be a serious problem as a result of the growth of large-scale data across a variety of industries, including the business, educational, and governmental sectors.

It is essential to correctly categorize the minority class to solve real-world issues. Finding effective solutions that may address the data imbalance and eliminate bias against the majority class is therefore necessary. SMOTE [19] offers a way to create

Table 5.1. An extensive literature survey on the existing methods for heart disease prediction.

Author	Year	Dataset	Method	Results	Observations
Sibo Prasad Patro *et al* [9]	2023	Cleveland database of UC Irvine repository	Hybridization of fusion-based ensemble model with ML	Enhances the prediction because of histogram-based gradient boosting model	Boosting techniques are mainly used to improve the accuracy of small datasets.
Chintan M. Bhatt *et al* [10]	2023	Kaggle CVD	Multiple models such as random K-mode, decision tree (DT), random forest (RF), XGBoost (XGB), multilayer perceptron (MP)	Accuracies reported for all algorithms 86% with the highest MP as 87.23%	(1) Single dataset (2) Lifestyle factors (3) Held-out test performance not evaluated (4) The interpretability not evaluated
Mohamed G. El-Shafiey *et al* [11]	2022	Cleveland and Statlog database	Hybrid genetic algorithm and PSO with RF. GAPSO-RF	95.6% (Cleveland) and 91.4% (Statlog)	(1) More classifiers should be evaluated (2) High computational cost and temporal complexity
Vikas Chaurasia *et al* [12]	2022	Kaggle dataset	DT, SVM, gradient boosting (GBC), Gradient boosting-based sequential feature selection (GBSFS), RF, Multilayer perceptron (MLP), Extra Tree(ET), Linear Regression (LR), KNN	11 features accuracy reported at 98.78%	
Abid Ishaq *et al* [13]	2021	UC Irvine ML repository has 299 records with 13 features	SVM, gradient boosting classifier, extra tree classifier, Gaussian naive Bayes (GNB), LR, stochastic gradient classifier, DT, and AdaBoost	92.62% with SMOTE	

(*Continued*)

Table 5.1. (*Continued*)

Author	Year	Dataset	Method	Results	Observations
Aqsa Rahim *et al* [14]	2021	Framingham and Cleveland	GBSFS features, data imbalance SMOTE, DT, RF, MLP, SVM, ET, GBC, LR, and stacking-based cardiovascular disease diagnosis missing values are used	Framingham, Heart Disease, and Cleveland), with respective accuracy levels of 99.1, 98.0, and 95.5%.	(1) More classifiers can be used in the ensemble technique. (2) Also real-time evaluation is needed rather than on state-of-the-art datasets.
Devansh Shal. *et al* [15]	2020	Cleveland database	NB, DT, KNN, and RF	90.78% (KNN)	To get greater accuracy, a more complicated model combination must be used.
Mustafa Jan *et al* [16]	2018	Cleveland and Hungarian	SVM, artificial neural network, GNB, regression analysis, and RF are five classifier model techniques that are combined in ensemble learning. Tenfold cross-validations are also employed.		Analysis on different datasets can be done.
Nida Khateeb *et al* [17]	2017	Cleveland and Statlog	IBK (KNN), J48, and bagging classifiers/ML techniques, as well as ens naive Bayes. SMOTE	79.20% by using resampling	
Ismail Babaoglu *et al* [18]	2010	Exercise stress testing	SVM classification model based on Feature selection technique using BPSO (BPSO–FST) or Feature selection technique using GA (GA–FST) Architecture	81.46% using BPSO-FST with 11 features	Too many tuning parameters to find accuracy for such a small data

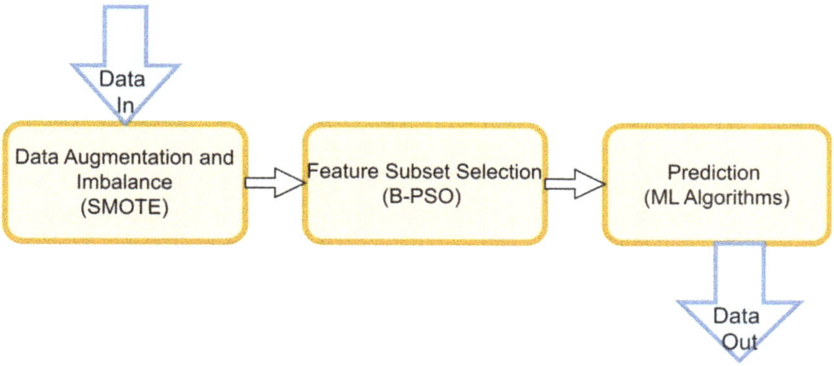

Figure 5.2. Basic outlines of the major steps carried out.

synthetic samples for the underrepresented class, balancing the distribution of classes and improving their representation. Researchers and practitioners can overcome the difficulties caused by class imbalance and guarantee more precise categorization in real-world circumstances by using SMOTE.

Several methods, which can be broadly categorized into three approaches [20], can be used to classify unbalanced data.

1. Data-level approach: The data-level approach involves adjusting or changing the original dataset to create a balanced dataset that enables the use of common ML algorithms. This entails methods like undersampling, oversampling, or a hybrid strategy that combines the two. The objective is to achieve a more equal representation of the classes by either decreasing the instances of the majority class (undersampling) or increasing the instances of the minority class (oversampling).
2. The algorithm-level approach: This approach modifies or creates new methods or algorithms expressly to deal with unbalanced data. These changes improve the algorithm's ability to manage class imbalance effectively without requiring considerable data manipulation. The goal is to develop algorithms that can automatically manage unbalanced datasets and make accurate predictions.
3. Cost-sensitive method: To increase classification accuracy while taking into account the costs associated with misclassification, the cost-sensitive method incorporates both data-level and algorithm-level strategies. This strategy seeks to lessen the negative effects of misclassifying the minority class, which are frequently more severe in terms of repercussions or expenses. The cost-sensitive technique aims to optimize the trade-off between accuracy and the costs associated with misclassification by taking the imbalanced nature of the data into account during the learning process.

These strategies can be used to apply a variety of solutions to the problems that unbalanced datasets provide, ultimately improving classification performance and reducing the effects of class imbalance. There are also three sub-approaches within the data level approach: undersampling, oversampling, and a hybrid method.

1. Undersampling: To lessen the dominance of the majority class in the sample, instances from that class are randomly deleted. The dataset is made more balanced by deleting examples from the dominant class.
2. Oversampling: To boost the representation of the minority class in the dataset, this method includes randomly duplicating examples from the minority class. The dataset becomes more balanced by increasing the number of cases from the minority class.
3. Hybrid approach: To produce a balanced dataset, the hybrid approach incorporates both undersampling and oversampling methods. To boost the number of samples from minority classes, the oversampling technique is typically used initially. To ensure an equal representation of both classes, the undersampling technique is then used to remove samples from the dominant class.

Both undersampling and oversampling methods, however, have some drawbacks. SMOTE can be used to overcome these restrictions. SMOTE is an oversampling technique that generates artificial samples only for the minority class. It avoids the overfitting problem that could result from arbitrary oversampling. SMOTE uses interpolation between positive examples that are close to one another and concentrates on the feature space rather than just replicating instances. To create a balanced dataset [21], this algorithm creates new instances that are more representative of the minority class.

In this study, we have applied SMOTE in both cases, the data augmentation as well as to resolve the issue of data or class imbalance. The in-detail steps are as follows:

Algorithm 1. SMOTE (data augmentation and class imbalance)

Input: Preprocessed data cube: $\boldsymbol{D}\{\boldsymbol{M} \times \boldsymbol{N}\}$, where $\boldsymbol{M}$ is the observations and $\boldsymbol{N}$ features
Output: Augmented and balanced dataset: $\widetilde{\boldsymbol{D}}\{\text{-}\widetilde{\boldsymbol{M}} \times \widetilde{\boldsymbol{N}}\}$
Step 1: Choose random data subset $\widetilde{\boldsymbol{D}} \epsilon \boldsymbol{S}$
Step 2: Find Euclidean distance from KNN $\boldsymbol{X}$
Step 3: Select a random number $\delta \epsilon (0, 1)$
Step 4: Multiple the difference with δ and add to generate the new samples
$\widetilde{\boldsymbol{D}_i} = \boldsymbol{Z}_i + \delta(\boldsymbol{Z}_i - \boldsymbol{X}_i)$

5.3.2 Feature subset selection (BPSO)

The dimensionality of data plays a crucial role in the performance of classification algorithms [22]. When a dataset contains numerous features, it increases the likelihood of confusion for learning algorithms such as classifiers, leading to poorer performance. To address this challenge, feature selection (FS) is a basic preprocessing step that is used to improve the performance of classification methods [23]. FS is the process of getting rid of duplicate, useless, and noisy features in a dataset.

The benefits of FS extend beyond improving algorithm efficiency; it also reduces the computational time required for calculations. However, when developing an FS algorithm, two key considerations need to be taken into consideration [24]. Firstly, it is important to find the most suitable subset of traits that will give the best results. This involves identifying the most relevant and informative features while discarding redundant or less significant ones. Secondly, the quality of a feature subset needs to be evaluated, ensuring that the selected features effectively capture the discriminative patterns in the data.

By addressing these concerns and implementing effective FS techniques, the dimensionality of the data can be reduced, leading to improved performance of classification algorithms and more efficient computation times.

In the research on FS, filter models and wrapper models have been used most often to judge the quality of a group of features. Filter models mostly evaluate the group of features based on how they relate to and depend on the class. Principal component analysis, information gain, and F-score are all examples of filtering methods [25]. These methods rate the usefulness and ability to tell one feature from another without looking at how a specific learning algorithm works. Wrapper models, on the other hand, evaluate the effectiveness of a chosen learning algorithm (e.g., a classification algorithm) during the assessment process. The evaluation is performed by incorporating the learning algorithm as a part of the feature subset evaluation. Wrapper techniques typically require more time for optimization compared to filter approaches because each feature subset is evaluated using the selected learning algorithm. However, wrapper methods are better when the main objective is to improve the performance of a specific learning technique.

The look for the best set of features is another important part of FS methods. There are three basic approaches commonly used: complete (exhaustive) search, random search, and heuristic search. The exhaustive search involves evaluating every possible feature subset and selecting the best one. However, this approach becomes impractical and time-consuming for large datasets, as it would require evaluating 2^N subsets for a dataset having N number of features.

A random search is an alternative approach that explores feature subsets randomly. While random search may not be guaranteed to find the absolute best subset, it can still be effective in practice, and it avoids the computational burden associated with exhaustive search. Heuristic search methods use specific rules or strategies to help find the best subset of features. These approaches try to achieve a balance between the quality of a selected subset and computational effectiveness.

To deal with complex and computationally challenging conditions, a class of optimization methods called metaheuristic algorithms is developed. These algorithms are especially helpful when conventional exact approaches fall short of delivering workable solutions in a reasonable amount of time. Metaheuristic can more efficiently search the solution space because they are motivated by natural phenomena, social behaviors, or evolutionary processes. Metaheuristics offer approximations that are frequently of good quality, in contrast to exact approaches that guarantee optimal answers. They are adaptable and can be used to address a wide range of issues in a variety of fields, including engineering, logistics, and finance. Metaheuristic algorithms provide a strong and adaptable approach to optimization by utilizing their ability to intelligently explore and exploit the search space, enabling the finding of close to optimal solutions for challenging real-world issues.

There are several taxonomies used to categorize metaheuristic algorithms, and one of them is based on the source of inspiration [26]. Among population-based metaheuristic algorithms, swarm intelligence and evolutionary algorithms are two of the most widely used. Evolutionary algorithms select an initial population at random at the beginning of the optimization process. Once a stopping point is reached, the population is then modified repeatedly in an evolutionary manner. The genetic algorithm [27] and differential evolution [28] are two well-known examples of evolutionary algorithms.

A well-known metaheuristic technique called particle swarm optimization (PSO) [29] is based on how fish school and flocks of birds interact. In PSO, the optimum solution is identified by navigating a collection of particles, each of which stands in for a potential solution. Each particle adjusts its position based on the best positions it has already identified as well as the best positions each other particle in the swarm has discovered. By exchanging information, particles can communicate with one another and grow closer to the ideal state of the universe. PSO is renowned for its simplicity, speed of convergence, and use. It has been employed to address optimization issues like function optimization, grouping, and neural network training. Because the method can swiftly adjust to changing conditions and easily search the search space, PSO is a powerful tool for addressing complicated optimization issues in a variety of fields.

An expansion of the standard PSO method created expressly to address binary or discrete optimization issues is BPSO [30]. Decision variables are frequently expressed in real-world situations as binary values, in which each variable may only take on one of two potential states, such as on/off or presence/absence. By modifying PSO's guiding principles for binary search spaces, BPSO tackles these kinds of issues. BPSO utilizes the collective intelligence of a number of particles to search the binary solution space for the best set of binary variables to maximize or minimize the objective function.

In this article, the authors have implemented BPSO; the detailed algorithm is shown in Algorithm 2. The different BPSO parameters are initialized as described in the table 5.2 below.

The algorithm is designed to iterate until the highest number of repetitions, which is set to 1 000. The dataset's dimensionality is set to 16, which corresponds to the number of features, and the population size is set to 30. Each particle in the algorithm is represented as a string consisting of 0s and 1s, where 0 indicates

Algorithm 2. BPSO.

Input: Preprocessed data cube: $\widetilde{D}\{\widetilde{\text{-}M} \times \widetilde{N}\}$
Output: Feature subset : $\widetilde{S} = \{\widetilde{\text{-}M} \times \tilde{n} \ll \widetilde{N}\}$
Initial parameters: $c_1 = 0.5$, $c_2 = 0.7$, $w = 0.3$, $\alpha \in [0, 1]$, $V\epsilon\,[0, 1]$
Step 1: Encoding : $P\{\widetilde{\text{-}M} \times \widetilde{N}\} = (1 \ldots 0\,0 \cdot\!\cdot\!. 1\,0 \ldots 0)$
$V\{\widetilde{\text{-}M} \times \widetilde{N}\} = (0.\,..11 \cdot\!\cdot\!.01\ldots1)$
Step 2: Cost function:
$gbest = RMSE$
$\text{RMSE} = \sqrt{(y_t - y_{svm})^2}$
Step 3: Position update:
$\widetilde{V_{i+1}} = w \times V + c_1 \times \alpha \times (pbest - P_i) + c_2 \times \alpha \times (nbest - P_i)$
$\widetilde{P}_{i+1} = \widetilde{V_{i+1}} + \widetilde{P}_i$
Step 4: Mutation:
If $\widetilde{P}_{i+1}(k) = \{0\,1\ if\ \alpha > 0.5$
Step 5: Stopping condition: if $T = 5000$
Step 6: Subset $\widetilde{S} = gBest$

Table 5.2. Different BPSO parameters.

Parameters	Values
Population size (P)	30
Maximum iterations (T)	500
Acceleration constant (c_1)	0.5
Acceleration constant (c_2)	0.7
Initial weight (w)	0.3
Dimension (d)	$\widetilde{N} = 16$

the absence of a feature, and 1 represents the selection of a feature. The initial population and velocity are randomly selected matrices of the same size.

Root mean square error (RMSE) is the cost function used in the algorithm. It measures the difference between the labels projected by the SVM and the actual labels. The particle locations are modified following the cost function evaluation. The algorithm keeps track of two optimal positions: the personal best (***pbest***), which represents the best position a particle has reached so far, and the global best (***gBest***), which is the best position achieved by any particle in the neighborhood. Additionally, each particle also considers the best position reached by its neighboring particle, known as the local favorite (***nbest***).

When a particle considers the entire population as its neighborhood, the best performance of the neighborhood becomes the global best (***gBest***). This global best position serves as the selected feature subset achieved by the algorithm.

5.3.3 Prediction (various ML algorithms)

In this analysis, we have compared five ML algorithms. In the following segment, the details of each are provided with the required parameter setting.

1. Logistic regression [31]: Using the statistical technique known as logistic regression, a categorical dependent variable is related to one or more independent variables. When the dependent variable simply has two alternative values, such as true or false, yes or no, 0 or 1, etc., it can be applied to a wide range of binary classification issues. In logistic regression, a sigmoid function, also referred to as the logistic function, is used to describe the dependent variable as a function of the independent variables. The S-shaped logistic function effectively converts input values to a range between 0 and 1. Depending on whether the patient's heart is present or absent, this range indicates the likelihood that the dependent variable will be 1.
2. Decision tree [32]: Decision trees are commonly used as a visual representation of the decision-making process in ML, data mining, and other related fields. Its structure is designed to resemble a tree, with leaf nodes standing in for class names, internal nodes for attribute testing, and branches for test results. Both classification and regression applications use decision trees. In contrast to regression, which aims to forecast a continuous value, the goal of classification is to determine a new instance's class based on its features. By periodically splitting the data into subgroups based on feature values, the decision tree algorithm builds the tree. This process continues until either a predetermined halting threshold is met or each subset becomes homogeneous. A decision tree is all about how evenly the classes are distributed.
3. Random forest classifier [33]: Using decision trees, the random forest classifier is a potent ensemble learning technique. It is frequently used in ML applications involving binary and multiple class categorization. On several random selections of the training data, the algorithm constructs decision trees, integrating their predictions to get the final prediction. The random forest classifier uses a set of randomly chosen features to train each tree, which significantly decreases overfitting and improves the model's capacity to generalize to new data. The algorithm applies each decision tree to the input data and establishes the anticipated class label for each tree in order to produce predictions using the random forest classifier.
4. Naive Bayes [34]: A probabilistic approach used in ML for classification is called the naive Bayes classifier. The theoretical underpinning for this reasoning is the Bayes theorem, which asserts that the likelihood of a hypothesis given the proof corresponds to the probability of the proof given the hypothesis, multiplied by the prior probability of the hypothesis. After computing the possibility of each class label given the input features, the naive Bayes classifier selects the class label with the highest probability. The Bayes theorem is used to determine this probability on the presumption that the features are conditionally independent. By assuming that the presence or

absence of one specific attribute has no bearing on the presence or absence of any other, the computation is made simpler and easier.

5. KNN [35]: ML applications for classification and regression frequently use the flexible nonparametric KNN method. It provides a simple but effective method for classifying or predicting the label of a specific data point by taking into account its near neighbors. The KNN algorithm uses a distance metric, such as Euclidean distance or cosine similarity, to determine from the training data the K data points that are a new data point's nearest neighbors. These neighbors are chosen based on how close they are to the most recent data point added to the feature space. In classification tasks utilizing the method, the label that appears the most frequently among the K neighbors is then applied to the most recent data point. When performing regression tasks, the computer is capable of predicting the value of the upcoming data point.

5.4 Parameter settings for the simulation study

This section contains comprehensive information on the dataset used, the platform used, and the numerous analysis measures used. Various analysis indicators are used to evaluate the algorithm's usefulness and performance. Based on their applicability to the issue area and their precision in describing the algorithm's performance, these metrics were selected after careful consideration. In general, this part gives a thorough discussion of the dataset used, the analysis platform used, and the several metrics used to assess the algorithm's success. These aspects are taken into account in order to have a thorough grasp of the experimental design, which results in insightful observations and trustworthy findings.

5.4.1 Experimental dataset

For the dataset required to forecast heart illness, the Kaggle archives were mined [36]. Four distinct datasets, the Cleveland, Hungary, Switzerland, and Long Beach V databases, were used to compile the data. A summary of the features of the dataset, including the number of patient records and the features is present. The dataset, which contains a total of 1025 patient records, is a significant and comprehensive database of data for the prognosis of cardiac disease. It includes 13 separate components that each captures a different aspect of the patient's health and medical background. One dependent feature is also part of the goal variable for determining whether heart disease will exist or not. The dataset provides a comprehensive and representative sample for training and assessing heart disease prediction models by merging data from various sources. The robustness and generalizability of the dataset are improved by the inclusion of many datasets from various regions.

5.4.2 Experimentation platform

For evaluating our ML models, we utilized Google Colab. All that is required is a web browser to access and work within the Colab environment. These hardware specifications ensured that our experiments were conducted efficiently, allowing for

Table 5.3. Different parameters of the heart disease database.

Metric	Formula
Accuracy	$\frac{(TP + TN)}{(TP + FN + TN + FP)}$
Precision	$\frac{TP}{(TP + FP)}$
Recall	$\frac{TP}{(TP + FN)}$
F1-score	$\frac{2 \times (\text{Precision} \times \text{Recall})}{(\text{Precision} + \text{Recall})}$

faster processing and optimal performance of the ML algorithms utilized. To conduct our experiments, we performed them on a computer system with the following specifications: the system with 8 GB of RAM with an Intel(R) Core(TM) i7–3630QM processor running at 2.40 GHz.

5.4.3 Evaluation metric

The main issue is figuring out whether the model can distinguish between the original and derived label. We used a number of indicators to assess the model's performance in this difficult assignment. It is crucial to remember that other project elements should not be prioritized above the choice and implementation of the model. To evaluate the model's capacity to identify bogus news, the data was exposed to a comprehensive set of evaluation criteria. ML models were assessed using a variety of metrics, such as classification reports, confusion measures, and others. A true positive denotes an accurate forecast of a positive result when both the actual and anticipated class values are positive. In contrast, true negative designates appear in table 5.3.

5.5 Results and analysis

The approach used in this study included a number of vital phases. Data augmentation and class balancing were the first steps. The synthetic minority over-sampling technique was used to increase the dataset size and take care of the class imbalance. This method, which ensured that each class had an equal number of samples, effectively doubled the amount of observations in the dataset. The preprocessing of features was done after the data was augmented and classes were balanced. It required a number of significant actions. Firstly, normalization was applied to scale the features and bring them within a consistent range. Secondly, missing values were filled using appropriate techniques to ensure a complete dataset. Lastly, categorical features were encoded into numerical representations, enabling their inclusion in subsequent analysis steps. The next step in the methodology was the selection of a smaller group of features from the original set. The BPSO method was used to make this decision. The RMSE in conjunction with SVM was the defined cost function used by the BPSO method, as it iteratively sought the best feature subset. The optimization curve for the BPSO algorithm, shown in figure 5.3,

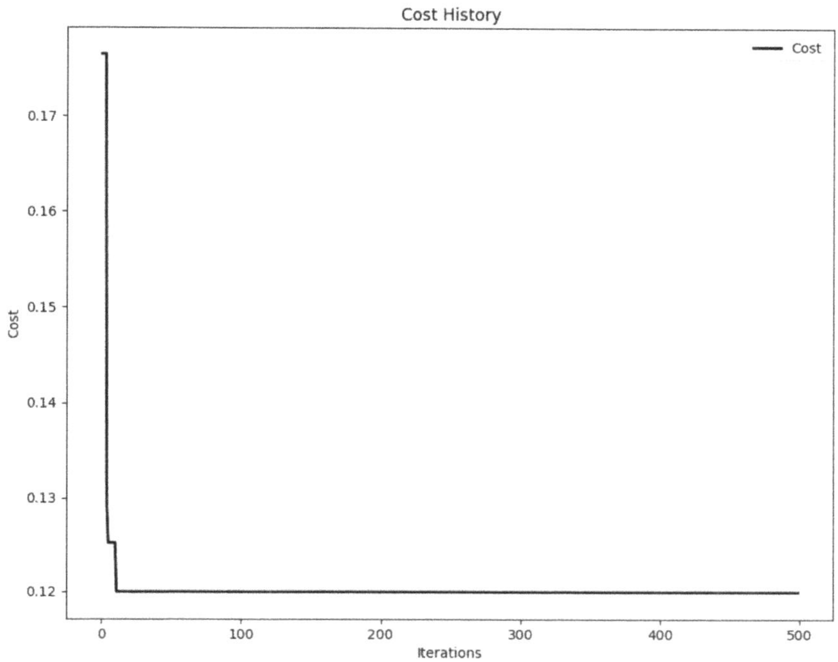

Figure 5.3. Optimization curve for BPSO.

demonstrated its convergence within 75 iterations, indicating its efficiency in finding a suitable feature subset.

A learning curve depicts the correlation between a ML model's efficacy and the volume of training data used. It explains how the functionality of the model changes as more training data become accessible. The training set and the test set are the two main elements that make up a learning curve. The training set's size, or the number of training examples, is represented by the x-axis, and accuracy as a performance indicator is shown by the y-axis. The learning curve displays how the model performs as the size of the training and testing sets increase. It raises awareness of potential problems with underfitting or overfitting and offers useful information on the behavior and effectiveness of the system.

In figure 5.4, we can observe that in the Navy Bayesian method, the learning and testing curves are highly converged and there are also many gaps between them. Thus, it is almost an ideal fit, and the model is able to generalize to new data effectively. In the case of the KNN, the curve is on the rising curve. Thus, initially, the model is not able to generalize.

An illustration that displays the overall classification criteria for a classification model is called a ROC curve (receiver operating characteristic curve). On the ROC curve, with the false positive rate on the x-axis and the true positive rate on the y-axis, the true positive rate is displayed against the false positive rate for various categorization thresholds. The performance of the model is shown by the ROC

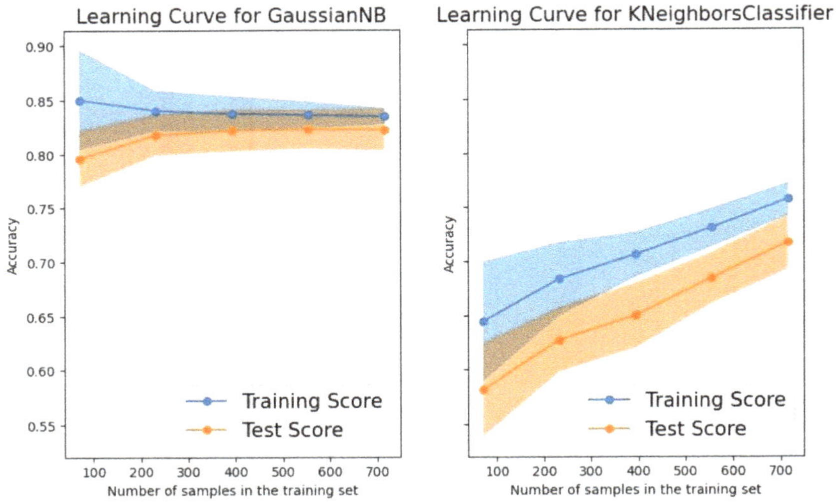

Figure 5.4. Curves for two ML algorithms.

curve's area under the curve (AUC) value. The ROC curve for our analysis is shown in figure 5.5. The observation for different AUCs is noted. The LR model has strong discriminatory power in separating the positive and negative groups when the AUC is 0.904, which shows. This AUC value's ROC curve will be clearly separated from the diagonal line, indicating that the model has a high rate of true positives and a low rate of false positives. The model's performance is comparable to the prior scenario, with an AUC of 0.905. It nevertheless has strong selective ability, performing somewhat better than an AUC of 0.904. The model will achieve a high rate of true positives and a low rate of false positives, as shown by the ROC curve, which will follow a similar pattern. The model's performance is still respectable but slightly worse than in the earlier examples in which the AUC is 0.884. An indication of a considerably larger false positive rate compared to the real positive rate can be seen in the ROC curve associated with this AUC number, which may exhibit a less distinct separation between the curve and the diagonal line.

The model exhibits high discriminatory power with an AUC of 0.970. A significant difference between the diagonal line and the ROC curve indicates a very low rate of false positives and a high rate of true positives. This suggests a model that is very accurate at differentiating between the classifications of the positive and the negative. An ideal model that achieves the best possible discrimination between the positive and negative classes is one in which the AUC is 1. With an AUC of 1, the ROC curve will be a step function, showing no false positives and an ideal true positive rate.

We sought to examine the link between the number of neighbors and its effect on both training and testing accuracy using figure 5.5. The graph illustrates how the model's accuracy varies as the number of neighbors rises. By looking at the graph, it is obvious that accuracy clearly decreases as the number of neighbors rises. For

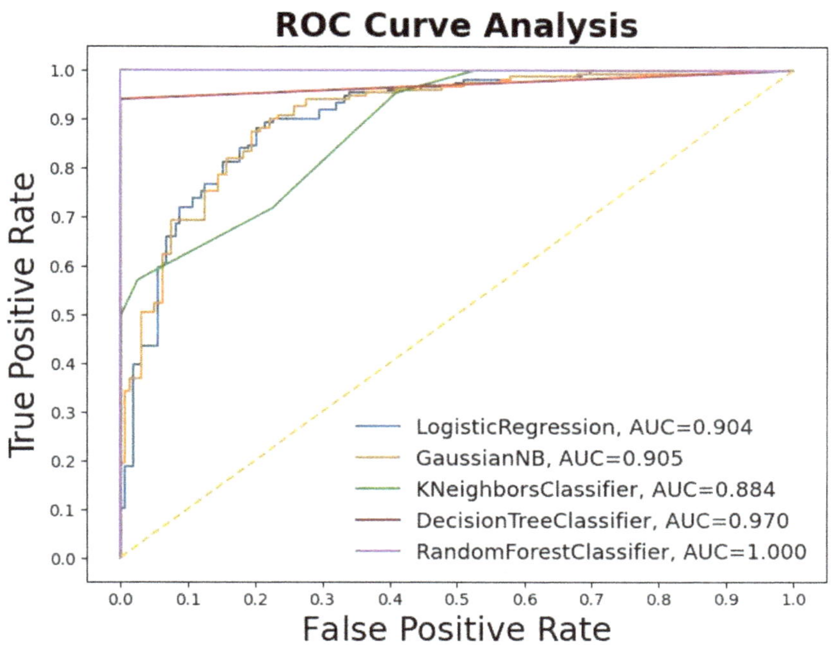

Figure 5.5. ROC curve for the various ML applied.

classification or regression tasks, the model tends to demonstrate lesser accuracy in both the training and testing phases, as it takes into account a larger number of neighbors. The results from figure 5.6 imply that adding neighbors may cause the model's decision-making process to become more variable or noisy. As the model is more influenced by the potentially antagonistic or unrepresentative neighbors, accuracy may suffer as a result. Making educated selections when choosing the right value for this parameter requires an understanding of the relationship between the number of neighbors and model correctness. By achieving a balance between including enough neighboring cases for precise predictions and preventing over-fitting or excessive influence from irrelevant neighbors, it enables the model's performance to be optimized.

Figure 5.7 can be used to analyze the degree of accuracy attained by various ML techniques. It is clear that among the employed techniques, the random forest technique and decision tree technique exhibit the greatest accuracy. In contrast, when compared to the other models, logistic regression shows the lowest accuracy. The random forest and decision tree models' better accuracy leads to the conclusion that these algorithms are ideally adapted for the task at hand and successfully identify the underlying patterns in the data. However, logistic regression's lower accuracy shows that it is not always capable of correctly predicting the target variable in a given situation.

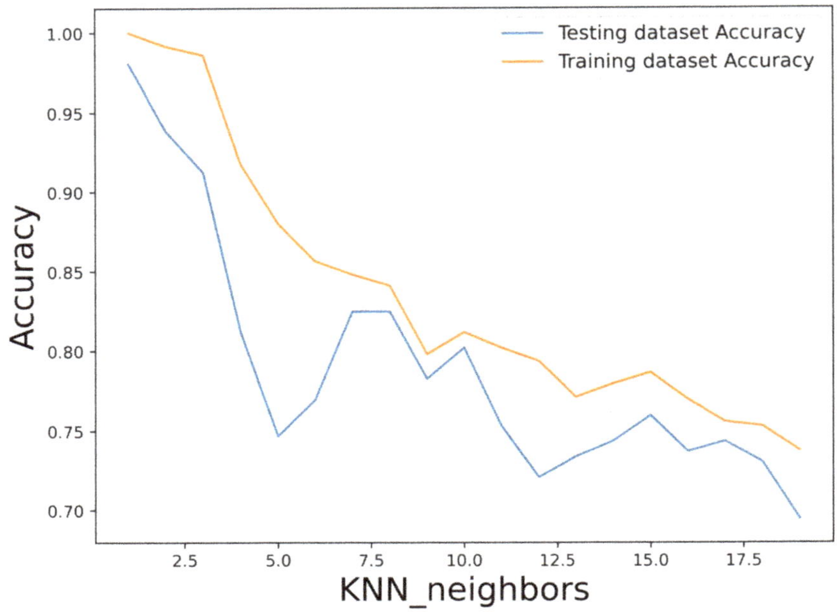

Figure 5.6. KNN neighbor analysis against accuracy of testing and training.

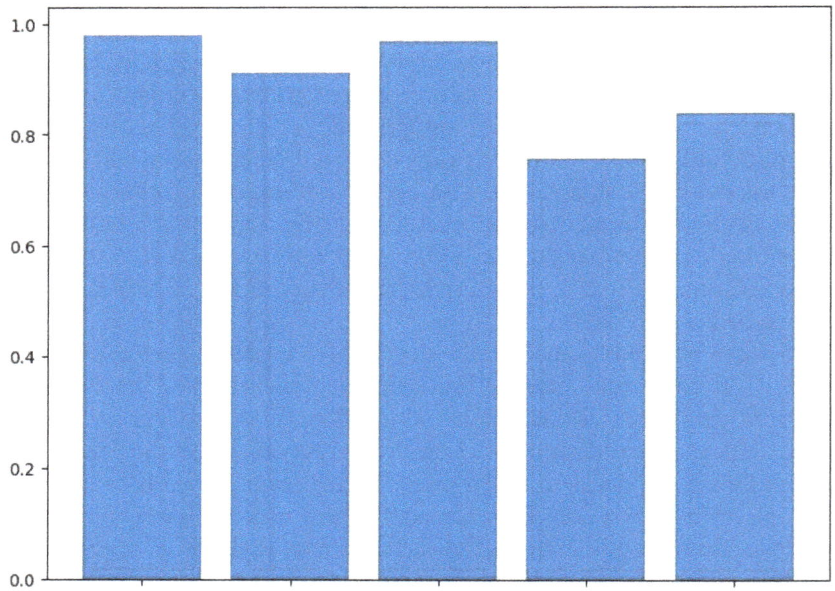

Figure 5.7. Accuracy analysis for the various models designed.

Table 5.4 contrasts the current research with earlier studies that used the same datasets. The work from [12] uses the stacking method to reach a remarkable accuracy of 98.78% while employing GBSFS for feature selection. Another work, cited as [37], combines multiple ML techniques with genetic algorithm with recursive feature reduction. Similar research using stacked SVM with an accuracy of 92.22%

Table 5.4. Comparative analysis of the various works done on heart disease prediction.

Paper	Year	ML algorithms	Accuracy (%)
Rasmy [41]	2018	• Recurrent neural network	82.2
Nirschi [43]	2018	• Random forest (RF) • Deep learning	• 95.2 • 97.4
Zhang [42]	2018	• RF • n-Gram features SVM	RF → 98.2 SVM → 95.2
Amin *et al* [40]	2019	• Feature selection • Voting method	87.4
Ali *et al* [39]	2019	• Stacked SVMs	92.22
Fitriyani *et al* [38]	2020	Heart disease prediction model • SVM • Decision tree • RF • Naive Bayes (NB) • LR • MLP	Heart disease prediction model-> • 95.90 (staglog) • 98.40 (Cleveland)
Rani *et al* [37]	2021	Genetic algorithm and recursive feature elimination • RF • Adaboost • SVM • NB • LR	• RF-86.6
Vikas Chaurasia *et al* [12]	2022	GBSFS features and • GBC • LR • SVM • DT • KNN • RF • MLP	Stacking-98.78

(*Continued*)

Table 5.4. (*Continued*)

Paper	Year	ML algorithms	Accuracy (%)
		• ET • stacking	
Proposed method	2023	BPSO Feature selection and • DT • RF • KNN • NB	RF-98

can be found in [39]. By combining feature selection with a voting system, the author of [40] reports an accuracy of 87.4%. Recurrent neural networks are used in the study cited as [41]; it reports an accuracy of 82.22%. The analysis in the last two publications, [42] and [43], is based on random forest approaches. These comparative studies are included in table 5.4 to provide a thorough assessment of the current work in comparison to previously published research. It shows how accurate and competitive the suggested strategy is as well as how well the strategies for feature selection and classification are used. These comparisons help researchers understand how various procedures perform and help them pinpoint the advantages and disadvantages of particular strategies.

5.6 Conclusion

This article explores different types of ML methods for the prediction of heart disease and classification accuracy improvement, especially when dealing with issues like high sample size and an unbalanced dataset. Data augmentation will be carried out using the SMOTE technique to overcome these issues. SMOTE contributes to the dataset's balancing and growth, which improves the models' performance. Additionally, feature selection will be used as an essential preprocessing step in the classification process. The dataset will be cleaned up by removing unnecessary, redundant, and noisy characteristics. The model's performance can be improved, computational costs may be decreased, and the problem of the curse of dimensionality can be alleviated by performing feature selection. The efficient feature selection method to be used is BPSO. Due to its capacity to efficiently explore huge search spaces, computational efficiency, simplicity of use, and low parameter requirements, BPSO is a good choice for this purpose. This project will make use of the Kaggle dataset on cardiac illness. The goal is to increase accuracy and decrease the incidence of false positives in cardiac disease prediction by utilizing the aforementioned methods and models. The models have performed well, according to the data so far, showing improved accuracy and a decrease in false positives. These encouraging results demonstrate the potency of the employed methods and serve as inspiration for additional research and model development in order to forecast cardiac illness.

In order to increase the prediction accuracy of heart disease, this article explores various ML models. This is especially important when dealing with issues like large sample sizes and an uneven dataset. In the future, we intend to look for better dataset and feature selection methods.

Bibliography

[1] World Health Organization 2017 *World Health Statistics 2017: Monitoring Health for the SDGs*

[2] Ordonez C 2006 Association rule discovery with the train and test approach for heart disease prediction *IEEE Trans. Inf. Technol. Biomed.* **10** 334–43

[3] Kusiak A *et al* 2006 Hypoplastic left heart syndrome: knowledge discovery with a data mining approach *Comput. Biol. Med.* **36** 21–40

[4] Poirier P 2008 Healthy lifestyle: even if you are doing everything right, extra weight carries an excess risk of acute coronary events *Circulation* **117** 3057–9

[5] Murphy K P 2012 *Machine Learning: A Probabilistic Perspective* (Cambridge: MIT Press) Kanimozhi V A and Thirunavu K 2016 A survey on machine learning algorithms in data mining for prediction of heart disease *Int. J. Adv. Res. Comput. Commun. Eng.* **5** 552–7

[6] Tripoliti E E, Papadopoulos T G, Karanasiou G S, Naka K K and Fotiadis D I 2017 Heart failure: diagnosis, severity estimation and prediction of adverse events through machine learning techniques *Comput. Struct. Biotechnol. J.* **15** 26–47

[7] Anbarasi M, Anupriya E and Iyengar N C H S N 2010 Enhanced prediction of heart disease with feature subset selection using genetic algorithm *Int. J. Eng. Sci. Technol.* **2** 5370–6

[8] Florence S, BhuvaneswariAmma N G, Annapoorani G and Malathi K 2014 Predicting the risk of heart attacks using neural network and decision tree *Int. J. Innov. Res. Comput. Commun. Eng.* **2** 2320–9798

[9] Patro S P, Padhy N and Sah R D 2022 An ensemble approach for prediction of cardiovascular disease using meta classifier boosting algorithms *Int. J. Data Warehous. Min. (IJDWM)* **18** 1–29

[10] Bhatt C M, Patel P, Ghetia T and Mazzeo P L 2023 Effective heart disease prediction using machine learning techniques *Algorithms* **16** 88

[11] El-Shafiey M G, Hagag A, El-Dahshan E S A and Ismail M A 2022 A hybrid GA and PSO optimized approach for heart-disease prediction based on random forest *Multimedia Tools Appl.* **81** 18155–79

[12] Chaurasia V and Chaurasia A 2023 Novel method of characterization of heart disease prediction using sequential feature selection-based ensemble technique *Biomed. Mater. Devices*

[13] Ishaq A, Sadiq S, Umer M, Ullah S, Mirjalili S, Rupapara V and Nappi M 2021 Improving the prediction of heart failure patients' survival using SMOTE and effective data mining techniques *IEEE Access* **9** 39707–16

[14] Rahim A, Rasheed Y, Azam F, Anwar M W, Rahim M A and Muzaffar A W 2021 An integrated machine learning framework for effective prediction of cardiovascular diseases *IEEE Access* **9** 106575–88

[15] Shah D, Patel S and Bharti S K 2020 Heart disease prediction using machine learning techniques *SN Compu. Sci.* **1** 345

[16] Mustafa J, Awan A A, Khalid M S and Nisar S 2018 Ensemble approach for developing a smart heart disease prediction system using classification algorithms *Res. Rep. Clin. Cardiol.* **9** 33–44
[17] Khateeb N and Usman M 2017 *Proc. of the Int. Conf. on Big Data and Internet of Thing Efficient heart disease prediction system using K-nearest neighbor classification technique* 21–6
[18] Babaoglu İ, Findik O and Ülker E 2010 A comparison of feature selection models utilizing binary particle swarm optimization and genetic algorithm in determining coronary artery disease using support vector machine *Expert Syst. Appl.* **37** 3177–83
[19] Fernández A, Garcia S, Herrera F and Chawla N 2018 SMOTE for learning from imbalanced data: progress and challenges, marking the 15-year anniversary *J. Artif. Intell. Res.* **61** 863–905
[20] Chawla N, Bowyer K W, Hall L O and Kegelmeyer W P 2002 SMOTE: synthetic minority over-sampling technique 1106.1813
[21] Chawla N, Lazarevic A, Hall L O and Bowyer K W 2003 *Knowledge Discovery in Databases: PKDD 2003, 7th European Conf. on Principles and Practice of Knowledge Discovery in Databases SMOTEBoost: improving prediction of the minority class in boosting (Cavtat-Dubrovnik, Croatia)* 107–19
[22] Kwon O and Sim. J M 2013 Effects of data set features on the performances of classification algorithms *Expert Syst. Appl.* **40** 1847–57
[23] Dash M and Liu H 1997 Feature selection for classification *Intell. Data Anal.* **1** 131–56
[24] Liu H and Motoda H 1998 *Feature Selection for Knowledge Discovery and Data Mining* (New York: Springer) The Springer International Series in Engineering and Computer Science 454
[25] Katrutsa A and Strijov V 2017 Comprehensive study of feature selection methods to solve multicollinearity problems according to evaluation criteria *Expert Syst. Appl.* **76** 1–11
[26] Talbi E-G 2009 *Metaheuristics: From Design to Implementation* (New York: Wiley)
[27] Holland J H 1992 *Adaptation in Natural and Artificial Systems* (Cambridge, MA: MIT Press)
[28] Storn R and Price K 1997 Differential evolution–a simple and efficient heuristic for global optimization over continuous spaces *J. Global Optim. 11* **4** 341–59
[29] Fernández Martínez J L and García Gonzalo E 2008 The generalized PSO: a new door to PSO evolution *J. Artif. Evol. Appl.* **2008** 861275
[30] Chuang L Y, Chang H W, Tu C J and Yang C H 2008 Improved binary PSO for feature selection using gene expression data *Comput. Biol. Chem.* **32** 29–38
[31] Hilbe J M 2011 Logistic regression *International Encyclopedia of Statistical Science* ed M Lovric (Berlin: Springer) 755–8
[32] Kingsford C and Salzberg S L 2008 What are decision trees? *Nat. Biotechnol.* **26** 1011–3
[33] Biau G and Scornet E 2016 A random forest-guided tour *Test* **25** 197–227
[34] Rish I 2001 An empirical study of the naive Bayes classifier *IJCAI 2001 Workshop on Empirical Methods in Artificial Intelligence* **3** 41–6
[35] Guo G, Wang H, Bell D, Bi Y and Greer K 2003 *On the Move to Meaningful Internet Systems 2003: CoopIS, DOA, and ODBASE KNN model-based approach in classification (Catania, Sicily, Italy)* 986–96
[36] https://www.kaggle.com/datasets/johnsmith88/heart-disease-dataset
[37] Rani P, Kumar R, Ahmed N M and Jain A 2021 A decision support system for heart disease prediction based upon machine learning *J. Reliab. Intell. Environ.* **7** 263–75

[38] Fitriyani N L, Syafrudin M, Alfian G and Rhee J 2020 HDPM: an effective heart disease prediction model for a clinical decision support system *IEEE Access* **8** 133034–50
[39] Ali L, Niamat A, Khan J A, Golilarz N A, Xingzhong X and Noor A *et al* 2019 An optimized stacked support vector machine based expert system for the effective prediction of heart failure *IEEE Access* **7** 54007–14
[40] Amin M S, Chiam Y K and Varathan K D 2019 Identification of significant features and data mining techniques in predicting heart disease *Telemat. Inform.* **36** 82–93
[41] Rasmy L, Wu Y, Wang N, Geng X, Zheng W J and Wang F *et al* 2018 A study of generalizability of recurrent neural network-based predictive models for heart failure onset risk using a large and heterogeneous EHR data set *J. Biomed. Inform.* **84** 11–6
[42] Zhang R, Ma S, Shanahan L, Munroe J, Horn S and Speedie S 2018 Discovering and identifying New York heart association classification from electronic health records *BMC Med. Inform. Decis. Mak.* **18** 48
[43] Nirschl J J, Janowczyk A, Peyster E G, Frank R, Margulies K B, Feldman M D and Madabhushi A 2018 A deep-learning classifier identifies patients with clinical heart failure using whole-slide images of H&E tissue *PLoS One* **13** e0192726

Chapter 6

Python based bio-signal processing: mitigation of baseline wandering in pre-recorded electrooculogram

K K Mujeeb Rahman

Baseline wandering is a common artefact observed in bio-signals, particularly in electrocardiogram (ECG), electrooculogram (EOG), and electroencephalogram (EEG) signals. It refers to the slow drift or fluctuation of the baseline of the signal, which can obscure the underlying physiological information. Baseline wandering can be caused by various factors, which introduce low-frequency noise into the signal, resulting in a shifting baseline. One of the biggest challenges in bio-signal processing is mitigating baseline wandering without compromising the signal quality. In this chapter, we introduce baseline wandering and its origin, using a pre-recorded EOG signal as an example. We discuss techniques such as the normalisation and standardisation of an EOG, wavelet decomposition of EOG signals, and estimations of signal power at different levels. We also demonstrate the filtering of the baseline drift on EOG samples using a threshold set based on wavelet energy. The chapter is complemented with a hands-on Python code that covers loading the signal, wavelet decomposition, and baseline correction.

6.1 Introduction

Bio-medical signals, often known as bio-signals, are signals that come from various human body organs and reveal the state of those organs' health. According to their initial form, bio-electric signals and non-electric bio-signals are the two main categories into which bio-signals fall. Bio-electric signals include, but are not limited to, ECGs, EOGs, EEGs, and electromyograms (EMGs) [1]. Some examples of non-electric bio-signals are blood pressure, body temperature, heart rate, $Sp0_2$, etc. Bio-potential electrodes must be attached to a patient in order to collect bio-electric signals, whereas bio-sensors or transducers must be used to receive non-electric bio-signals [2]. Additionally, a bio-data acquisition (Bio-DAQ) module is also required to record

doi:10.1088/978-0-7503-5929-0ch6

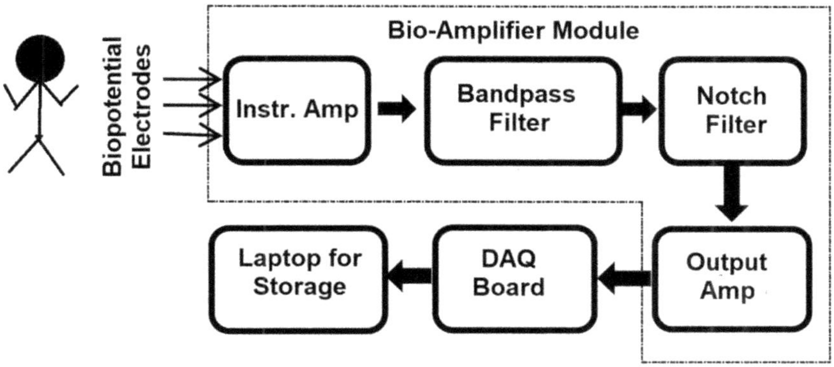

Figure 6.1. The structure of the bio-electric signal acquisition module.

these bio-signals. These procedures are non-invasive, which means they do not cut or pierce the human body. The structure of a general-purpose Bio-DAQ module is in figure 6.1 [3]. The figure consists of five different sub-modules: an instrumentation amplifier, a bandpass filter, a notch filter, an output amplifier, and a DAQ module. A Bio-DAQ module's performance, including the amount of noise it introduces, can certainly change depending on the particular device being used, its design, and the setting in which it is being evaluated [4]. It is recommended to refer to the manufacturer's specifications to learn more precisely about the noise levels or performance parameters of Bio-DAQ.

Generally, bio-signals are highly susceptible to contamination due to their characteristics and various kinds of noise and artefacts [5]. 'Noise' refers to random signals that are superimposed on bio-signals, while 'artefacts' are disturbances or distortions that have identifiable causes and patterns. Both noise and artefacts can impact the accuracy and quality of bio-signals. Therefore, pre-processing is a crucial step in bio-signal analysis or interpretation to reduce the effect of noises and artefacts present in the raw data without damaging the pertinent data, making the signals more suitable for further analysis. The most common types of noise present in the bio-signals are baseline wander, muscle tremor noise, and powerline interferences [6].

6.1.1 Baseline wander noise

Baseline wander is a common occurrence, particularly when ECG, EEG, and EOG signals are being recorded. The cause of baseline wander could be any of the following factors. Baseline fluctuations may result from the heart's modest movement as a result of breathing in and out. Even the tiniest of a patient's movement or muscular contraction can cause the electrodes to move and affect the baseline. Baseline fluctuations can also be caused by issues with the electrical contact between the electrodes if the electrical contact is sporadically broken. The baseline may fluctuate as a result of uneven electrode–skin contact caused by sweat, lubricants, or other impurities on the skin [7]. Furthermore, over time, a minor voltage potential

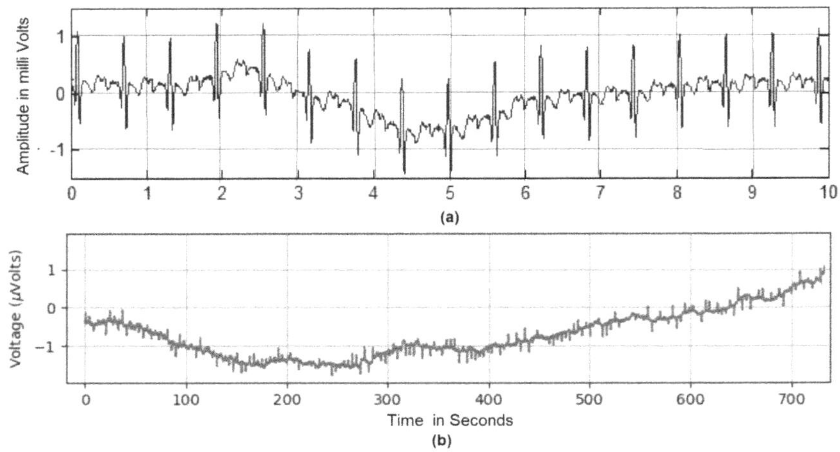

Figure 6.2. The effect of baseline wandering on: (a) an ECG signal and (b) an EOG signal.

forms at the electrode–skin contact as a result of polarisation; the voltage potential may cause the baseline to drift in a slowly shifting way. Baseline drift can be seen on the ECG and EOG signals shown in figure 6.2(a) and (b), respectively. Because baseline wander can distort the original underlying waveform and make it challenging to extract relevant characteristics or detect minute changes in the bio-signal, it must be eliminated. The bio-signal is effectively aligned to a constant reference level by removing baseline drift, enabling subsequent analysis techniques, like signal processing or pattern recognition algorithms, to operate more precisely and successfully [8].

6.1.2 Muscle tremor noise

When muscles contract or move unintentionally during recording of a bio-electric signal, unwanted electrical impulses are produced. These signals are referred to as 'muscle tremor noise.' These tremors can inject noise into bio-signals, which makes it difficult to accurately extract and analyse the needed information [9]. Bio-signals like EMGs, ECGs, EOGs, and EEGs are frequently impacted by muscle tremor noise. These signals can be contaminated by tremor artefacts, especially if electrodes are positioned over muscles or if the subject moves their muscles too much. Figure 6.3 shows the effect of involuntary muscle activity in an ECG signal.

6.1.3 Powerline interference

Bio-signal recordings, such as those of ECG, EEG, and EMG signals, frequently encounter interference from powerline noise. Electromagnetic coupling between the powerlines and the electrodes or recording apparatus is the main contributor to the interference. Depending on the country, either 50 Hz or 60 Hz is the common most powerline frequency [9]. Due to its close frequency proximity to the physiological

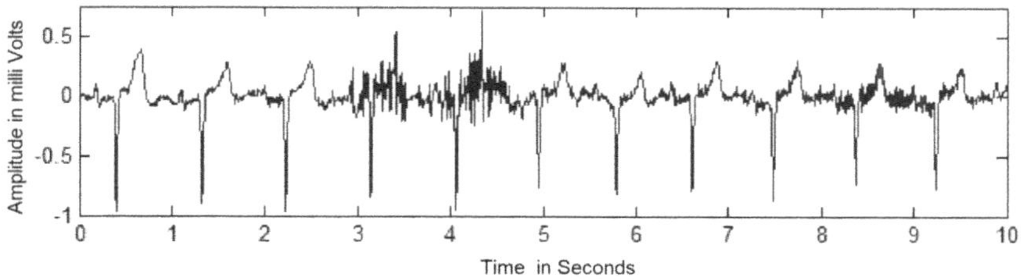

Figure 6.3. The muscle tremor noise in an ECG signal.

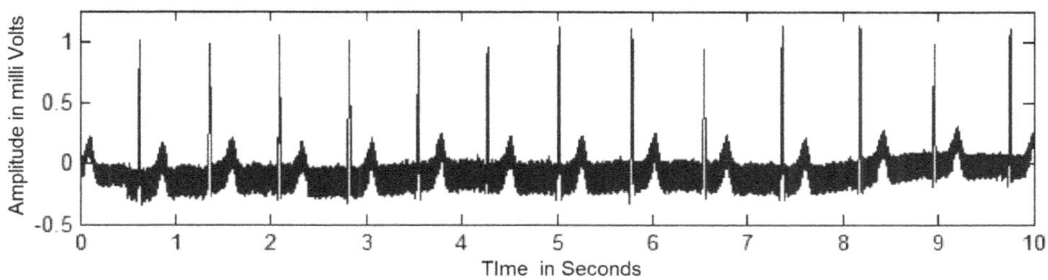

Figure 6.4. The powerline interference with ECG.

signals of interest, powerline noise can significantly affect bio-signal recordings, as shown in figure 6.4. Underlying physiological data may be concealed or altered by these disturbances, which are periodic in nature.

6.1.4 EOGs and their clinical use

Electrooculography, a method for measuring and capturing the electrical potential generated by eye movements, is based on the concept that the cornea (the front surface of the eye) and the retina (the rear surface of the eye) have distinct electrical charges. An electrooculography uses electrodes placed around the eyes to detect this dipole [10, 11]. EOG signals are useful tools for a variety of clinical and scientific applications relating to eye movements and eye-related illnesses because they are non-invasive and reasonably simple to capture. With an emphasis on minimising baseline drift brought on by EOG signals, in this chapter, we provide a range of bio-signal processing techniques to enhance the signal-to-noise ratio of previously collected raw EOG data. An EOG serves a number of crucial purposes: an EOG aids in the evaluation of smooth pursuits, fixation, and saccades in eye movement analysis. By examining the rapid-eye-movement period, a stage of sleep marked by elevated brain activity and vivid dreams, an EOG aids in the understanding of various sleep phases [12]. Eye movements are used by human–computer interface systems to enable persons with disabilities to interact with tools and technology [13]. EOG and infra-red cameras offer devices for tracking the users' gaze, enhancing assistive technology. Nystagmus, oculomotor nerve palsy, and myasthenia gravis are also diagnosed and monitored by EOGs [14]. Aviation, automotive, and human factors studies all benefit from using EOGs to assess visual attention, effort, and eye tiredness. These assessments then enhance safety and human–machine interactions. Users may simply interact with virtual items in virtual reality (VR) and augmented reality (AR) systems by moving their eyes, which creates an immersive experience [15]. An EOG aids in the comprehension of visual perception, attentional mechanisms, and cognitive load. For patients with eye movement deficits, EOG-based therapy regimens enhance eye movement coordination and quality of life, effort, and eye tiredness, all of which enhance safety and human–machine interactions [16]. Users may simply interact with virtual items in VR and AR systems by moving their eyes, which creates an immersive experience. An EOG aids in the comprehension of visual perception, attentional mechanisms, and cognitive load [17]. For patients with eye movement deficits, EOG-based therapy regimens enhance eye movement coordination and quality of life [18].

On the Google Colab platform, we use Python programming to show how these signal processing techniques are implemented. Additionally, this chapter provides instructions on using Google Colab for Python coding.

6.2 Introduction to Google Colab

Google Colab® (https://colab.google/) is an online application that allows users to write, run, and collaborate on Python code in a group context. The application offers a variety of tools and resources for Python programming as well as for data analysis and machine learning (ML). It is based on the well-liked Jupyter Notebook user interface. One can write and execute Python code with Google Colab without having to install or configure anything locally. With pre-installed libraries and dependencies, including well-known ones like pandas, NumPy, and TensorFlow, Google Colab offers a virtual machine environment that makes it simple for both newcomers and experienced coders to get started straight away [19].

The integration of Google Colab with Google Drive (Gdrive) is one of the major benefits of using Google Colab. The processes of accessing files and datasets on Gdrive, importing files into Colab notebooks, and saving work are all done simply. This makes it possible for simple notebook sharing and seamless collaboration. Additionally, Google Colab offers strong hardware resources like tensor processing units (TPUs), graphical processing units (GPUs), and central processing units (CPUs). These tools are free and dramatically increase the pace of computationally demanding operations, such as developing ML models.

The markdown cells available in Colab notebooks allow the user to write formatted text, add photos, and add interactive components, making them ideal for papers that combine code, visualisations, and explanations.

6.2.1 How to use Google Colab

The procedures and illustrations below explain how to use Google Colab:

(a) Open Google Colab: Open your web browser and navigate to Google Colab at https://colab.research.google.com/.
(b) Sign in: Sign into your Google account. If you don't have one, you can create a new account for free.
(c) Create a new notebook: Once you're signed in, click on 'New Notebook' to create a new Colab notebook. Figure 6.5 shows the screenshot of a new notebook page. It has mainly two types of cells: a code cell and a text cell. As the names of the cells imply, users can enter Python code in the code cell and text input in the text cell. The main menu options are: 'File,' 'Edit,' 'View,' 'Insert,' 'Tools,' and 'Help,' with the usual meanings. The 'Runtime' tab is used to control the runtime environment for the notebook with commands such as 'start,' 'halt', and 'restart.' Additionally, user can change the runtime type, which offers choices for CPU, GPU, or TPU accelerators.
(d) Add code cells: In the notebook, you'll see an empty code cell where you can write and execute Python code. Click on the '+ Code' button to add a new code cell, as shown in figure 6.5.
(e) Write and execute code: In the code cell, you can write your Python code. To execute the code, either click on the 'Play' button next to the code cell or use the keyboard shortcut 'Shift + Enter.' The code will be executed, and the output, if any, will be displayed below the cell. The code example given below produces the sum of two numbers after receiving two numbers from the user. Figure 6.6 shows the output of the sample Python code.

```
************************************
# A sample program in python
a=input('Enter first number: ')
b=input('Enter first number: ')
c=int(a)+int(b)
print(a, " + ",b," = ",c)
************************************
```

(f) Add markdown cells: You can add markdown cells to provide explanations, add headings, or include formatted text. To add a markdown cell, click on

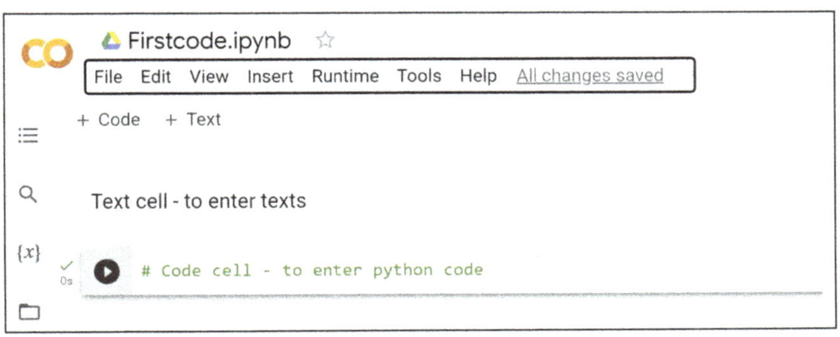

Figure 6.5. The text cell and code cells of a Python Integrated Development Environments (IDE).

```
Enter first input:32
Enter second input:46
32 + 46  =  78
```

Figure 6.6. The output of the sample Python code.

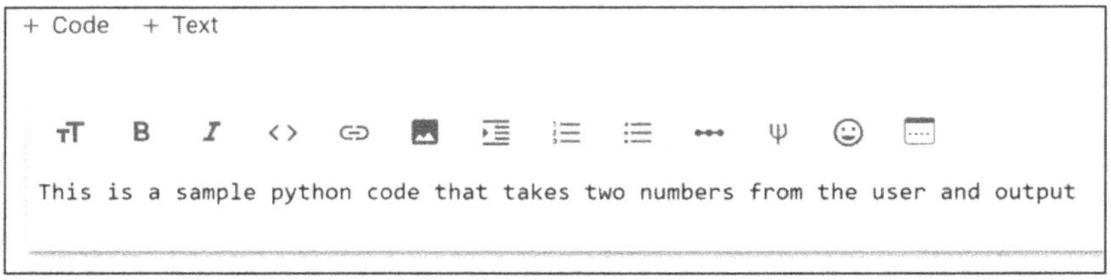

Figure 6.7. The markdown cells in Python IDE.

the '+ Text' button or select 'Insert' from the menu and choose 'Text cell,' as illustrated in figure 6.7.

(g) Save and load notebooks: Google Colab saves your notebook to your Gdrive automatically. Alternatively, you may use the 'File' option to upload an existing notebook from your local computer or download the notebook as an .ipynb file.

(h) Access files and datasets: Google Colab provides easy access to files and datasets stored in your Gdrive. You can mount your Gdrive using the provided code snippet, and then access files using the file path. The first code cell, shown in figure 6.7, is used to mount Google access of your data from Gdrive. The execution of this code prompts the user to provide a valid Gmail identification and password to grant access.

(i) Share and collaborate: You can share your Colab notebooks with others by clicking on the 'Share' button. You can also invite collaborators to view or edit your notebook, and they can make changes in real-time.

(j) Install and import libraries: You can install and import new libraries (e.g., pydicom) as explained in the second and third cells of figure 6.8.

(k) Share and collaborate: You can share your Colab notebooks with others by clicking on the 'Share' button. You can invite collaborators to view or edit your notebook, and they can make changes in real-time.

```
+ Code   + Text

[1]  from google.colab import drive
     drive.mount('/content/drive/')

     Mounted at /content/drive/

[2]  !pip install pydicom

     Collecting pydicom
       Downloading pydicom-2.4.2-py3-none-any.whl (1.8 MB)
                                    ━━━━━━━━━━━━━━━━━━━━━━━━ 1.8/1.8 MB 10.4 MB/s eta 0:00:00
     Installing collected packages: pydicom
     Successfully installed pydicom-2.4.2

[3]  import pydicom
```

Figure 6.8. Mounting GDrive in Google Colab and installation and import of new libraries.

6.3 Algorithm for correction of baseline wandering in pre-recorded EOG signals

This section describes Python implementation of the correction of baseline wandering associated with EOG signals. The signals used in this notebook are available upon request. Any signal processing process begins by importing data from a source folder, which might be at a specified place on a personal computer (e.g., desktop). Google Colab, however, has access to the data stored in Gdrive. As a result, the initial step is to store the data in Gdrive. Hence the first step is to mount Gdrive to Colab, as described in section 6.2.1, and then load the data from the source folder. We created a folder with a name 'Sample signals' in the 'My Drive' folder that contains two-channels (left channel, horizontal eye movements; right channel, vertical eye movements) of EOG signals employed in our study. Figure 6.9 illustrates an approach for baseline correction. Starting with loading two-channel pre-recorded EOG signals from the source file, this approach separates the EOG data into horizontal and vertical directions before using wavelet decomposition to remove the baseline wandering noise that corrupted the supplied EOG data. Each step of the algorithm is described in the following sections with the Python codes and relevant outputs.

6.3.1 Steps 1–3

To begin with, the 'os' built-in library function needs to be imported, which offers a means of interacting with the operating system. The function provides a number of methods and functions for carrying out operations on files and directories, managing environment variables, and other related activities. Python instructions for Steps 1–3

Step 1→ Import the required dependencies
Step 2→ Mount Google drive in Colab to fetch the data
Step 3→ Get path of the source folder
Step 4→ Load the EOG data from the source folder
Step 5→ Convert the data into a dataframe using pandas
Step 6→ Segregate the data into vertical and horizontal channels
Step 7→ Visualize the data
Step 8→ Standardize the data for each channel
Step 9→ Perform frequency spectrum analysis using the periodogram
Step 10→ Use discrete wavelet transform (DWT) to break up the EOG signal until the energy of the detailed coefficients reach a local minimum
Step 11→ Obtain the baseline by reconstructing the approximate coefficients at the above level and using inverse DWT.
Step 12→ Obtain the output signal with minimal baseline wandering.

Figure 6.9. The algorithm used for baseline correction.

```
+ Code   + Text

[1]  from google.colab import drive
     drive.mount('/content/drive/')

     Mounted at /content/drive/

[2]  # To get the path of current directory
     import os
     path=os.getcwd()
     print(path)

     /content

[3]  # To change the directory where the source folder is located
     os.chdir('/content/drive/My Drive/Sample signals/')
     path=os.getcwd()
     os.listdir()

     ['100m.mat', 'EOG.mat']
```

Figure 6.10. Mounting Gdrive and setting up the source directory's path.

are given below, and the corresponding screen view is shown in figure 6.10. As previously mentioned, cell-1 is used to mount Gdrive to Colab, cell-2 is used to retrieve the path of the current working directory, and cell-3 is used to find the necessary file by modifying the current directory.

```
**************************************************
# To mount the Gdrive
from google.colab import drive
drive.mount("/content/drive/")

# To get the path of working directory
import os
path = os.getcwd()
print(path)

# To change the directory where the source folder is located
os.chdir("/content/drive/My Drive/Sample signals/")
path = os.getcwd()
os.listdir()
**************************************************
```

6.3.2 Step 4

In this step, the data from the source directories is loaded by using the Python library 'SciPy.' SciPy is a free and open-source Python library used for technical and scientific computing. For typical tasks in science and engineering, SciPy includes modules for optimisation, linear algebra, integration, interpolation, special functions, fast Fourier transform (FFT), signal and image processing, ordinary differential equation solvers, and other related topics.

```
*********************************************
import scipy.io
# Load the .mat file
data = scipy.io.loadmat("EOG.mat")
# To discard the header and retain EOG data only
EOG_data = data["EOG"]
# To show dimensions od the data
print("Dimension of EOG data:", EOG_data.shape)
# To print a few samples of the loaded data
print(EOG_data)
*********************************************
```

In figure 6.11, the screen view of the above-mentioned codes is displayed. Data and headers are loaded into memory by the first cell's code. The code in the second cell specifies the EOG data dimensions. The loaded data contains two channels: left of which represents horizontal eye movement data and right of which indicates vertical eye movement data. To print a few samples of the loaded data, use the code in cell-3.

```
[4]  import scipy.io
     # Load the .mat file
     data = scipy.io.loadmat('EOG.mat')
     # Access the variables in the .mat file
     data

    {'__header__': b'MATLAB 5.0 MAT-file, Platform: PCWIN64, Created on: Mon Nov 18 08:38:30 2019',
     '__version__': '1.0',
     '__globals__': [],
     'EOG': array([[ 0.00000000e+00, -2.53125000e+00, -3.39062500e+00, ...,
             -1.76262598e+03, -1.76262305e+03,  0.00000000e+00],
            [ 0.00000000e+00,  1.34619141e+00,  1.07324219e+00, ...,
              1.45675195e+03,  1.45539575e+03,  0.00000000e+00]])}

[5]  EOG_data=data["EOG"]
     print('Dimension of EOG data:',EOG_data.shape)

    Dimension of EOG data: (2, 312190)

[6]  print(EOG_data)

    [[ 0.00000000e+00 -2.53125000e+00 -3.39062500e+00 ... -1.76262598e+03
      -1.76262305e+03  0.00000000e+00]
     [ 0.00000000e+00  1.34619141e+00  1.07324219e+00 ...  1.45675195e+03
       1.45539575e+03  0.00000000e+00]]
```

Figure 6.11. Loading EOG data from the source directory and its pre-processing.

6.3.3 Steps 5–6

To make things simpler, the imported data was used to generate a dataframe. According to the Python code provided below, this calls for importing the pandas package. The pandas module is a well-known open-source tool for data organisation and analysis of the Python programming language. Data loading and saving from various file formats (CSV, Excel, SQL databases, etc.), handling erroneous data, merging and combining datasets, data cleaning and transformation, data selection, filtering, sorting, data summarisation, statistical computations, etc. are characteristics of this module. Use the given Python code's instruction 1 to import pandas, and instruction 2 to convert the EOG data into a dataframe. Instruction 3 must be used to create dataframes for the left and right channels. Lines 6–8 can be used to show the size of the loaded EOG signals as well as their left and right channels.

```
****************************************************
import pandas as pd
EOG_pd=pd.DataFrame(EOG_data)
left =EOG_pd.iloc[0]
right=EOG_pd.iloc[1]
# To know the dimnesions of eacg channel
print('Dimensions of EOG data:', EOG_pd.shape)
print('Dimensions of Left channel:', left.shape)
print('Dimensions of Right channel:', right.shape)
****************************************************
```

6.3.4 Step 7

This stage's objective is to plot the left and right channel of the EOG signals. The whole length (322 190 samples) of the left and right channels is shown in a single plot using the given code. The import of the 'matplotlib.pyplot' package is necessary for this step. The package contains a matplotlib sub-module that offers a practical and simple interface for making different plots and visualisations. Line 1 of the code imports the matplotlib library. The commands on lines 2 and 3 result in the red and blue left and right channels of the EOG signals. Line 4 shows grid lines on the plot. The operations on lines 5 and 6 provide the x and y labels, and the commands on line 6 return the appropriate legend.

```
****************************************************
import matplotlib.pyplot as plt
plt.plot(left, color='red',label='Left channel')
plt.plot(right, color='blue',label='Right channel')
plt.grid() # to show grid lines
plt.xlabel('Samples')
plt.ylabel('Magnitude')
plt.legend()
****************************************************
```

Figure 6.12 displays the horizontal and vertical graphs for the whole pre-recorded EOG, with samples on the x-axis and EOG magnitude on the y-axis, totalling in 312 190 EOG samples. The plots reveal that there is considerable baseline wandering

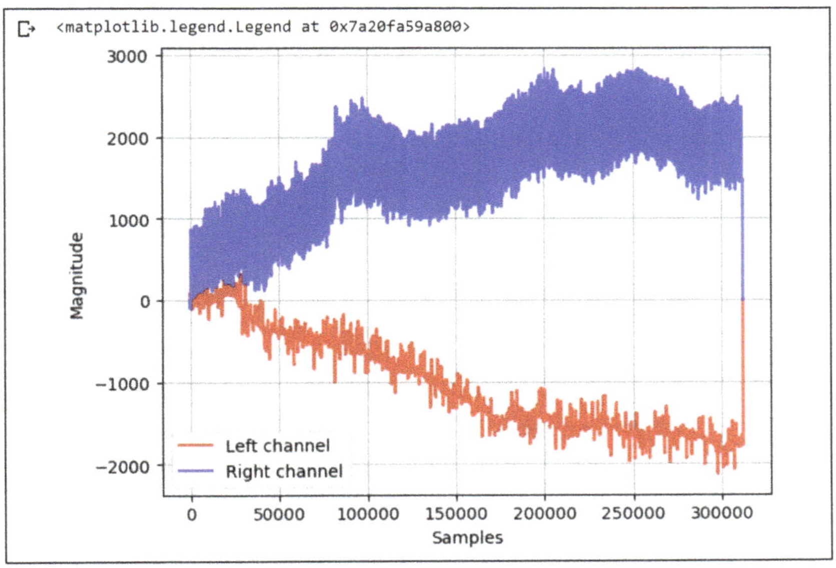

Figure 6.12. Visualisation of the two channels of EOG signals.

noise associated with EOGs, which is obvious on both charts. The baseline wandering should be corrected before further processing or analysis.

The EOG's temporal data are essential in the vast majority of applications. The time intervals may be calculated using equation (6.1) with the number of samples ($n = 312\ 190$) and the sampling rate (256 samples/s for the supplied EOG data). The 'NumPy' Python library, which offers numerical computation, must be included at this point in order to use the Python code that is supplied below to compute time intervals. For effectively manipulating these arrays, NumPy provides a wide range of mathematical operations as well as large, multi-dimensional arrays and matrices. The time series and plots are constructed in lines 4 and 5 of the provided Python code, respectively. The whole duration of the EOG signal is 1219 s.

$$\text{Time in seconds} = \frac{\text{Number of samples}}{\text{Sampling rate}} \tag{6.1}$$

```
*****************************************************************
import numpy as np
number_of_samples = EOG_pd.shape[1]
sampling_rate = 256
time = np.arange(0, number_of_samples / sampling_rate, 1 / sampling_rate)
print('Duration of the signal: ', max(np.round(time)),' in seconds')
plt.plot(time[1:10240],left[1:10240], color='red',label='Left channel')
plt.plot(time[1:10240],right[1:10240], color='blue',label='Right channel')
plt.grid()
plt.xlabel('Time in Seconds ')
plt.ylabel('Magnitude')
plt.legend()
*****************************************************************
```

Figure 6.13 shows the initial 40 s EOG segment. A forward saccade lasting one second (rising edge), a backward saccade lasting one second (falling edge), and a blink are all seen on both EOG channels. Blinks appear better on vertical channels than on horizontal channels, as seen in figure 6.13, though forward and reverse saccades are more distinct on horizontal channels.

6.3.5 Step 8

Standardisation of data points, also known as '*Z*-score normalisation' or 'feature scaling,' is a data preparation method used to transform numerical data such that it has a mean of 0 and a standard deviation of 1. This process aims to bring all characteristics or variables to a similar level for fair comparisons and analysis. The *Z*-score transformation offers several benefits, making the data scale-independent and removing the original data's unit of measurement for equitable comparisons

```
Duration of the signal:  1219.0  in seconds
<matplotlib.legend.Legend at 0x7af0fa36eda0>
```

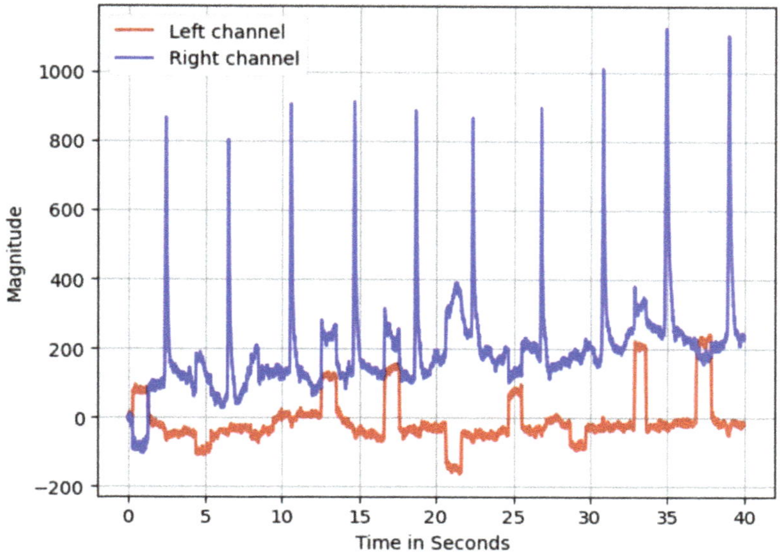

Figure 6.13. A short segment of EOG data with temporal details.

and analysis of variables with different scales. Additionally, the *Z*-score transformation makes data outliers more noticeable, aiding in data purification and benefiting statistical investigations and modelling strategies that rely on the assumption of normality. Furthermore, standardisation can enhance the stability and convergence of several ML techniques, resulting in more accurate and reliable models. However, it is not a must to apply standardisation to all data. Whether to do so depends on the data's properties and the type of modelling. The *Z*-score of a one-dimensional variable, x, may be obtained using equation (6.2), where $\overline{x}$ is the mean value and $\sigma(x)$ is the standard deviation of x.

$$Z_{score}(x) = \frac{x - \overline{x}}{\sigma(x)} \tag{6.2}$$

The following Python script may be used to find the *Z*-score. Line 1 imports the 'scipy.stats' module, which is a crucial tool for many data analysis and statistical modelling tasks because it provides a wide range of statistical functions and probability distributions. Lines 3 and 4 calculate the *Z*-scores for the left and right channels, respectively.

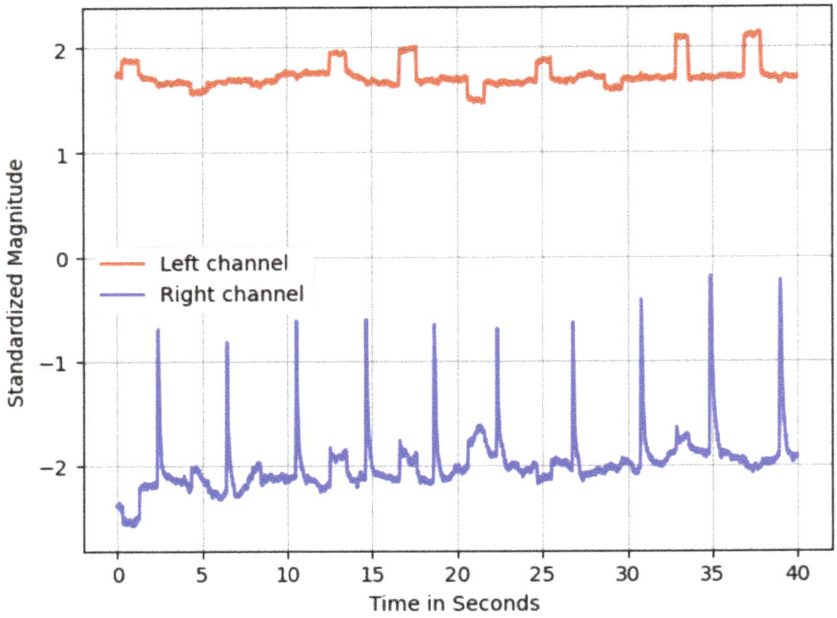

Figure 6.14. EOGs after the *Z*-score standardisation.

```
************************************************************************
from scipy import stats
# Calculate z-scores
left_z = stats.zscore(left)
right_z = stats.zscore(right)
plt.plot(time[1:10240],left_z[1:10240], color='red',label='Left channel')
plt.plot(time[1:10240],right_z[1:10240], color='blue',label='Right channel')
plt.grid()
plt.xlabel('Time in Seconds ')
plt.ylabel('Standardized Magnitude')
plt.legend()
************************************************************************
```

Figure 6.14 displays the plots of normalised EOGs; however, the first 40 seconds are displayed to provide a comprehensive overview. While *Z*-score standardisation does not alter the breadth of data storage, it does alter the scale and units of the data, which is a crucial distinction to make. The waveforms appear to be comparable to those in the graphs' raw version despite the addition of *Z*-scoring to the magnitudes (figure 6.13).

6.3.6 Step 9

The power spectral density (PSD) of a signal (figure 6.15) may be determined using a periodogram, a technique used in spectral analysis and signal processing. The PSD gives a spectrum that shows how much power each frequency in the signal contributed. Fundamentally, equation (6.3) is employed to obtain a periodogram

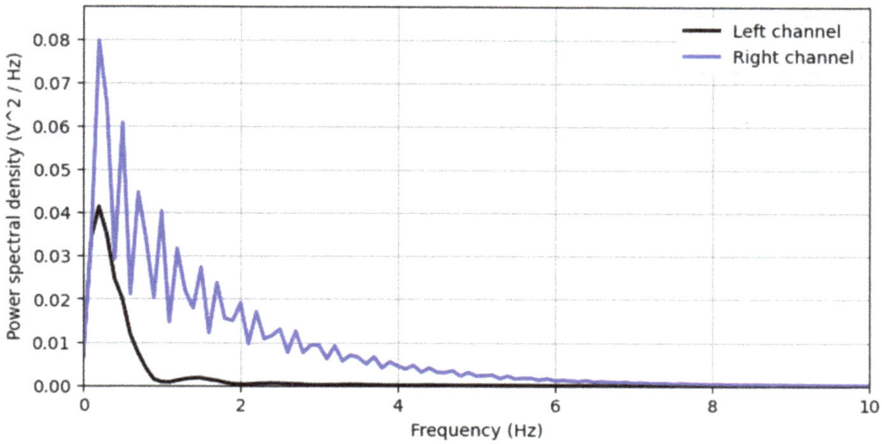

Figure 6.15. The estmated power spectral densities (PSDs) of the left and right channels.

of a discrete signal, where |FFT(x)| is the FFT's absolute value, N is the number of samples, and SR is the sampling rate [20].

$$\text{PSD}(x) = \frac{|\ \text{FFT}(x)\ |^2}{N*SR} \tag{6.3}$$

The basic periodogram, however, has certain drawbacks, including excessive volatility and low-frequency resolution. Improvements have been made to solve these problems, including windowing the signal, averaging numerous periodograms, and employing a more sophisticated spectrum estimate method, called the 'Welch method.' In comparison with the simple periodogram, this technique provides a better PSD estimate. The Welch method is the favoured option in many signal processing applications because it offers superior frequency resolution, less spectral leakage, and more consistent results.

The major steps in Welch's method are as follows [20]:

(a) Divide the original signal into 'n' number of overlapping segments.
(b) Apply a window function to each segment.
(c) Compute a periodogram for each windowed segment by taking its FFT.
(d) Average the individual periodograms to lessen variance and smooth the estimated PSD.

The PSD of the EOG signal is now assessed to better comprehend the link between EOG frequency and signal intensity. The scipy.signal module from the SciPy Python library, which is used for technical and scientific computing, is required for this step. This library offers a broad variety of signal processing capabilities that are beneficial in many applications. The main applications of scipy.signal include signal creation, spectrum analysis, filter design, convolution, frequency response, wavelet transforms, and traditional signal processing methods. Here is a snapshot of the Python script for the Welch's periodogram. The sampling frequency (line 2) and window size (line 3) for this phase must be supplied. The signal.welch function (line 4) produces a list of frequencies and the PSDs that go along with them. The PSDs for each channel must typically be calculated.

```
*****************************************************
from scipy import signal
sf=256 # Sampling rate
win = 10 * sf # choose the winodow size

freqs, psd = signal.welch(left_z, sf, nperseg=win)
plt.figure(figsize=(8, 4))
plt.plot(freqs, psd, color='k', lw=2,label='Left channel')
plt.ylim([0, psd.max() * 1.1])

freqs, psd = signal.welch(right_z, sf, nperseg=win)
plt.plot(freqs, psd, color='b', lw=2,label='Right channel')
plt.xlim([0, 10])
plt.ylim([0, psd.max() * 1.1])

plt.xlabel('Frequency (Hz)')
plt.ylabel('Power spectral density (V^2 / Hz)')
plt.title("Welch's periodogram")
plt.grid()
plt.legend()
*****************************************************
```

Figure 6.19 displays the PSDs of the horizontal (left channel) and vertical (right channel) EOGs.

6.3.7 Step 10

The discrete wavelet transform (DWT) is a powerful signal processing technique used to analyse signals and images. The DWT decomposes a signal into a time–frequency representation by using a set of wavelet functions as the basis functions [21]. To run the DWT, a signal is divided into approximation and detail coefficients at different levels based on a variety of wavelet functions. The detail coefficients capture the signal's high-frequency or minute details, whereas the approximation coefficients reflect the signal's low-frequency components. The selection of wavelet functions is often based on their characteristics and usefulness for specific applications. Examples of wavelet functions employed in the DWT include the Haar, Daubechies, Symlet, Coiflet, and Biorthogonal families [22, 23]. There are a variety of additional wavelet families out there, each with unique features and benefits. Iteratively repeating this decomposition process results in a multi-resolution examination of the signal. The DWT provides a variety of advantages over other signal decomposition techniques, such the Fourier transform. The DWT's primary benefits lie in its ability to effectively capture localised features and transient activity in signals [23]. Its capacity to localise in both time and frequency domains allows for precise analysis, while the multi-resolution analysis ensures a comprehensive representation of the signal's frequency content across various scales or levels of detail. Figure 6.20 illustrates the breakdown of a one-dimensional signal using DWT. The input signal is divided into approximation (A1) and detailed (D1) components at level 1. A1 breaks down into D2 and A2 at level 2, and this process is repeated until level 4. D1, D2, D3, D4, and A4 provide the coefficients as output at the conclusion [24, 25].

A Python script that performs a level-4 decomposition of the EOG signal is presented below. In this work, the Daubechies 2 wavelet (db2) is the best fit for the employed EOG signal. The wavelet decomposition is carried out using the PyWavelets (pywt) library, a popular and useful tool for wavelet research. The pywt library offers a wide range of routines that handle both DWTs and continuous wavelet transforms, enabling tasks such as signal and image denoising, feature extraction, compression, and more. Figure 6.16 shows the five coefficients obtained from the decomposition: the approximate coefficient A4 and the detailed coefficients D4, D3, D2, and D1. These coefficients are plotted using a sub-plot arrangement of 5 × 1, as shown in figure 6.17.

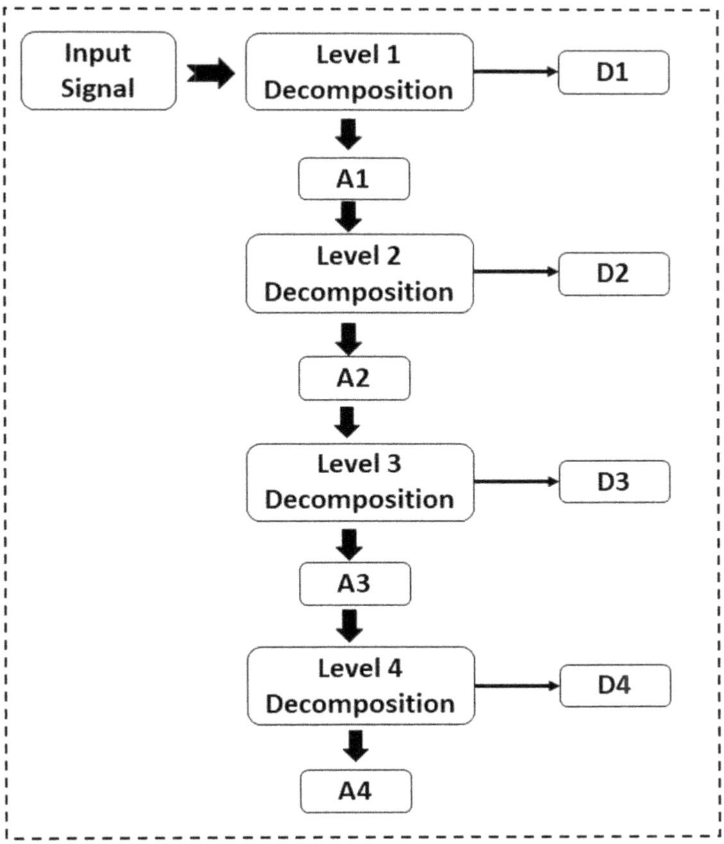

Figure 6.16. The process of a four-level decomposition of a 1d-signal using DWT.

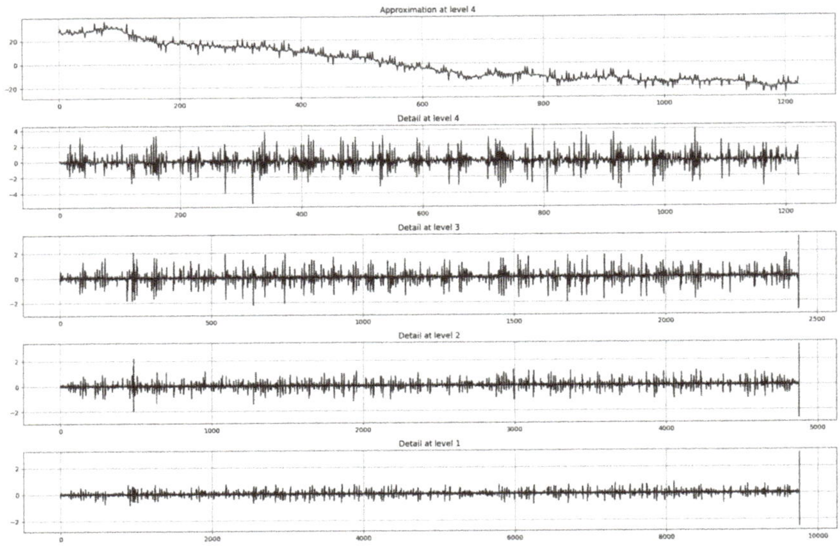

Figure 6.17. Plots of the approximate (A4) and detailed coefficients (D4, D3, D2, and D1).

****

```
import pywt
# Perform four-level decomposition
coeffs = pywt.wavedec(signal, 'db2', level=4)

# Extract the approximation and detail coefficients at each level
A4, D4, D3, D2, D1 = coeffs

# Plot the coefficients (five subplots)
plt.figure(figsize=(12, 6))
plt.subplot(5, 1, 1)
plt.plot(A4)
plt.grid()
plt.title("Approximation at level 4")

plt.subplot(5, 1, 2)
plt.plot(D4)
plt.grid()
plt.title("Detail at level 4")

plt.subplot(5, 1, 3)
plt.plot(D3)
plt.grid()
plt.title("Detail at level 3")
```

```
plt.subplot(5, 1, 4)
plt.plot(D2)
plt.grid()
plt.title("Detail at level 2")

plt.subplot(5, 1, 5)
plt.plot(D1)
plt.grid()
plt.title("Detail at level 1")
plt.tight_layout()
plt.show()
************************************************************
```

Specific thresholding approaches, depending on the particular application, are adopted to omit some of the detailed coefficients. The goal of applying thresholding approaches to the detailed coefficients is to selectively remove certain components of the signal, such as noise or fine-grained details, while retaining the most relevant information. As portrayed in figure 6.16, we use DWT to repeatedly decompose the EOG signal until the energy of the detailed coefficients reaches a local minimum. To find local minima we take three consecutive energy points, and if the middle point is the smallest among the three, it represents a local minimum. Up until the first local minimum is attained, the decomposition process continues. The Python code given below explains how to repeatedly deconstruct the left channel EOG signal (n = 15) and calculate the energy of the detailed coefficients at each level.

```
************************************************
import pywt
arr=[]
ssds = np.zeros((3))
sig = np.copy(left_z)
iterations = 0
i=1
while (i<15):
  lp, hp = pywt.dwt(sig, "db2")
  #print('Energy at level ',i,np.sum(hp ** 2))
  sig=lp
  i=i+1
  arr.append(np.sum(hp ** 2))
plt.plot(arr,label='Left channel')
plt.xlabel('Levels of decomposition')
plt.ylabel('Energy of the detailed coefficients')
plt.legend()
plt.grid()
****************************************************
```

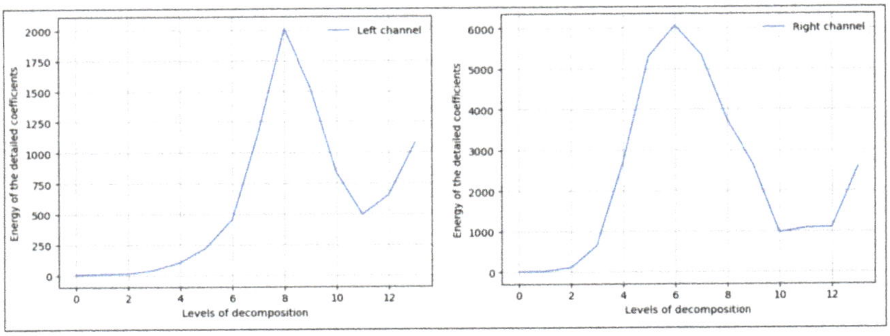

Figure 6.18. Plots of the energries of the detailed coefficeints at different levels of decomposition.

According to the energy curves in figure 6.18, the first local minimum, which is assumed to be where the breakdown comes to an end, is observed at level 12 for the left EOG and at level 11 for the right EOG.

6.3.8 Steps 11 and 12

As stated earlier, the decomposition operation must be repeated until the least energy requirement is satisfied, yielding approximative and precise coefficients. Local minima in the energy distribution of the detailed coefficient, which may be regions with relatively low energy or weak noise components, may be used to detect baseline wandering [12, 13]. By omitting the detailed coefficient at this stage, the baseline may be determined by using the approximation coefficient that corresponds to the lowest energy level and the inverse DWT with the same number of upsampling rounds as in the decomposition. Finally, by deducting the estimated baseline from the original input, we may lessen or eliminate the baseline wandering noise that is present in the pre-recorded EOG. The Python scripts for calculating the left channel EOG's baseline are provided below.

```
*************************************************************************
# Copy left channel EOG
signal = np.copy(left_z)
# to store three concecutive energy values
pnts = np.zeros((3))
latest_lp = np.copy(signal)
iterations = 0
while True:
# Repeated single level decomposition
  lp, hp = pywt.dwt(latest_lp, "db2")
```

```
# Shift and calculate the energy of detail coefficient
  pnts = np.concatenate(([np.sum(hp ** 2)], pnts[:-1]))
# Check if we are in the local minimum of energy function of detailed coefficient
  if pnts[2] > pnts[1] and pnts[1] < pnts[0]:
    break
  latest_lp = lp[:] # to save the lp when energy is min
  iterations += 1
  baseline = latest_lp[:]

# Reconstruct the baseline from the lp iteratively up to the original length
for _ in range(iterations):
   baseline = pywt.idwt(baseline, np.zeros((len(baseline))), "db2")
baseline=baseline[: len(signal)]
# Remove baseline from original signal
output=signal - baseline

# To plot the results
plt.subplot(3, 1, 1)
plt.plot(signal, "b-", label="Z-scored version of Original EOG signal with baseline
wandering")
plt.legend(loc='upper right')
plt.grid()
plt.figure(figsize=(12,8))
plt.subplot(3, 1, 2)
plt.plot(baseline, "r-", label="Estimated baseline")
plt.legend(loc='upper right')
plt.grid()
plt.figure(figsize=(12,8))
plt.subplot(3, 1, 3)
plt.plot(output, "g-", label="Baseline Corrected EOG signal")
plt.legend(loc='upper right')
plt.grid()
plt.show()
****************************************************************************
```

Figure 6.19(a) displays the original (standardised) left channel EOG, (b) the estimated baseline, and (c) the baseline corrected left channel EOG. In figure 6.20, the corresponding graphs (a–c) for the right channel EOG are shown.

The above outcomes clearly demonstrate the efficacy of the suggested method in calculating the baselines of the left channel and right channel EOGs. Additionally, the results show that the algorithm adequately corrects the baseline drifting.

Furthermore, a single-channel ECG signal affected by baseline wandering can be used to evaluate the effectiveness of the proposed algorithm in correcting the baseline wandering artefact. After making the necessary adjustments, the algorithm is applied to a pre-recorded ECG signal (loaded from ecg_sample1.csv) for baseline correction. Similar to EOG signals, the 'db2' wavelet function proves to be an optimal choice. The modified Python code is found below, which includes scripts for loading the ECG

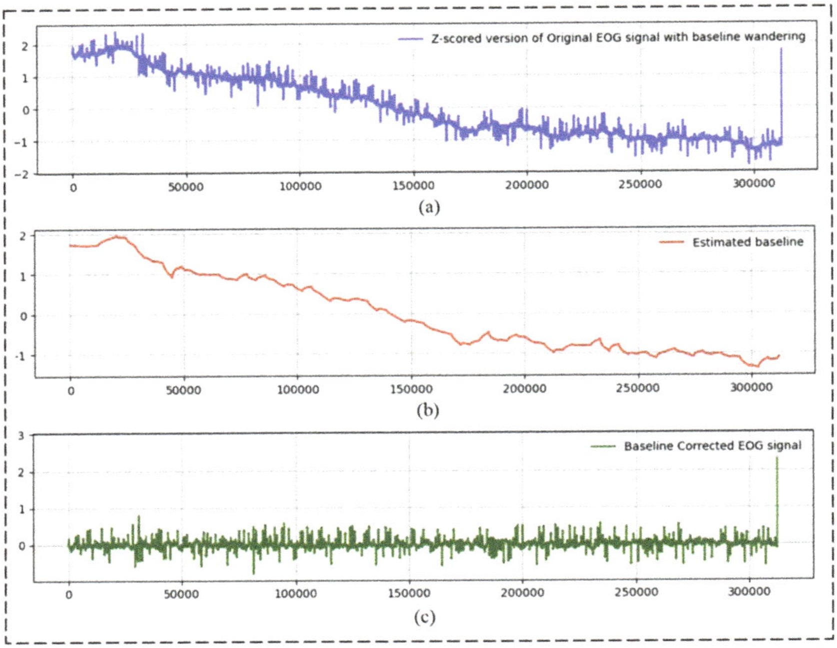

Figure 6.19. Plots of the original, estimated baseline and baseline corrected left channel EOG.

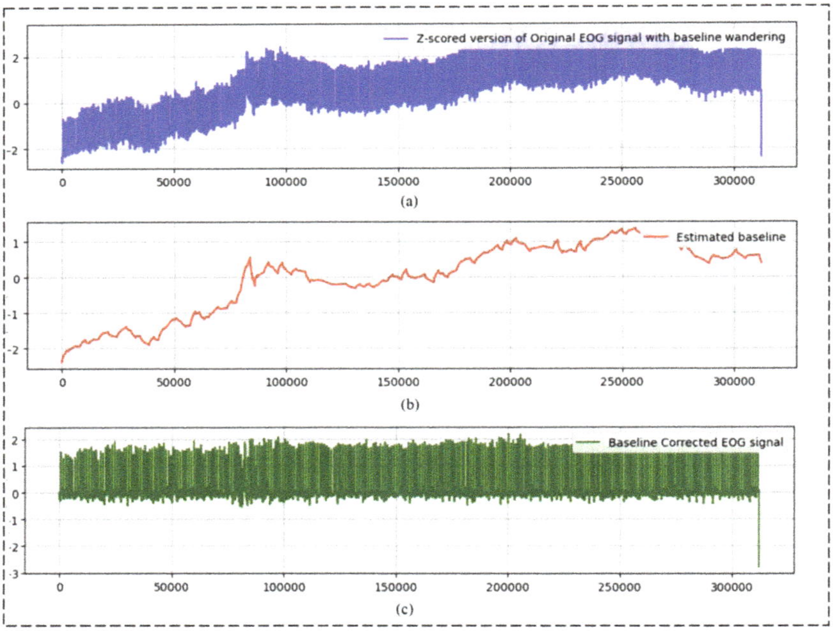

Figure 6.20. Plots of the original, estimated baseline and baseline corrected right channel EOG.

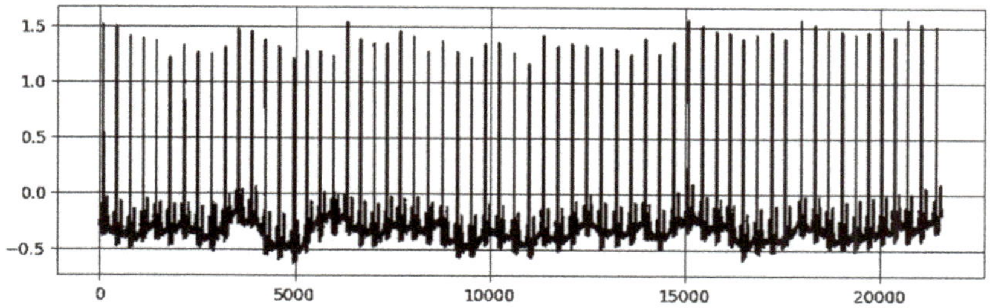

Figure 6.21. An ECG signal (full length) used for the experimental work.

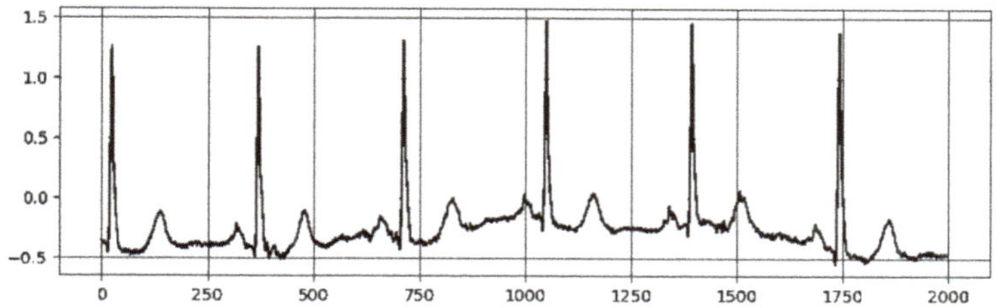

Figure 6.22. A short segment of the ECG signal.

signal, discarding its header lines, and plotting the loaded signals in full length and short duration. The remaining parts of the code remain unchanged. The plots of the ECG for full length and short interval are shown in figures 6.21 and 6.22, respectively.

```
*************************************************************************
import matplotlib.pyplot as plt
# Read ecg_sample1.csv file
f = open('ecg_sample1.csv', 'r')
lines = f.readlines()
f.close()

# Discard the first two header lines.
signal = [0]*(len(lines)-2)
for i in range(len(signal)):
  signal[i] = float(lines[i+2].split(',')[1])

# To show the ECG signal for the entire length
plt.figure(figsize=(10,3))
plt.plot(signal)
plt.grid()

# To show a small segment of the ECG signal
plt.figure(figsize=(10,3))
plt.plot(signal[2500:4500])
plt.grid()
*************************************************************************
```

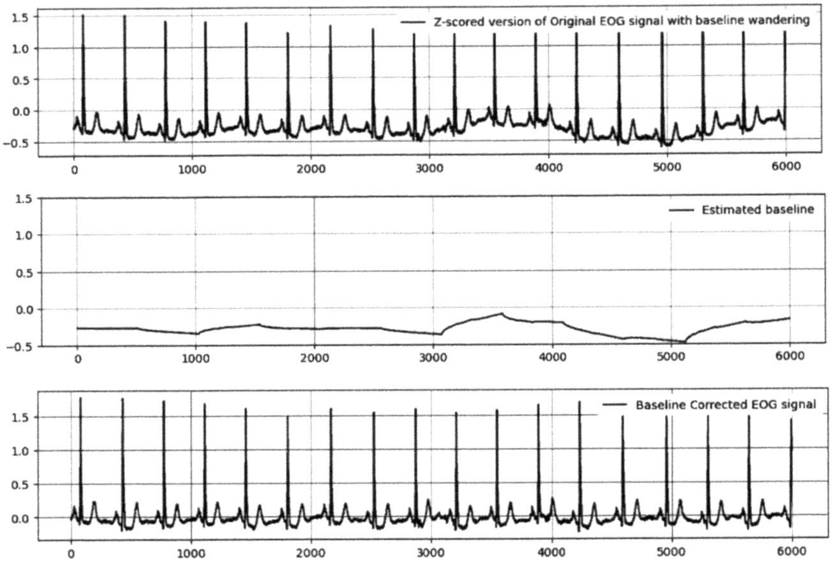

Figure 6.23. Plots of the ECG before and after baseline correction along with the estimated baseline.

The results presented above conclusively show the capacity of the suggested technique to handle baseline artefacts in bio-electric signals, such as an EEG signal, efficiently and reliably (as demonstrated in figure 6.23).

6.4 Conclusion

This chapter covers a comprehensive overview of the origin of bio-electric signals, the methods of acquiring such signals, and the use of Python for pre-processing pre-recorded EOG signals. Additionally, the chapter has explored algorithms and techniques for baseline wander correction, accompanied by detailed Python code and explanations. The chapter began by introducing the fundamental concepts of bio-electric signals and their significance in various physiological processes. To address baseline wander in EOG signals, the chapter introduced the wavelet transform as a powerful tool for signal analysis and decomposition. The chapter then delved into the practical implementation of baseline wander correction using wavelet transform in Python. It provided a step-by-step guide with Python code, illustrating the process of loading pre-recorded EOG data, performing wavelet decomposition, identifying the baseline component, and reconstructing the EOG signal without baseline wander. The code examples and explanations offered a clear understanding of the algorithms and techniques employed for baseline wander correction.

Bibliography

[1] Najarian K and Splinter R 2012 *Biomedical Signal and Image Processing* (Boca Raton: CRC Press) 2nd edn

[2] Sörnmo L and Laguna P 2005 *Bioelectrical Signal Processing in Cardiac and Neurological Applications* (Amsterdam: Elsevier Academic)
[3] Rahman K M and Nasor M 2015 *2015 Int. Conf. Inf. Commun. Technol. Res. (ICTRC) Multipurpose low-cost bio-daq system for real time biomedical applications (Piscataway, NJ)* (IEEE) 286–9
[4] Roncagliolo P, Arredondo L and González A 2007 Biomedical signal acquisition, processing, and transmission using smartphone *J. Phys. Conf. Ser.* **90** 012028
[5] Huigen E, Peper A and Grimbergen C A 2002 Investigation into the origin of the noise of surface electrodes *Med. Biol. Eng. Comput.* **40** 332–8
[6] 1998 *The Measurement, Instrumentation, and Sensors Handbook* ed J G Webster (Boca Raton: CRC Press)
[7] 2011 *Advanced Methods of Biomedical Signal Process.* ed S Cerutti and C Marchesi (Hoboken: Wiley-IEEE Press)
[8] Reilly R B and Lee T C 2010 II.3. Electrograms (ECG, EEG, EMG, EOG) *Stud. Health Technol. Inform.* **152** 90–108
[9] Islam M K, Rastegarnia A and Sanei S 2021 Signal artifacts and techniques for artifacts and noise removal ed M A R Ahad and M U Ahmed *Signal Processing Techniques for Computational Health Informatics. Intelligent Systems Reference Library* **Vol 192** (Cham: Springer)
[10] Bulling A, Ward J A, Gellersen H and Tröster G 2011 Eye movement analysis for activity recognition using electrooculography *IEEE Trans. Pattern Anal. Mach. Intell.* **33** 741–53
[11] Krill A E 1970 The electroretinogram and electro-oculogram: clinical applications *Invest. Ophthalmol* **9** 600–17
[12] Brown M, Marmor M, Vaegan, Zrenner E, Brigell M, Bach M and ISCEV 2006 ISCEV standard for clinical electro-oculography EOG 2006 *Doc. Ophthalmol.* **113** 205–12
[13] Gopal I S and Haddad G G 1981 Automatic detection of eye movements in REM sleep using the electrooculogram *Am. J. Physiol.* **241** R217–21
[14] Belkhiria C, Boudir A, Hurter C and Peysakhovich V 2022 EOG-based human–computer interface: 2000–2020 review *Sensors (Basel)* **22** 4914
[15] Barbara N, Camilleri T A and Camilleri K P 2019 EOG-based eye movement detection and gaze estimation for an asynchronous virtual keyboard *Biomed. Signal Process. Control* **47** 159–67
[16] Zao J K *et al* 2016 Augmenting VR/AR applications with EEG/EOG monitoring and oculo-vestibular recoupling ed D Schmorrow and C Fidopiastis *Foundations of Augmented Cognition: Neuroergonomics and Operational Neuroscience. AC 2016. Lecture Notes in Computer Science* **Vol 9743** (Cham: Springer)
[17] Němcová A, Janoušek O, Vítek M and Provazník I 2017 Testing of features for fatigue detection in EOG *Biomed Mater Eng* **28** 379–92
[18] Dibble J M and Teters C K 2004 Monitoring eye movement with a computer based electro-oculogram (EOG) *Biomed. Sci. Instrum.* **40** 463–8
[19] Bisong E 2019 Google Colaboratory *Building Machine Learning and Deep Learning Models on Google Cloud Platform: A Comprehensive Guide for Beginners* (Berkeley: Apress) 59–64
[20] Solomon O M 1991 *PSD computations using Welch's method [Power Spectral Density (PSD)]* US Dep. Energy Off. Sci. Tech. Inf.
[21] Daubechies I 2004 *An Introduction to Wavelet Analysis* (Berlin: Springer)

[22] Mallat S 1999 *Wavelets: Theory, Algorithms, and Applications* 1st edn (New York: Academic)

[23] Rioul O and Vetterli M 1991 Wavelets and signal processing *IEEE Signal Process Mag.* **8** 14–38

[24] Gandhi T, Panigrahi B K and Anand S 2011 A comparative study of wavelet families for EEG signal classification *Neurocomputing* **74** 3051–7

[25] Zhang D 2006 *2005 IEEE Eng. Med. Biol. 27th Ann. Conf. Wavelet approach for ECG baseline wander correction and noise reduction (Piscataway)* (IEEE) 1212–5

Chapter 7

Efficient nanoscale device modeling using artificial neural networks with TensorFlow and Keras libraries in Python

Abhishek Kumar Singh, Rohini Palanisamy and Aakash Kumar Jain

This chapter presents the potential approach of developing artificial neural networks (ANNs) for the efficient modeling framework of multigate Field Effect Transistors (MuGFETs). The ANNs are developed using Python-based deep learning framework TensorFlow and Keras libraries, allowing for hypertuning, efficient training, implementation, and easy integration into existing workflows. The performance of the ANNs was evaluated and compared against the test data from Technology Computer-Aided Design (TCAD) simulations, demonstrating a prominent level of accuracy in predicting the electrical characteristics of MuGFET devices. The results demonstrate the potential of ANNs created by a framework based on TensorFlow for fast and accurate modeling of MuGFETs. Furthermore, optimization techniques for improving performance and reducing resource consumption are discussed. ANN-based device models required for the very large scale integration industry are fast, accurate and promise a new compact modeling paradigm with pioneer advancements.

7.1 Introduction

7.1.1 Motivation and background

The semiconductor industry is witnessing unprecedented growth in the complexity of emerging nanoscale devices, with leading manufacturers predicting that more than three dozen distinct types of materials would be required for the fabrication of advanced sub-7 nm nodes. As an illustration, sub-10 nm FinFET nodes are already composed of more than two dozen varied materials, making the situation even more intricate with the introduction of organic materials. This poses significant challenges for modeling the electrical behavior of these materials, which must be done using

doi:10.1088/978-0-7503-5929-0ch7

concurrent SPICE models. The presence of multiple SPICE simulators and models from numerous vendors requires significant collaboration between semiconductor companies, research and development institutions, and electronic design automation vendors. Developing and standardizing new SPICE models to facilitate this process requires running extremely expensive test wafers through fabs, spending copious amounts of time and capital on development and characterization of foundation intellectual property (IP), and generating and maintaining mountains of data while performing repetitive tasks. Therefore, there is a critical need for more efficient approaches to modeling these complex nanoscale devices that can reduce the cost and time required for their development and characterization.

7.1.2 Problem statement and objectives

The development and characterization of complex nanoscale devices pose significant challenges in terms of the cost and time required to generate and maintain the data needed for their modeling. Conventional solutions using compact models and simulators from numerous vendors require significant collaboration and are often time-consuming. Therefore, there is a critical need for more efficient approaches to modeling these devices.

Artificial intelligence (AI)-based device modeling solutions using ANNs have been shown to be effective in handling complex behaviors of nanoscale devices. These models are designed to learn patterns and relationships from large datasets and make accurate predictions. Neural networks are especially good at capturing non-linear behavior, which is important in modeling complex nanoscale devices. Compared to conventional solutions using SPICE models and simulators, AI-based device modeling solutions offer a more efficient and accurate approach. They can reduce the cost as well as time to market for developing and characterizing nanoscale devices while still maintaining accuracy and performance.

7.2 Brief literature survey

7.2.1 MuGFETs and their characteristics

MuGFETs are a promising proven alternative to traditional planar FETs for sub-10 nm technology nodes due to their improved electrostatic control and reduced short-channel effects [1]. MuGFETs have multiple gate electrodes surrounding the channel region, allowing for better gate control and reduced channel length. Some examples of MuGFETs include gate-all-around (GAA) FETs, FinFETs, and nanowire FETs. Numerous studies have demonstrated the superior performance of MuGFETs over traditional planar FETs, including improved subthreshold swing, higher drive currents, and reduced leakage [2].

This chapter primarily focuses on the transfer characteristics (I_d–V_{gs}) and output characteristics (I_d–V_{ds}) of a GAA Metal Oxide Semiconductor Field Effect Transistor (MOSFET) devices, as they provide a comprehensive understanding of the devices' direct current behavior and performance. The transfer characteristics give insight into the FET's threshold voltage, subthreshold swing, leakage and transconductance, among other parameters. These parameters are important in designing and optimizing FET circuits for low power and high performance technology. The output characteristics provide information on the devices' output resistance (r_o), saturation voltage (V_{Dsat}), channel length modulation, velocity saturation and breakdown voltage. These properties are crucial for understanding the devices' critical operating limits and for designing FET circuits that meet desired performance specs.

7.2.2 ANNs and their application in compact modeling

ANNs have gained significant attention in the field of compact modeling due to their ability to capture complex and nonlinear device behaviors. ANNs offer several advantages over traditional compact models, including increased accuracy, reduced computational time, and the ability to model a wide range of operating conditions [3–7]. ANNs have been successfully applied to model various semiconductor devices such as MOSFETs, Bipolar Junction Transistors (BJTs), and diodes, among others. These models have been shown to accurately capture device behavior under a range of operating conditions, making ANNs a promising tool for the design and optimization of modern semiconductor devices.

To summarize, ANN-based models perform better than other approaches in the following ways:

- Scalability: deep neural networks (DNNs) can be easily scaled to achieve the desired level of accuracy and precision without exponentially increasing the solution space complexity with an increasing number of variables.
- Generalization: Both look up tables (LUTs) [8] and physics-inspired neural networks (Pi-NNs) [9] cannot be generalized due to being quite customized and physics-inspired modifications, respectively. Unlike them, a wide range of learning algorithms gives DNN the flexibility to be well generalized.
- Smoothness: To achieve an ideal model, it is essential for it to possess infinite differentiability. LUT-based models cannot be made differentiable beyond a certain limit, as including higher order polynomials for curve fitting will exponentially increase the computational complexity [8].

7.2.3 Previous work using ANNs for modeling MuGFETs

With the expeditious evolution of emerging semiconductor devices, to enable device–circuit co-design, compact models are quite crucial and inevitable [10]. Most of the compact models are device-physics driven, and hence these physics-based interactions on a device level are quite complex to comprehend in addition to being computationally expensive [3]. This creates a hurdle in making purely physics-based compact models like Berkeley Short-channel IGFET Model (BSIM) or surface potential based models [4, 5]. Also, there are proposed LUT-based methods as alternatives to conventional compact modeling, but they suffer from lengthier simulation time and other convergence issues for large-scale circuits where precision is required. Also, they are sort of rigid, as they lack tuners that could be employed to alter output characteristics [7]. For RF (radio frequency) device applications, ANNs have a well-documented history of being employed as efficient compact models for semiconductor devices [11]. With the surge of machine learning (ML) applications and supporting parallelized hardware for training and popular software packages, it has become efficacious to use ANNs for compact modeling of advanced FET nodes. In some recent studies, long short-term memory (LSTM) based approaches were also suggested [12] but later proved to be too much computationally expensive [13].

In one of the explored approaches, a comprehensive evaluation was conducted to assess the effectiveness of the ANN-based compact modeling methodology for advanced FET modeling [8]. The evaluation focused on several key aspects, including model fitting capability (accuracy), model generation (ANN training) time, SPICE simulation time, model retargeting, and feasibility of variability modeling. By introducing essential elements to the ANN-based compact modeling methodology, remarkable achievements were made in terms of high model accuracy, fast ANN training time, and efficient SPICE simulation time. The ANN was employed as a compact model for the FET, with the input layer neurons representing voltage biases applied to the FET terminals and instance parameters such as L, W, and T. While the number of hidden layers can be adjusted, it was determined using hypertuning parameters. The ANN weights were represented in a double-precision floating-point format, and the output variables of the ANN were transformed into terminal currents (I) or terminal charges (Q) using a conversion function. For applications involving RF distortion simulations, it was ensured that the I–V model passed the Gummel symmetry test at $V_{ds} = 0$ V to maintain accuracy [14].

Another study introduced a general purpose, unified process-aware ML-based compact model targeting resistive random-access memory (RRAM) and similar devices with hysteresis [15]. Here, an LSTM model was deployed to accurately fit the RRAM's I–V characteristics. Due to its unique memorization capability, LSTM effectively recaptures both the low resistance and high resistance state of the RRAM. The model's training data was obtained from fabricated RRAM samples featuring a $TaN/HfO_2/Pt/Ti/SiO_2/Si$ structure. Notably, the resulting fitting error is remarkably low, measuring and 0.0096 mA for the sinusoidal wave input voltage and 0.0148 mA for the random walk voltage sequences. To demonstrate the process awareness of the model, a post-oxide annealing dataset spanning temperatures from 300°C–500°C and the root mean squared error (RMSE) was impressively small, measuring at 0.0028.

7.3 Methodology

7.3.1 Overview of proposed approach

1. The choice of device for this study is a GAA MOSFET device based on specifications from [16]. Additional device parameters, such as doping distribution and work function, are obtained from the BSIM manual [BSIM-CMG 111.0.0 (09/12/2019)] [17] (figure 7.1).

2. Using the Synopsys Sentaurus Structure Editor, the device is constructed in accordance with the specifications.

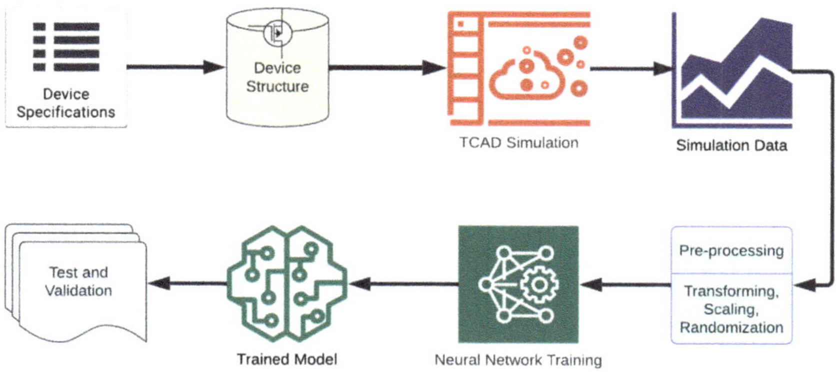

Figure 7.1. Representative workflow.

3. To generate training data, the device model is simulated for a range of voltage values (V_{ds} and V_{gs}) with one bias kept constant and the other sweeped.
4. The generated data are stored in *PLT* files, from which I_{ds}, V_{ds}, and V_{gs} values are extracted and stored in a *CSV* format. This is done either through writing a shell or Python script or manually by plotting and exporting curves in Synopsys Sentaurus Visual.
5. The extracted data are then pre-processed by arranging them in a tuple format (V_{gs}, V_{ds}, I_{ds}), storing them in a numpy ndarray or pandas DataFrame, shuffling, scaling, and dividing them into test–train–validation sets, and converting them to NPZ format to make them TensorFlow compatible.
6. A hyperband optimization algorithm from the Keras tuner package is used to determine the most optimum hyperparameters (number of neurons, number of layers, learning rate, activation function etc) for the neural network design. This was preferred over Bayesian optimization for its significant speedup.
7. After determining the hyperparameters, the DNN model is configured and trained on the pre-processed training data. The training time depends on the size of the dataset, complexity of the neural network, and the hardware.
8. The effective performance of the trained model is assessed by determining the MSE (mean square error) and regression coefficient. New data are generated from TCAD, and the model is validated by comparing the predicted and actual drain current values.

7.3.2 Implementation details

7.3.2.1 Dataset preparation

To create the dataset, the device structure is constructed using the Synopsys Sentaurus Structure Editor, as illustrated in figure 7.2. Once the structure design is completed, details such as doping profiles, material properties, mobility, and

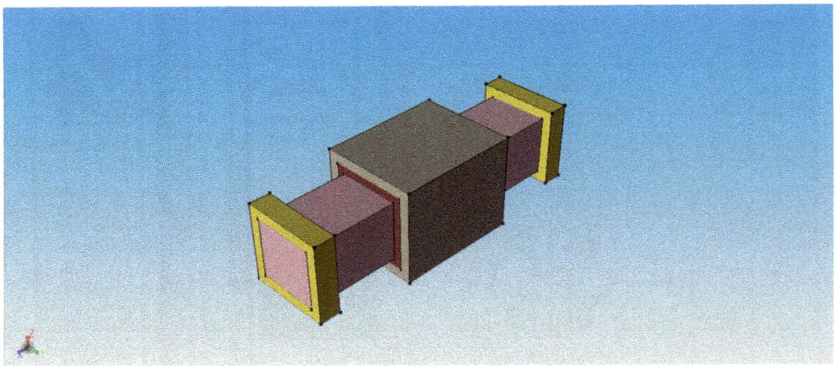

Figure 7.2. A three-dimensional view of the device designed using the Synopsys Sentaurus Structure Editor.

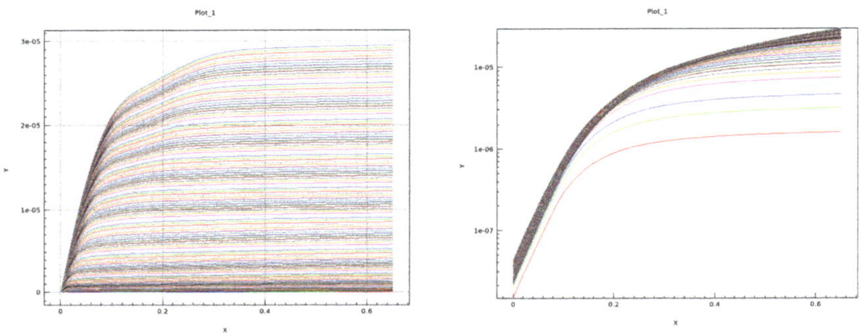

(a) Simulated Output Characteristics

(b) Simulated Transfer Characteristics

Figure 7.3. Simulated characteristics of the constructed device.

effective mass are established, and mesh is generated. Transport models and quantum mechanics models are included in a *.cmd* file for numerical simulation solvers. The device structure is then simulated using the Synopsys Sentaurus Device '*sdevice*' command, and a sweep is performed for a range of 0–0.65 V with 5 mV increments for drain and gate biases separately. This yields a total of 130 *PLT* files for both sweeps. Visual representation of the output data is obtained using the Synopsys Sentaurus Visual, as shown below in figures 7.3(a) and (b).

There are two methods for extracting data: The first option involves using a script to extract values from the *PLT* files, while the second option involves loading all *PLT* files and then consolidating them into a single *CSV* file. In this scenario, the second approach is taken to extract the data, and the resulting data appears as depicted in figure 7.4.

However, pre-processing is required to make it usable in the workflow. The dataset generated from Synopsys Sentaurus Visual is not structured in a suitable format to be used as inputs for the DNN model. Therefore, a pre-processing step is required to transform the dataset into a suitable structure that can be fed into the DNN model.

	A	B	C	D	E	F	G
1	drain TotalCurrent(init_3_1_plo	drain TotalCurrent(init_3_1_	drain TotalCurrent(init_3_2_p	drain TotalCurrent(init_3_2_plot	drain TotalCurrent(init_3_3	drain TotalCurrent(init_3_	drain
2	0	-4.53E-19	0	-3.96E-19	0	-2.05E-20	
3	0.00065	3.16E-09	0.00065	3.71E-09	0.00065	4.36E-09	0.0
4	0.001776667	7.81E-09	0.001776667	9.19E-09	0.001776667	1.08E-08	0.00
5	0.003929852	1.43E-08	0.003929852	1.69E-08	0.003929852	1.98E-08	0.0
6	0.007179852	2.00E-08	0.007179852	2.37E-08	0.007179852	2.79E-08	0.0
7	0.010429852	2.30E-08	0.010429852	2.72E-08	0.010429852	3.21E-08	0.0
8	0.013679852	2.44E-08	0.013679852	2.89E-08	0.013679852	3.42E-08	0.0
9	0.016929852	2.51E-08	0.016929852	2.97E-08	0.016929852	3.52E-08	0.0
10	0.020179852	2.54E-08	0.020179852	3.01E-08	0.020179852	3.56E-08	0.0
11	0.023429852	2.55E-08	0.023429852	3.03E-08	0.023429852	3.58E-08	0.0
12	0.026679852	2.56E-08	0.026679852	3.04E-08	0.026679852	3.60E-08	0.0
13	0.029929852	2.57E-08	0.029929852	3.05E-08	0.029929852	3.61E-08	0.0
14	0.033179852	2.59E-08	0.033179852	3.07E-08	0.033179852	3.63E-08	0.0
15	0.036429852	2.61E-08	0.036429852	3.09E-08	0.036429852	3.66E-08	0.0
16	0.039679852	2.63E-08	0.039679852	3.12E-08	0.039679852	3.69E-08	0.0
17	0.042929852	2.65E-08	0.042929852	3.14E-08	0.042929852	3.72E-08	0.0
18	0.046179852	2.67E-08	0.046179852	3.17E-08	0.046179852	3.75E-08	0.0
19	0.049429852	2.69E-08	0.049429852	3.19E-08	0.049429852	3.78E-08	0.0
20	0.052679852	2.71E-08	0.052679852	3.21E-08	0.052679852	3.80E-08	0.0
21	0.055929852	2.73E-08	0.055929852	3.24E-08	0.055929852	3.83E-08	0.0
22	0.059179852	2.75E-08	0.059179852	3.26E-08	0.059179852	3.85E-08	0.0
23	0.062429852	2.77E-08	0.062429852	3.28E-08	0.062429852	3.88E-08	0.0

Figure 7.4. Extracted data from Sentaurus Visual in *CSV* format.

1 to 5 of 5 entries Filter

index	vds(V)	ids(A)	vgs(V)
0	0.0	-4.53e-19	0.005
1	0.00065	3.16e-09	0.005
2	0.001776667	7.81e-09	0.005
3	0.003929852	1.43e-08	0.005
4	0.007179852	2e-08	0.005

Show 25 per page

Figure 7.5. Post-processed input dataset.

In the below snippet, the raw data is imported as a DataFrame using pandas library.

```
#Import Pandas and tcad data from SVisual
import pandas as pd
raw_df_svisual = pd.read_csv('gaafet_idvd_all.csv')
```

The following code generates V_{gs} values and structures all into a DataFrame, renames its columns, processes it, and returns a new DataFrame that contains the V_{gs}(V), V_{ds}(V), and I_{ds}(A) data for each measured data point in the raw data.

```
vgs = 0
newdf = pd.DataFrame()
for i in range(int(raw_df_svisual.shape[1]/2)):
  tempdf = raw_df_svisual.iloc[:,[2*i,2*i+1]].copy()
  #Renaming the columns

  tempdf.rename(columns = {list(raw_df_svisual)[2*i]:'vds(V)', list(
        raw_df_svisual)[2*i+1]:'ids(A)'}, inplace = True)

#Adding Vgs column
  vgs = vgs+0.005 #Incrementing vgs by 0.005 in each step
  tempdf['vgs(V)'] = vgs
  #Adding to overall dataframe

newdf = newdf.append(tempdf, ignore_index = True)
```

After the pre-processing, the dataset looks like figure 7.5.

Furthermore, a copy in *CSV* format is saved as '*dataset_stcad.csv*' to be used in a later stage.

7.3.3 Hypertuning for optimizing performance

Now, we discuss performing the preliminary operations on the dataset: shuffling and scaling.

- Shuffling ensures that the model does not learn the order of the data and prevents the model from overfitting to the sequence of the data. Secondly, scaling is done to ensure that all features of the data are on a similar scale.
- Scaling is done to normalize the data and bring all features onto the same scale. This is important because atypical features in the dataset may have different ranges, units, and distributions, which can lead to difficulties in training the neural network. Scaling can also improve the convergence rate and the performance of the neural network.

The following snippet performs the same operation. For that, the *preprocessing.scale* method is designed to scale and preprocess input data, and since the data is transformed by scaling, the variance and mean values of the original dataset are preserved for future use during the testing process when analyzing real-world data. Finally, using *numpy*, the indices are shuffled, essentially creating a shuffled index array representing the rows of the dataset in a new random order. The shuffled indices are then used to shuffle the input and target datasets.

```
unscaled_dataset = pd.read_csv('dataset_stcad.csv',delimiter=',')
#Saving the data as inputs containing vgs and vds and targets conta
ining ids
unscaled_inputs = unscaled_dataset[['vgs(V)', 'vds(V)']].values
unscaled_targets = unscaled_dataset[['ids(A)']].values
##Scaling or standardizing the inputs for better accuracy
scaled_inputs = preprocessing.scale(unscaled_inputs)
scaled_targets = preprocessing.scale(unscaled_targets)
#scaled_targets = unscaled_targets
#Extracting mean and std deviation for UNSCALING the output later
vgs_mean, vds_mean = unscaled_inputs.mean(axis=0)
vgs_stddev, vds_stddev = unscaled_inputs.std(axis=0)
ids_mean = unscaled_targets.mean(axis=0)
ids_stddev = unscaled_targets.mean(axis=0)
#Shuffling the inputs and targets
#Make changes for SCALED and UNSCALED datasets
shuffled_indexes = np.arange(unscaled_inputs.shape[0])
np.random.shuffle(shuffled_indexes)

scaled_inputs = scaled_inputs[shuffled_indexes]
scaled_targets = scaled_targets[shuffled_indexes]
```

```
print(train_samples_count, test_samples_count, validate_samples_count)

21112 2639 2639
```

Figure 7.6. Data bifurcation for the DNN.

Normally, splitting a dataset into three sets—train, test, and validation sets—is an essential step in deep learning model development, and the standard 8:1:1 ratio is commonly used. Using this ratio ensures that there are enough samples in the training set for the model to learn from while still having enough samples in the validation and test sets for accurate assessments of the model performance. Now the dataset is apportioned into train, test and validate sets in an 8:1:1 ratio:

```
NUM_DATAPOINTS = 203*130
train_samples_count = int(0.8*NUM_DATAPOINTS)
test_samples_count = int(0.1*NUM_DATAPOINTS)
validate_samples_count = int(0.1*NUM_DATAPOINTS)

train_inputs = scaled_inputs[:train_samples_count]
train_targets = scaled_targets[:train_samples_count]

validation_inputs = scaled_inputs[train_samples_count:train_samples_
count+validate_samples_count]
validation_targets = scaled_targets[train_samples_count:train_sample
s_count+validate_samples_count]

test_inputs = scaled_inputs[train_samples_count+validate_samples_cou
nt:]
test_targets = scaled_targets[train_samples_count+validate_samples_c
ount:]
```

Following the aforementioned process, a distributed dataset was created and appears as illustrated in figure 7.6. The dataset includes a total of 26 390 data points, out of which 21 112 data points were assigned for training, while 2 639 data points were each used for testing and validation purposes, respectively.

Now using the *NPZ* file format, which is a practical choice for storing and handling large datasets in TensorFlow, a copy for hyperband tuning is created. Hence, the *savez* method from *numpy* is used to save the arrays into a compressed *NPZ* file format.

```
np.savez('gaafet_dataset_train', inputs=train_inputs, targets=train
_targets)
np.savez('gaafet_dataset_validation', inputs=validation_inputs, tar
gets=validation_targets)
np.savez('gaafet_dataset_test', inputs = test_inputs, targets=test_
targets)
```

In this study, the Keras tuner was modeled in and ran as a separate program for more flow control. But it could be integrated into the main program as well. Prior to further operation, it is necessary to have the required packages for tuning of the hyperparameters.

```
!pip install -q -U keras-tuner
import keras_tuner as kt
```

A function is required to create and optimize a DNN for this task using the Keras Tuner package. The hyperparameters are adjusted to discover the most effective combination of values that maximize the model's performance for the specified task.

The model_builder function takes a hyperparameter object as input and returns a compiled neural network model. The input layer is transformed into a flat structure, and the number of hidden layers as well as the number of units in each layer can be fine-tuned using the *hp.Int* function. The activation function for each hidden layer can be configured to either *relu* or *elu* using the *hp.Choice* function. The output layer consists of a solitary neuron. The learning rate for the optimizer is also tuned using the *hp.Choice* function. The compiled model uses the Adam (adaptive moment estimation) optimizer, *MeanSquaredError* loss function, and *RootMeanSquaredError* metric.

The Adam optimizer is an enhanced version of the stochastic gradient descent (SGD) optimization algorithm, which is specifically engineered to enhance the convergence rate and enhance the accuracy of the model during training. This optimizer is widely used in deep learning models, especially for models involving large datasets and complex architectures. It essentially consists of two main components: moving averages of the gradient and moving averages of the gradient squared. It combines these moving averages to compute adaptive learning rates and update the model parameters in an iterative manner.

The adaptive learning rate is derived by dividing the moving average of the gradient by the square root of the moving averages of the squared gradients. This effectively adjusts the learning rate for each weight in the network by a factor that takes into account the gradient history for that weight. This helps to overcome the disadvantage of traditional SGD algorithms, where the learning rate is constant throughout the training process.

The loss function used is the MSE, which calculates the mean squared difference between the actual and predicted values in the output layer of the network. This function quantifies the disparity between the predicted and actual values, serving as a measure of the model's performance with respect to the given task. Lastly, the metrics list contains RMSE as the evaluation metric, which is the square root of the MSE. It offers an estimation of the mean deviation between the predicted values and the actual ones. By using this metric, the model's performance and accuracy can be monitored during the training process.

```
def model_builder(hp):
  comp_model = keras.Sequential()
  #Adding the input layer
  comp_model.add(keras.layers.Flatten(input_shape=(2,)))

  for i in range(hp.Int("num_layers", 1, 5)):
      comp_model.add(
          keras.layers.Dense(
              # Tune number of units separately.
              units=hp.Int(f"units_{i}", min_value=32, max_value=51
2, step=32),
              activation=hp.Choice("activation", ["relu", "elu"]),
          )
      )
  #Add the output layer
  comp_model.add(keras.layers.Dense(1))

  # Tuning the learning rate for the optimizer
  # Choosing an optimal value from 0.0001, 0.001, or 0.01
  hp_learning_rate = hp.Choice('learning_rate', values=[1e-4, 1e-
3, 1e-2])

  comp_model.compile(optimizer=keras.optimizers.Adam(learning_rate=
hp_learning_rate),
                loss=tf.keras.losses.MeanSquaredError(),
                metrics=[tf.keras.metrics.RootMeanSquaredError()])

   return comp_model
```

The *hyperband* method from *keras_tuner* is used to optimize hyperparameters of the Keras-based model architecture to enhance performance in the following code. Also, the *optimizer* parameter specifies that the tuner will be minimizing the value of the validation set loss function. The *max_epochs* parameter determines the number of epochs that the tuner will train the model for. The *factor* parameter scales subsequent epochs and resource allocations for each band of hyperparameter configurations.

```
#Setting up the tuner now
tuner = kt.Hyperband(model_builder,
                     objective='val_loss',
                     max_epochs=20,
                     factor=3,
                     directory='my_dir',
                     project_name='dnn_for_device_modeling')
stop_early = tf.keras.callbacks.EarlyStopping(monitor='val_loss', p
atience=5)
```

Among many available optimizers, hyperband optimization is used, the reason being that this hyperparameter optimization algorithm outperforms traditional methods like grid search, random search, and Bayesian optimization. This is due to its ability to explore the hyperparameter space effectively through a process called successive halving, which discards underperforming models and prioritizes computation towards promising models. It is also less memory intensive and more effective, finding optimal combinations of hyperparameters quickly and accurately. Its

```
The hyperparameter search is complete.
Optimal number of layers is 3.
The optimal number of units in the first densely-connected layer is 20.
The optimal number of units in the second densely-connected layer is 24.
The optimal number of units in the third densely-connected layer is 48.
The optimal learning rate for the optimizer is 0.001.
The optimal activation function is relu.
```

Figure 7.7. Hyperparameter search result.

intuitive design makes it easy to implement for complex DNN architectures, making it an ideal choice for hyperparameter tuning in deep learning applications.

The search method is invoked to explore the hyperparameter space in the following code snippet. This step may take a long time according to constraints set in the model builder.

```
tuner.search(train_inputs, train_targets, epochs=50, validation_spli
t=0.2, callbacks=[stop_early])

# Getting the optimal hyperparameters now
best_hps=tuner.get_best_hyperparameters(num_trials=1)[0]
```

To display the results of hyperparameter optimization process, do the following:

```
print(f"""
The hyperparameter search is complete.
Optimal number of layers is {best_hps.get('num_layers')}.
The optimal number of units in the first densely-
connected layer is {best_hps.get('units_1')}.
The optimal number of units in the second densely-
connected layer is {best_hps.get('units_2')}.
The optimal number of units in the third densely-
connected layer is {best_hps.get('units_3')}.

The optimal learning rate for the optimizer is {best_hps.get('learni
ng_rate')}.

The optimal activation function is {best_hps.get('activation')}.

""")
```

This outputs the following result (figure 7.7).

7.3.4 DNN modeling and training

In this section, the model is defined with hyperparameters obtained from the optimization step. In this context, the variables *input_size* and *output_size* represent the count of input and output neurons in the neural network, which is determined by the number of features and desired output dimensions, respectively. There are three hidden layers responsible for extracting the relevant features from the input data. Each hidden layer consists of a rectified linear unit (ReLU) activation function. ReLU is a simple, computationally efficient, and nonlinear function that provides the network with the ability to model complex, nonlinear relationships in data by replacing the negative input values with zero and leaving positive values unchanged. All of these are combined to form a sequential model. A learning rate of 0.001 is set for the Adam optimizer. The optimizer is responsible for minimizing the loss function and updating the weights in the neural network during the training process. Also, the '*EarlyStopping*' callback is used to stop the process of training if there is no predetermined improvement in the validation loss for five epochs. It is used to prevent overfitting to the training data and to reduce training time. The training is done using the fit method, and '*verbose = 1*' gives information on the progress of the training. The trained model and history variables are stored for subsequent use.

```
inputs_size = 2
output_size = 1
hidden_layer_1_size = 20
hidden_layer_2_size = 24
hidden_layer_3_size = 48
comp_model = tf.keras.Sequential([
                               tf.keras.layers.Input(inputs_size),
     tf.keras.layers.Dense(hidden_layer_1_size, activation='relu'),
     tf.keras.layers.Dense(hidden_layer_2_size, activation='relu'),
     tf.keras.layers.Dense(hidden_layer_3_size, activation='relu'),
                               tf.keras.layers.Dense(output_size)
])

#Optimizer
opt = tf.keras.optimizers.Adam(learning_rate=0.001)
comp_model.compile(optimizer=opt, loss='mean_squared_error')

early_stopping = tf.keras.callbacks.EarlyStopping(monitor='val_loss
', patience=5, verbose=1, mode='min', restore_best_weights=True)

batch_size = 50
max_epochs = 100

history = comp_model.fit(train_inputs,
          train_targets,
          batch_size = batch_size,
          epochs = max_epochs,
          validation_data=(validation_inputs, validation_targets),
          callbacks=[early_stopping],
          verbose = 1)
```

7.4 Results and analysis

7.4.1 Model evaluation and comparison with TCAD simulations

After this, a function is created to plot the training and validation loss over the incremental epochs of the resultant model, as follows:

```
def plot_loss(history):
  plt.plot(history.history['loss'], label='Training loss')
  plt.plot(history.history['val_loss'], label='Validation loss')

  plt.xlabel('Epoch')
  plt.ylabel('Error')
  plt.legend()
  plt.grid(True)
plot_loss(history)
```

This produces the following results (figures 7.8 and 7.9).

RMSE was found to be 2.3763 × 10^{-5}, and the regression coefficient was calculated to be approximately 0.999 9762, which is remarkably close to unity. The regression coefficient is used to quantify the relationship between two or more variables i.e. the actual and predicted values in linear regression. It should be noticed that I_{ds}(target) values are normalized. For denormalization, predictions must be multiplied with a standard deviation (of I_{ds}) of training and added to the mean (of I_{ds}). This normalization procedure is followed by the pre-processing method of the Sklearn library.

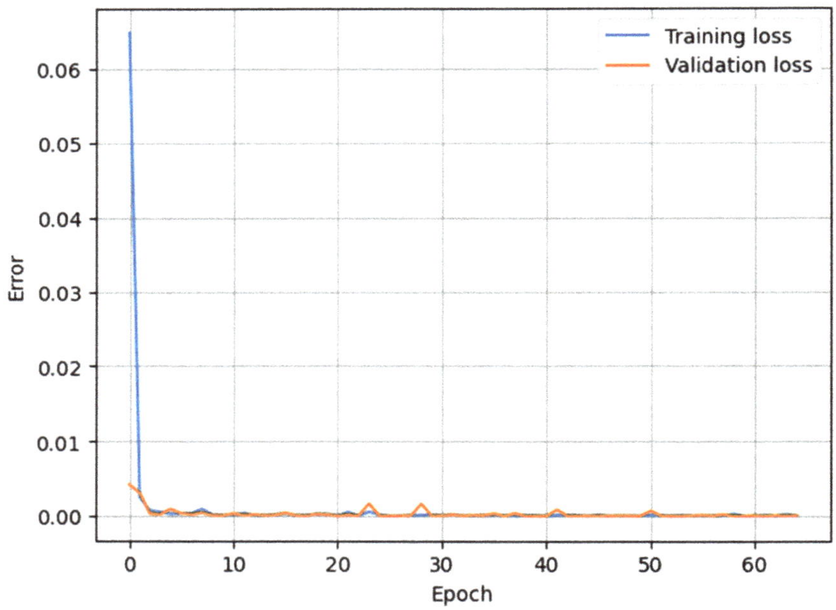

Figure 7.8. Loss curve for the DNN model.

```
[ ] print(model.evaluate(test_inputs, test_targets))

    83/83 [==============================] - 0s 2ms/step - loss: 5.6470e-10
    5.647037282230372e-10
```

Figure 7.9. Model MSE.

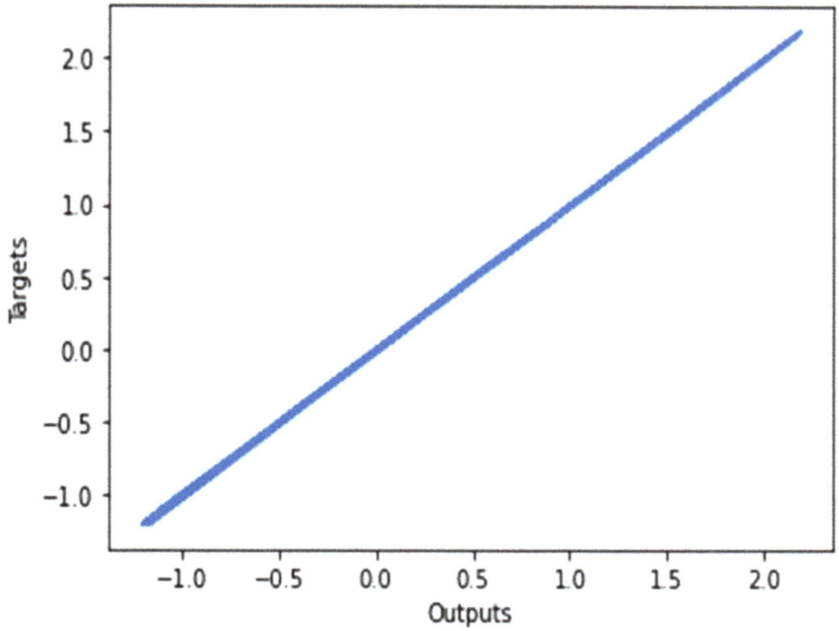

Figure 7.10. Performance on the test dataset.

Furthermore, figure 7.10 demonstrates the performance of the DNN model over the test dataset.

To validate the DNN model, for predetermined values, TCAD simulations were run and data were extracted. For the same voltage biases, predictions were taken out of the DNN model and denormalized. The following are results of the I_{ds}–V_{ds} curve (figure 7.11) and I_{ds}–V_{gs} curve (figure 7.12) for different input voltage biases.

The three-layer DNN can predict the output and transfer characteristics of the chosen device with reasonable accuracy. The following observations can be made based on the obtained results:

- The DNN model can predict I_{ds} over a large number of voltage biases (about 200) in a noticeably short amount of time (7 ms). This suggests that ANN-based models are highly responsive and have the potential to improve the runtime of true SPICE simulations.
- Compact model development can be completed in hours rather than weeks, resulting in significant improvements in turnaround time, which is one of the most critical factors for industrial designs.
- The ANN architecture used in this study for MuGFETs can be applied to a large number of similar devices, demonstrating the model's generalization capabilities.

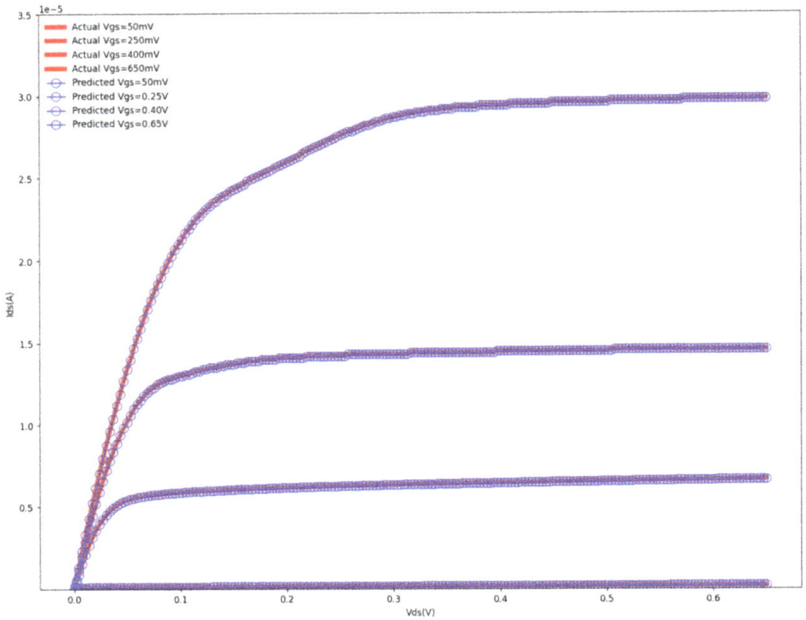

Figure 7.11. Output characteristics (*I*ds vs *V*ds) for 12 nm gate length and 5 nm radius.

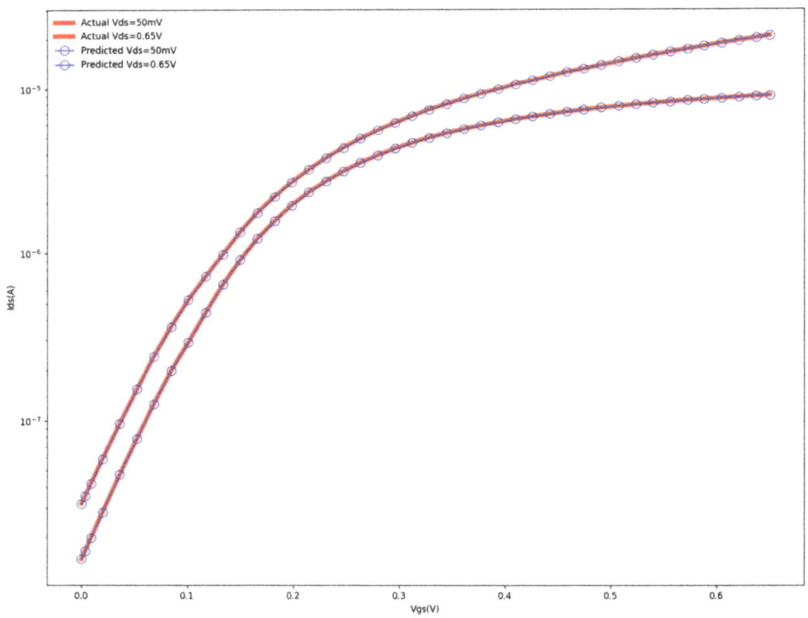

Figure 7.12. Transfer characteristics (I_{ds} vs Vgs) curve.

7.4.2 Potential and limitations of the proposed approach

The potential of the above project is vast. The application of a DNN in predicting the *I–V* characteristics of a device has several advantages. The developed compact model can be integrated with circuit simulation tools, which can then be used for circuit optimization. This can enable designers to quickly explore different circuit topologies and find the optimal design parameters, leading to better performing circuits in a shorter time.

The compact model can also be used as a building block in larger system-level simulations, where multiple devices and circuits are interconnected. This can enable simulations of complex systems such as integrated circuits, leading to better design decisions and more efficient systems. The fast prediction time of the deep neural network can be leveraged for real-time monitoring of the device characteristics. This can enable continuous monitoring of device performance, leading to early detection of faults and potential failures.

However, there are also some limitations to the ANN-based framework. One of the major limitations is that the DNNs' prediction accuracy heavily relies on the quality of data input. Furthermore, training a DNN requires a significant amount of data, which may be difficult to obtain for certain applications. Another limitation is that the DNN model is only as good as the data it is trained on. Therefore, if the device under study has anomalous behavior, it may not be detected by the DNN. Finally, the interpretation of results may be difficult, as DNNs are considered black box models and the relationship between the input and output is not always clear. Therefore, it is difficult to extract insight from the model to improve the device under study.

7.5 Conclusion

The continued shrinking of node size will inevitably lead to ANNs becoming the most optimal compact models. Such models will facilitate a smooth flow of bi-directional information between designers and foundries, further strengthening design/technology co-optimization. In addition, there are several actionable scopes, including extracting weights and biases to simulate circuits using behavioral Verilog-A models. Geometrical dimensions such as channel width (W), channel length (L), thickness (T) etc can also be swept, further extending the generalization capabilities of the DNN model. The DNN model can also be made aware of process, voltage, and temperature variation and mismatch, which can help predict variations resulting from Monte Carlo analyses, saving valuable time.

The need for AI in device modeling cannot be overstated. Its ability to provide accurate and efficient predictions, coupled with its potential for scalability and generalization, makes it a valuable tool for advancing technology and driving innovation. As the field continues to evolve, AI-based solutions will undoubtedly play a crucial role in meeting the challenges of the future.

Bibliography

[1] Soliman S A *et al* 2011 Multi-gate MOSFETs: a review *IEEE Trans. Electron Devices* **58** 20–38

[2] Appenzeller J 2010 Device physics of nanowire devices for logic and memory applications *Proc. IEEE* **98** 2201–15

[3] Hutchins J, Alam S, Zeumault A, Beckmann K, Cady N, Rose G S and Aziz A 2022 A generalized workflow for creating machine learning-powered compact models for multi-state devices *IEEE Access* **10** 513–51

[4] Xu J and Root D E 2015 Advances in artificial neural network models of active devices *2015 IEEE MTT-S Int. Conf. on Numerical Electromagnetic and Multiphysics Modeling and Optimization (NEMO)* pp 1–3

[5] Mehta K and Wong H Y 2020 Prediction of FinFET current-voltage and capacitance-voltage curves using machine learning with autoencoder *IEEE Electron Device Lett.* **42** 136–9

[6] Zhang L and Chan M 2017 Artificial neural network design for compact modeling of generic transistors *J. Comput. Electron.* **16** 825–32

[7] Wang J, Xu N, Choi W, Lee K-H and Park Y 2015 A generic approach for capturing process variations in lookup-table-based FET models *2015 Int. Conf. on Simulation of Semiconductor Processes and Devices (SISPAD) (Washington, DC)* pp 309–12

[8] Wang J, Kim Y-H, Ryu J, Jeong C, Choi W and Kim D 2021 Artificial neural network-based compact modeling methodology for advanced transistors *IEEE Trans. Electron Devices* **68** 1318–25

[9] Gildenblat G, Li X, Wu W, Wang H, Jha A and Van Langevelde R *et al* 2006 PSP: An advanced surface-potential-based MOSFET model for circuit simulation *IEEE Trans. Electron Devices* **53** 1979–93

[10] Root D E 2012 Future device modeling trends *IEEE Microwave Mag.* **13** 45–59

[11] Li M, İrsoy O, Cardie C and Xing H G 2016 Physics-inspired neural networks for efficient device compact modeling *IEEE J. Explor. Solid-State Comput. Devices Circuits* **2** 44–9

[12] Lin A S, Pratik S, Ota J, Rawat T S, Huang T-H, Hsu C-L, Su W-M and Tseng T-Y 2021 A process-aware memory compact-device model using long-short term memory *IEEE Access* **9** 3126–39

[13] Lin A S, Liu P-N, Pratik S, Yang Z-K, Rawat T and Tseng T-Y 2022 RRAM compact modeling using physics and machine learning hybridization *IEEE Trans. Electron Devices* **69** 1835–41

[14] Mcandrew C C 2006 Validation of MOSFET model source–drain symmetry *IEEE Trans. Electron Devices* **53** 2202–6

[15] Lin A S, Pratik S, Ota J, Rawat T S, Huang T H, Hsu C L, Su W M and Tseng T Y 2021 A process-aware memory compact-device model using long-short term memory *IEEE Access* **9** 3126–39

[16] Kao M-Y, Kam H and Hu C 2022 Deep-learning-assisted physics-driven mosfet current-voltage modeling *IEEE Electron Device Lett.* **43** 974–7

[17] Dunga M V, Lin C-H, Niknejad A and Hu C 2007 BSIM-CMG: a compact model for multi-gate transistors *FinFETs and Other Multi-Gate Transistors* ed J P Colinge (New York: Springer Science+Business Media, LLC) ch 3 pp 113–53

Chapter 8

A Python-based comparative study of convolutional neural network–based approaches for the early detection of breast cancer

Ishrat Jahan Mohima and Mosabber Uddin Ahmed

Breast cancer is a serious health problem for women worldwide, and timely identification is crucial for effective treatment. Convolutional neural networks (CNNs) have looked promising in the past few years for the timely identification of breast cancer. Here, a distinctive analysis of various CNN-based approaches for prior breast cancer detection is provided using histopathological image data. The proposed model comprises several layers, including CNN layers, which extract relevant features from the provided input data. Max pooling is included to down-sample extracted features and for reduction of the dimensionality of the input data. The flattened layer reshapes the pooled attributes into a one-dimensional (1D) vector. Then a long short-term memory (LSTM) layer is included, and the output generated by the LSTM layer is integrated to some dense layers, which apply non-linear transformations to the features. Finally, a classification layer is used to classify the input image. How well the model performs has been assessed using histopathological image dataset and achieved 88.13% accuracy, 88.15% precision, 88.14% recall, 88.14% F1 score, and 94.12% area under the curve (AUC) score. The results indicate how this suggested methodology may effectively identify breast cancer in its earliest stages.

8.1 Introduction

Breast cancer is considered as one of the key issues pertaining to public health globally. Also, this one is considered to be the next prime factor in women's cancer-related fatalities. Despite the fact that there has been a significant advancement in survival from this disorder in rich-resource countries, the danger is still rising, leading to high death rates in low income and middle income countries [1]. The World Health Organization approximated that there would be a total of 2.3

doi:10.1088/978-0-7503-5929-0ch8

million additional cases and 685 000 mortalities from breast cancer as a result of this disease in 2020 [2]. The second most familiar malignancy is breast cancer for women in the United States of America, right behind skin cancer [3]. According to estimates, this disease will affect one in eight women at some stage of their lives [4].

Around 500 000 newly discovered cases are identified every year, which makes it the most widespread cancer in Europe for women [5]. Age, family history, genetic abnormalities (such as BRCA1 and BRCA2), lifestyle choices, hormonal factors and exposure to specific environmental conditions are a few of the elements that affect the enlargement of breast cancer [6]. Breast cancer is frequently discovered later, when treatment options are more limited and survival rates are poorer, as in low- and middle-income nations. Breast cancer has a major financial impact, with annual treatment expenses and lost productivity estimated to be in the billions of dollars [8]. Mammograms and clinical breast exams are routine screenings that are essential for prior identification and better treatment outcomes in breast cancer.

But the mortality rate can be reduced and the chances of successful treatment can be improved significantly if breast cancer is detected early and diagnosed accurately [9]. Some imaging techniques are standard to identify breast cancer currently, such as ultrasound, mammography, magnetic resonance imaging (MRI) etc [10]. Still, there are some limitations to these methods such as low sensitivity, high false-positive rates, and high cost [7, 11]. Recent progress in deep learning and computer vision have been very promising for the betterment of accuracy and efficiency in detecting breast cancer [12, 13]. In particular, deep convolutional neural networks (CNNs) have been prosperously implemented in a variety of medical imaging procedures including breast cancer identification [14]. These networks are capable of automatic relevant feature extraction from images and of making accurate predictions.

The primary goal of this research is to introduce and thoroughly evaluate a novel deep CNN–based approach meticulously tailored to revolutionize breast cancer detection. This innovative strategy embraces the cutting-edge domain of deep learning techniques, aiming to transcend the prevailing limitations of current breast cancer diagnostic methodologies, which have historically hindered the quest for precision and early detection. To achieve this ambitious goal, a publicly available dataset of mammography images has been used extensively to train and assess the proposed approach on this renowned histopathology dataset, widely acknowledged for its pivotal role in breast cancer identification research. The richness and diversity of the dataset render it indispensable for not only constructing robust deep learning algorithms dedicated to breast cancer identification but also for facilitating the exploration and validation of computer-aided detection methods, further strengthening the diagnostic armamentarium in the battle against breast cancer. By examining histopathological data of images, which can also be called biomedical images of the biopsy, it is possible to detect these alterations in the structure of the cells [15]. However, due to the complexity of these images, malignancy must be determined by skilled pathologists. Deep learning diagnosis methods, on the other hand, may extract a lot of information from the photos and make a judgment call based on it [16]. The histopathology dataset includes multiple digital images of breast tissue samples, which include both healthy and malignant tissue. The advancement of image analysis systems that can precisely identify malignant tissue in breast biopsies depends on this dataset [17].

The scarcity of annotated data is one of the vital barriers to deep CNN–based breast cancer detection [18]. Acquiring annotated breast images requires both time and resources, and it is challenging to obtain a significant number of datasets to instruct a network model adequately [19]. The effectiveness of the proposed methodology will be measured with existing breast cancer detection modalities. This work has made a unique deep CNN–based strategy for identifying breast cancer. The suggested strategy holds promise in enhancing both the perfectness and timeliness of breast cancer recognition, leading to advanced clinical experiences and potential cost reductions in health care. This research will be highly valuable for health care researchers and professionals involved in the field of breast cancer identification, providing insights and methodologies to advance their work. By harnessing the capacity of deep CNNs and cutting-edge deep learning techniques, the aim is to transcend current limitations in breast cancer detection and empower health care practitioners with a state-of-the-art diagnostic tool. Earlier and more precise detection may lead to more effective treatment options, potentially improving patient outcomes and quality of life. The implications of the work extend beyond academia, with the potential to transform breast cancer management and contribute to more sustainable and equitable health care practices.

The article is structured into distinct sections for comprehensive coverage. Section 8.2 comprises an in-depth examination of the relevant publications. Section 8.3 meticulously expounds on the complete workflow of the proposed system, elucidating its intricate processes. The results obtained from the suggested method are critically assessed in section 8.4, providing valuable insights into the model's performance. Finally, section 8.5 houses the conclusion, presenting a concise summary of findings and implications for the future of breast cancer detection.

8.2 Related works

Rahman *et al* [20] developed an automated system that can analyze an entire stack of slides containing breast cancer imaging samples to determine exactly the exact locations of invasive ductal carcinoma (IDC) from the slides and make decisions in accordance with the results. The authors dealt with the class imbalance problem in their dataset of breast cancer images. They separated the images between the training and test sets after randomly choosing an equivalent number of images from each of the two classes (IDC (+) and IDC (−)). The network has six convolution layers with rectified linear unit (ReLU) activation, max pooling, and dropout, which are accompanied by a fully joined hidden layer and an output layer called softmax. The network is trained with an Adadelta optimizer, learning rate decay, early stopping, and data augmentation. The filter size used in each layer was 3×3. Max pooling is used to downsample the input shape from $50 \times 50 \times 64$ to $25 \times 25 \times 64$. The total number of parameters was 3 505 474. They got an accuracy of 89.34%.

Aiza M Romano and Alexander A Hernandez [21] have developed a deep learning framework to predict IDC. In this study, an upgraded CNN network was trained, and the model's performance on IDC patch-wise classification techniques was examined by the authors. Some experimental findings demonstrate that their methodology produces comparatively better performance on the IDC dataset. In comparison with the most recently published deep learning strategy that identifies IDC, it is obtained that the f-score is 85.28%, and the calculated stabilized accuracy is obtained at 85.41%. From their model, it is also obtained that the f-score was improved 11.5% and the model improved the balance accuracy to 0.86%.

Alghodhaifi *et al* [22] have quantitatively tested two CNN architectures employing depth-wise independent convolution and conventional convolution to improve evaluation matrices of their proposed CNN architecture. They scanned 162 slide images at 40×. Their dataset was also imbalanced. CNNs applied local feature detectors to the entire image to compare the resemblance amidst different image patches and training sets' signature patterns. Afterward, the dimension of the feature space was minimized through an aggregation or subsampling (pooling) function. They also examined many activation functions, including Tanh, Sigmoid, and ReLU, to check which function performs better. Gaussian noise is also used to test the two models' resilience. The outcomes demonstrate that CNN architecture beats the softmax classifier, where 87.5% classification accuracy was achieved, including 93.5% sensitivity and 71.5% specificity.

IDC in breast cancer whole slide images (WSI) is automatically detected and analyzed using a CNN approach in the research of Cruz-Roa *et al* [23]. In order to identify malignancy using traditional methods for identifying IDC, pathologists must scan substantial amounts of benign tissue. CNNs are trained on a broad dataset of WSI patches in the deep learning method to learn a hierarchical representation of IDC tissue areas. Using a dataset of 162 patients, the approach was assessed, and it performed better in terms of F-measure (71.80%) and overall accuracy (84.23%) than both conventional handmade characteristics and supervised classification. Additionally, the research contends that the restrictions in categorization errors were brought on by the challenge of getting extremely detailed annotations of the sick area from a seasoned pathologist.

Deep learning was utilized by Arevalo *et al* [24] to automate the identification of breast cancer in histopathology pictures. The region underneath receiver operating characteristic (ROC) curve to identify the breast cancer was 0.984 after the authors used a deep CNN with residual connections.

Roy *et al*'s alternative method [25] of classifying breast cancer via texture analysis was based on machine learning. From breast ultrasound pictures, they extracted several textural properties and used multiple machine learning techniques for classification. Their findings showed that they could distinguish between benign and malignant breast tumors with an exactness of 88.5%.

Additionally, Shen *et al* [26] utilized deep learning for breast cancer detection in mammography. Employing the CBIS-DDSM database, deep learning models were trained on mammogram patches, aiming to increase detection accuracy. This research highlighted the effectiveness of advanced image analysis in medical diagnostics, particularly for early breast cancer detection.

In addition, Li *et al* [27] proposed a cascaded CNN architecture-based deep learning approach for breast cancer recognition. To find problematic areas in mammograms, their model used a region proposal network, which was followed by a classification network to determine the likelihood of malignancy. The zone under the ROC curve for the suggested method's identification of breast cancer was 0.936.

Using radiomics, a method that extracts some quantitative information from medical pictures, Wang *et al* [28] suggested a unique method for detecting breast cancer. To categorize breast lesions as benign or cancerous, they used MRI data and machine learning algorithms. Their technique revealed the potential of radiomics in assisting with the detection of breast cancer and achieved a correctness of 87%.

The function of microRNAs (miRNAs) in the expansion of breast cancer was studied by Huang *et al* [30] Small non-coding RNA molecules called miRNAs control the expression of genes. The researchers pinpointed particular miRNAs linked to tumor formation, metastasis, and therapeutic response through their thorough examination of miRNA expression profiles in patients with breast cancer. These discoveries advance the knowledge of the molecular processes that underlie breast cancer and may help in the creation of specialized treatments.

A thorough analysis of the function of tumor-infiltrating lymphocytes (TILs) in breast cancer was carried out by Chen *et al* [31]. Immune cells known as TILs penetrate the tumor microenvironment and are essential for the host's defense against cancer. The authors focused on the potential of TILs as a biomarker for therapy response and patient outcome as they examined the prognostic and predictive importance of TILs in breast cancer.

Narayanan *et al* [32] suggested that their research represents a CNN model including deep learning for the correct interpretation of IDC in breast cancer histopathology images. Five CNN layers made up the suggested design, which came after a dense layer and a classifier called 'softmax' for the final classification. This model was conditioned on a publicly available dataset containing 277 524 patches, of which 78 786 contain IDC. The proposed method achieves high performance by using a batch of 27 753 photos, of which 7 879 had IDC. An AUC value of 0.935 was attained, which created a new standard for further research projects.

8.3 Methodology

The identification process of breast cancer is built on taking a dataset of histopathological images of breast tissue samples and predicting the chances of having breast cancer. The workflow diagram of the recommended technique is presented in figure 8.1.

8.3.1 Dataset

In this study, breast histopathology images were utilized from a dataset that is publicly available in Kaggle [33], which consists of microscopic images of breast tissue samples. The dataset was gained from the University of California, Irvine

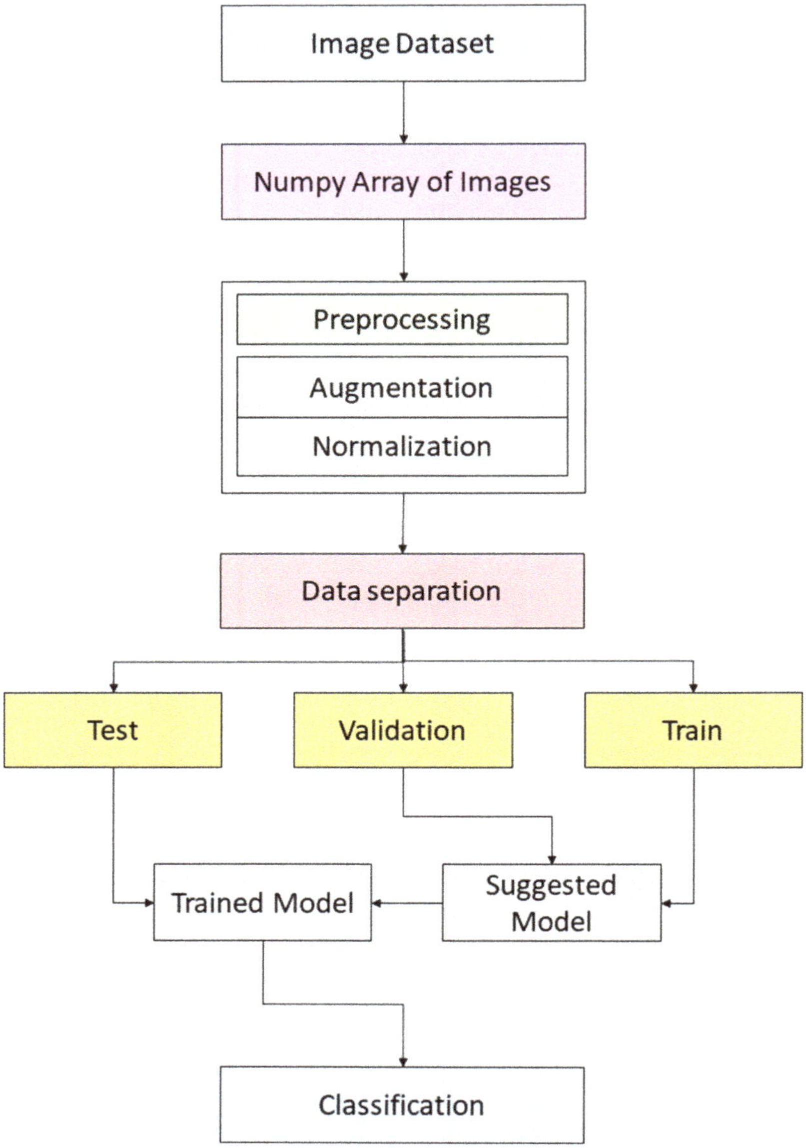

Figure 8.1. Workflow diagram of the recommended technique.

Machine Learning Repository. It includes a total of 277 524 image patches, taken at 40× magnification and digitized using a scanner, resulting in a resolution of 50 × 50 pixels. The dataset was heavily unbalanced with 198 738 negative and only 78 786 positive patches. In order to ensure a balanced dataset, 75 000 images have been collected from each class. So, the final dataset consisted of 150 000 images, which were split into three sets at random. These are the training set (80%), validation set

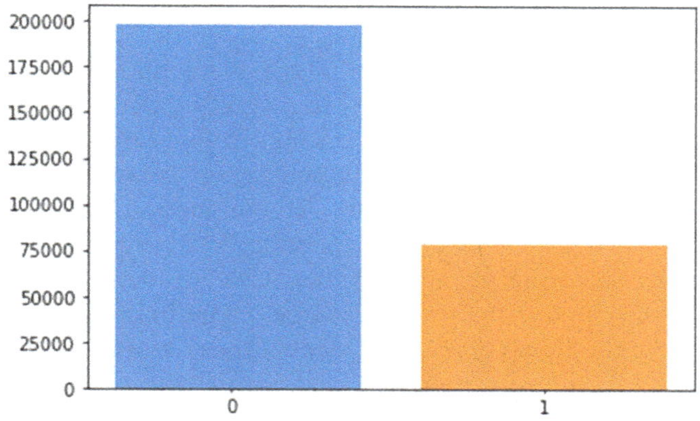

Figure 8.2. Unbalanced dataset.

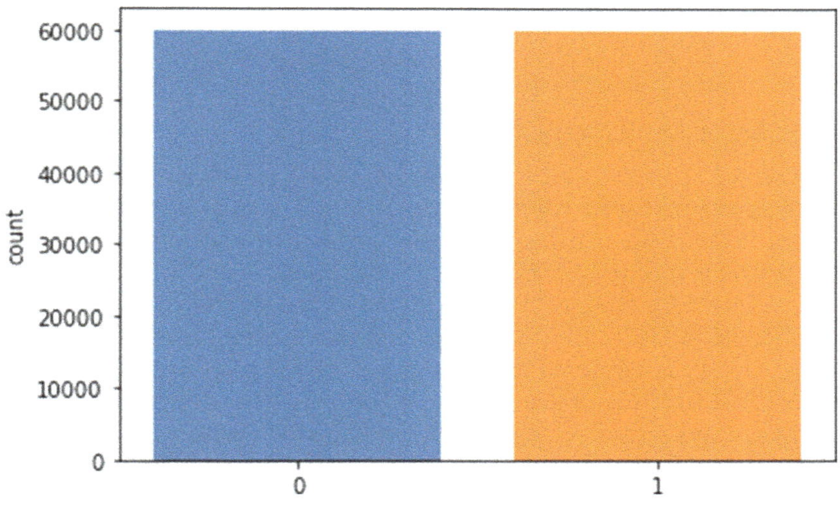

Figure 8.3. Balanced dataset.

(10%), and testing set (10%). Figure 8.2 represents the ratio of unbalanced data, and figure 8.3 represents the distribution of the balanced training dataset considered for this study. Each image in the dataset was labeled as benign or malignant based on expert annotations. Finally, the image data were preprocessed and saved as a numpy array or npy file to be used to train the model. This step was performed to ensure faster processing of data while running the model as well as to save the RAM space in the system. The study aspires to generate a machine learning model that can correctly classify breast tissue images as either benign or malignant, which can assist in the earlier identification of breast cancer.

8.3.2 Image preprocessing

The preprocessing of the dataset includes image augmentation and image normalization. Data augmentation is a prime way to magnify the function of deep learning models by addressing the problem of overfitting. In this case, to provide the model with a varied set of images, various data augmentation techniques were employed to improve the proposed model's scope [34]. A combination of techniques, such as height and width shifting, zooming, and shearing, has been used. To be specific, a 20% or 1.2× zoom was applied on randomly selected images, while shearing was performed on about 20% of the original image. In addition, the height and width were also shifted randomly up to 20% to augment the dataset. These techniques were used to generate additional synthetic data, thereby expanding the size of the dataset and providing the framework with a diversified set of images to learn from. The following code snippet illustrates the image preprocessing technique. All of the processing was accomplished using the *ImageDataGenerator* function.

```
datagen = ImageDataGenerator(
height_shift_range=0.2, width_shift_range=0.2, zoom_range=0.2, shear_range=0.2
)
```

8.3.3 Convolutional neural network

CNN stands for a distinctive neural network that particularly aims at processing image data as input, which is a grid-like architecture. An image is nothing but a two-dimensional (2D) array. An image consists of a matrix representation of pixels. These pixel values specify how much bright the image is and what color it reflects. There are mainly three types of building blocks or layers of CNN architecture. They are the convolution layer, the dense layer and the pooling layer.

The input image is subjected to filters or kernels in the convolution layer, which intertwines them over the image to extract features. These filters employ element-wise multiplications and summations to capture various patterns or features such as edges, textures, or forms. The network can learn and recognize visual patterns at various scales thanks to this mechanism.

The feature maps are compressed using the pooling layer after the convolution layer. The most important information is preserved while the spatial dimensions are reduced. Common pooling methods, such as maximum or average pooling, choose the highest or lowest value inside a particular area of the feature map. This lessens computing complexity while capturing the most important information.

The fully connected layer, sometimes referred to as the dense layer, translates the features that were extracted from the preceding layers to the appropriate output format such as classification labels. With adjustable weights and biases, it links each neuron in the lower layer to each neuron in the upper, dense layer. This layer integrates the features and uses non-linear transformations to create predictions or classifications.

Through a process known as backpropagation, the CNN learns the ideal weights and biases during training. The network improves its capacity to generate accurate predictions by minimizing the distance between its anticipated outputs and the actual labels through iteratively tweaking these parameters.

Convolution and pooling layers are generally used by CNNs to derive hierarchical representations of visual information from input images. These collected features are then used by the dense layer to classify or for other tasks. Due to their effective processing and comprehension of complicated visual data, CNNs are highly suited for tasks like semantic segmentation, object detection, and image recognition.

Input: The input layer puts the preprocessed image data from the dataset and consumes it for further processing by the following levels.

Convolution layer: Convolution layers are a fundamental building block in CNNs, where a deep learning model is generally implemented in computer vision functions like image identification and object classification. Kernels are learnable filters that a convolutional layer is made up of (e.g. 3×3, 5×5, 7×7), which are small matrices that slide over the input data to perform a mathematical operation called a convolution. A feature map is considered the ending result of the sequence of linear combinations from the source image as well as the learnable filter. Following each CNN layer, a ReLU modification is made to the feature map, providing non-linear characteristics to a model. ReLU introduces non-linearity by setting negative values to zero. This makes the model more resistant to changes in input data and enables the network to learn complicated correlations.

Pooling: Pooling layers are applied for downsampling. They reduce the amount of input parameters and perform dimensionality reduction. Pooling layers perform a significant role in capturing the most salient characteristics of an image in addition to downsampling and lowering the amount of input parameters. They work with segments or areas of the feature map [35]. Each segment of the image is subjected to the pooling procedure separately. In spite of the lack of learnable parameters in any of the pooling layers, the hyperparameters comprise the size of the filter, padding, and stride in pooling operations, which are more alike in convolutional operations. Max pooling is the most widely applied type of pooling operation. It takes a certain region from the feature map input, generates the most significant values in every region, and omits the rest of the values. The suggested approach applies a max-pooling layer where the filter size is 2×2, and it includes a stride of 2. This results in a further reduction in the spatial dimensions of the feature map as the pooling operation is applied to nonoverlapping 2×2 sections, and the stride of 2 assures that neighboring regions do not overlap.

Our strategy tries to extract and keep the most important features while lowering the computational burden of the succeeding layers by adding max pooling into the architecture. This aids in enhancing the general CNN's efficiency and efficacy for tasks like object categorization and image identification in computer vision applications.

LSTM: The benefit of employing a recurrent neural network layer called the LSTM layer is that it can help capture long-term dependencies in the sequence data. Unlike traditional neural network layers, which treat each input independently, LSTM layers have a memory mechanism that allows them to selectively remember or forget information over time. This is specifically helpful for tasks where the input sequence has a long-term structure or where certain elements of the sequence are more important than others. Complex temporal links in the data can be captured by using LSTM layers in a model, allowing for more precise predictions or classifications. In situations like natural language processing, speech identification, and time series forecasting, the LSTM layer's capacity to selectively retain information over protracted times is especially useful [36]. After the LSTM layer, a dense layer can also be implemented to map the LSTM output to the desired output format. The dense layer can also perform regression or classification tasks based on the type of the problem.

Dense layer: This is an essential component of the CNN's design to process the output feature maps from the final pooling layer [37]. The output feature maps out of the last pooling layer are typically planed, or reshaped to an array that is 1D, and joined to one or more fully connected layers where all the inputs and outputs are associated with a trainable weight. As a result, each neuron in the dense layer obtains input from every neuron in the layer below and bases its output on the acquired weights. These weights enable the dense layers to identify intricate links and patterns in the data. A subset of dense layers then maps the characteristics generated and supplied by the convolution layers and also maps the downsampling layers to the concluding outcome of the network. The output can be the probability of happening, either 0 or 1. In binary classification, for instance, the result may show the likelihood that the input is owned by one class or the other. The hindmost layer of these fully connected layers can perform the classification tasks for each class. The output of the network is translated into probabilities for each class using the appropriate activation functions, such as softmax. Based on the recognized patterns and features, the model may now assign the input data to the most appropriate class and make predictions.

CNNs can learn complex representations of the input data and take advantage of the strength of fully connected networks to successfully complete classification problems by using dense layers. The hierarchical features that were extracted by earlier levels and converted into predictions or classifications by these layers are crucial [38].

8.3.4 Model overview

```
model = keras.Sequential(
[
keras.layers.Conv2D(32, (3, 3), padding='same', activation='relu', input_shape=(50, 50, 3)),
keras.layers.MaxPooling2D(strides=2),
keras.layers.Conv2D(64, (3, 3), padding='same', activation='relu'),
keras.layers.MaxPooling2D((3, 3), strides=2),
keras.layers.Conv2D(128, (3, 3), padding='same', activation='relu'),
keras.layers.MaxPooling2D((3, 3), strides=2),
keras.layers.Conv2D(256, (3, 3), padding='same', activation='relu'),
```

```
10. keras.layers.MaxPooling2D((3, 3), strides=2),
11. keras.layers.Flatten(),
12. keras.layers.Reshape((−1, 256)),
13. keras.layers.LSTM(128, return_sequences=False),
14. keras.layers.Dense(512, activation='relu'),
15. keras.layers.Dense(128, activation='relu'),
16. keras.layers.Dense(1, activation='sigmoid')
17. # tf.keras.layers.Dense(2, activation = 'softmax')
18. ]

    )
```

The suggested approach has four convolutional layers followed by four pooling layers, consecutively. The above code snippet describes the model architecture. Convolution layers are a fundamental building block in CNNs [39]. Lines 3, 5, 7 and 9 of the code snippet illustrate the convolution layer. First, the images have been taken from the balanced dataset and then they have been passed to the first convolution layer, which includes a filter of number 32 and of size 3 × 3. The first convolution layer has an input shape of 50 × 50 × 3, and the output obtained was 50 × 50 × 32 after extracting the features once. Then, a max-pooling layer was applied to lessen the dimensionality of the output gained from the initial convolution layer. After applying the pooling layer once, the output shape was decreased to 25 × 25 × 32. Then, this approach was applied again three times. Filter numbers for the second, third, and fourth convolution layers were 64, 128, and 256, respectively. Each of them has a size of 3 × 3. After applying the last pooling layer, the output shape was found to be 2 × 2 × 256. In all four convolution layers, the ReLU activation function was applied due to its ability to produce sparse, non-linear, and computationally efficient activations while also helping to mitigate the vanishing gradient problem throughout the training. The output found after the feature extraction from the input data was a group of three-dimensional feature maps. Then, flattening was applied for these feature maps to be turned into a 1D vector. After flattening, the 1D output layer has the shape 1 024. Then the flattening layer was reshaped to 4 × 256 for compatibility, which improves the data representation, and it helped to enhance the model performance. By including an LSTM layer in the middle of the reshape and fully connected layers, the model can learn to better capture long-term dependencies and temporal patterns in the input data that may cause improved performance in image classification tasks.

Then, two fully attached dense layers were added with the ReLU activation function again, which includes 512 and 128 neurons, respectively. Lastly, the classification layer is formed by adding two neurons with a sigmoid activation function to finally predict whether a particular image contains benign or malignant cancer cells. The 'Adam' optimizer was used for training the model, and it can handle non-stationary and noisy objective functions. The loss function used here was 'binary cross entropy' because the main target was a binary classification task. Figure 8.4 represents the model architecture of the suggested model. The output shape and parameter numbers of each layer are presented clearly in table 8.1.

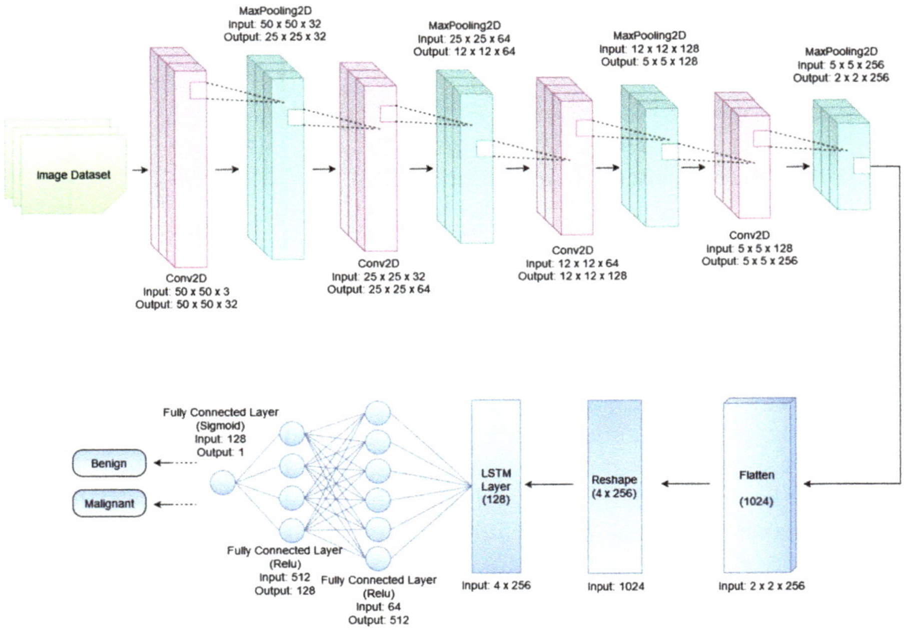

Figure 8.4. Model architecture of the suggested model.

Table 8.1. Numeric description of the suggested model architecture.

Backbone networks	Output shape	Number of parameters
Conv2D	50 × 50 × 32	896
MaxPooling2D	25 × 25 × 32	0
Conv2D	25 × 25 × 64	18 496
MaxPooling2D	12 × 12 × 64	0
Conv2D	12 × 12 × 128	73 856
MaxPooling2D	5 × 5 × 128	0
Conv2D	5 × 5 × 256	295 168
MaxPooling2D	2 × 2 × 256	0
Flatten	1 024	0
Reshape	4 × 256	0
LSTM	128	197 120
Dense	512	66 048
Dense	128	656 604
Dense	1	129

8.3.5 Evaluation metrics

These are the metrics included to judge the performance of proposed model. These metrics help to determine how well the model is performing and whether it is meeting the desired goals or not.

Accuracy: Accuracy signifies the percentage of predictions accurately done by the model from overall predictions. However, accuracy may not be the leading metric to apply in certain scenarios, particularly for the imbalanced dataset. Imbalanced datasets are those in which the distribution of classes is highly skewed, with samples from one class constituting the majority and samples from other classes constituting the minority. But the class imbalance problem of the dataset has been removed.

$$A = \frac{(\mathrm{TP} + \mathrm{TN})}{\mathrm{TP} + \mathrm{TN} + \mathrm{FN} + \mathrm{FP}} \tag{8.1}$$

Precision: Precision quantifies the percentage of true positive predictions (correctly identified samples) over the overall count of positive predictions (samples predicted as positive). It offers insightful information about the model's capacity to deliver precise positive predictions. This metric is useful in cases where false positives are costly. Since false positives are instances where the model wrongly labels a sample as positive when it should be negative, here priority has been given to avoid them by concentrating on precision. A lower rate of false positives indicates a model that is more dependable at correctly identifying positive cases, and a higher precision reflects this.

$$P = \frac{\mathrm{TP}}{\mathrm{TP} + \mathrm{FP}} \tag{8.2}$$

Recall: Recall quantifies the percentage of true positives over the overall count of actual positives (total samples belonging to the positive class). It sheds light on the model's capacity to correctly detect each positive instance. This metric is useful in cases where false negatives are costly or where there is a class imbalance. False negatives happen when the model is unable to detect positive instances, which, depending on the application, can have dire repercussions. For instance, failing to identify a patient's disease during a medical diagnostic could cause the therapy to be delayed or to cause further issues.

$$R = \frac{\mathrm{TP}}{\mathrm{FN} + \mathrm{TP}} \tag{8.3}$$

F1 score: The F1 score stands for the harmonic mean of two metrics, where one is known as precision and the other is recall. It works well when both of these metrics are important and the classes are imbalanced. For providing a thorough evaluation of the model's capacity to accurately identify positive cases while avoiding false positives and false negatives, the F1 score unites recall and precision into a single statistic. By using the harmonic mean, the F1 score ensures a balanced evaluation by giving precision and memory equal weight.

$$\mathrm{F1} - \mathrm{score} = \frac{2\mathrm{x}P\mathrm{x}R}{P + R} \tag{8.4}$$

ROC curve and AUC score: For binary classification tasks, evaluation metrics like the ROC curve and the AUC score are frequently utilized. The true positive rate (sensitivity) in opposition to the false positive rate (specificity) for several categorization thresholds is plotted on the ROC curve. How well the model performs can be examined when applied to various threshold values using the ROC curve, which graphically depicts the trade-off amid sensitivity and specificity. Greater classification performance is shown by a true positive rate that is greater and a false positive rate that is lower.

The region under the ROC curve is measured by the AUC score, which quantifies the model's overall efficacy. The model's discriminative capacity is expressed as a single numerical value in this statement. The AUC score is a numeric value between 0 and 1, where 1 indicates a perfect classifier and 0.5 denotes a random classifier.

The ROC curve and AUC score give us important information about how the model performs at various classification thresholds, enabling us to evaluate its discrimination capabilities and choose the classification threshold that is most appropriate for the given task. They are often-used evaluation metrics in many different fields such as fraud detection, anomaly detection, and medical diagnostics.

The area under the ROC curve provides a single score, and it represents the model's overall effectiveness.

8.4 Result analysis

Initially, there was no LSTM layer in the model. As a result, the accuracy obtained was not satisfying at all. So, an LSTM layer was added. Also, the model has been trained in different numbers of epochs such as 20 epochs, 25 epochs, and 65 epochs (table 8.2). For 20 epochs, the proposed model achieved an accuracy of 88.13%, whereas 87.23% and 87.22% accuracy were achieved for 25 and 65 epochs, respectively (figure 8.5).

```
plt.plot(history.history["loss"])
plt.plot(history.history["val_loss"])
plt.title("Loss vs Epoch Graph")
plt.ylabel("loss")
plt.xlabel("epoch")
plt.legend(["train", "validation"], loc="upper right")
plt.show()
```

Table 8.2. Evaluation matrices of the recommended approach.

Classes	Precision (%)	Recall (%)	F1 score (%)	Support
Benign	88.95	87.08	88.01	7 500
Malignant	87.35	89.19	88.26	7 500
Average	88.15	88.14	88.14	15 000

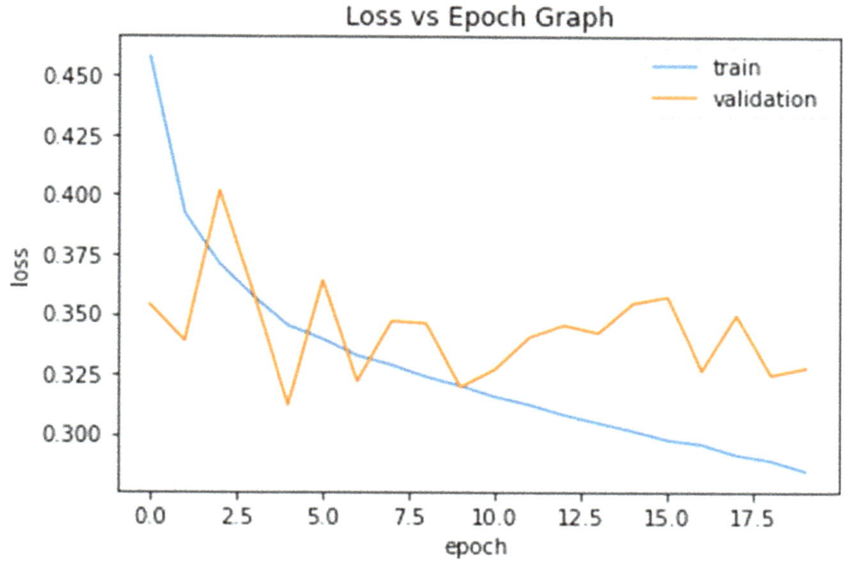

Figure 8.5. Loss over the number of epochs.

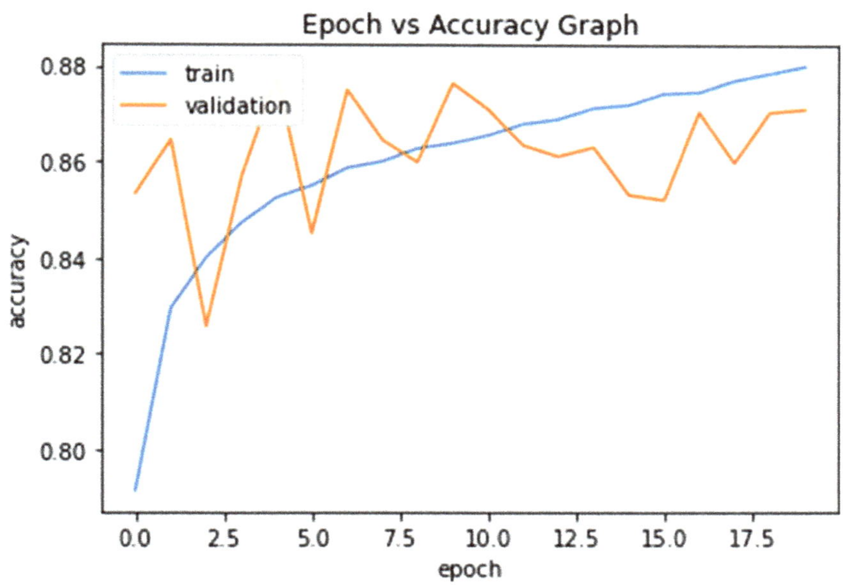

Figure 8.6. Accuracy over the number of epochs.

Here, matplotlib.pyplot was imported as plt. The recorded values of training and validation loss for the 20 epochs of training phase were plotted using plt.plot (figure 8.6).

```
plt.plot(history.history["accuracy"])
plt.plot(history.history["val_accuracy"])
plt.title("Epoch vs Accuracy Graph")
plt.ylabel("accuracy")
plt.xlabel("epoch")
plt.legend(["train", "validation"], loc="upper left")
plt.show()
```

The history records the training and validation accuracies throughout the training phase. These training and validation accuracies over the 20 epochs were plotted on the graph using matplotlib.pyplot.

So, the best result was found in 20 epochs. Table 8.3 denotes that, while comparing to earlier research works on the similar datasets, the suggested model achieves higher correctness. The model also outperforms some existing frameworks in terms of the rate of precision, F1 score, and AUC score. The average time of training each epoch of the proposed model was 123 seconds. In figure 8.7, it is shown that the number of true positive samples was 6 531 and the quantity of false positive samples was 969, respectively. It means this model has truly predicted 6 531 images as benign and that 969 images were predicted as malignant. On the other hand, the quantity of false negative samples was 811, and the number of true negative samples was 6 689. It means that 811 malignant images were predicted as benign and 6 689 malignant images were actually predicted as malignant. From this observation, the precision rate was calculated as 88.15%, the recall rate was achieved at 88.13%, and the F1 score obtained was 88.13%. The proposed framework has gained an exactness of 88.13%. The final results of the recommended model are represented in table 8.2. According to figures 8.5 and 8.6 the model's training loss was reducing and training accuracy was improving as the number of epochs increased. As figure 8.8 shows, these two classes have attained an 88% zone on the ROC curve, and the macro average was received at 88%. Additionally, the benign class has an 83.9% region on the precision–recall curve, and the malignant class has an 83.6% region, as shown in figure 8.9. The AUC score of the suggested model was 94.14%,

Table 8.3. Performance comparison among different existing models.

Existing methods	Precision (%)	Recall (%)	F1 score (%)	Accuracy	AUC score (%)
Romano and Hernandez [21]	88.95	87.08	88.01	7 500	
Alghodhaifi *et al* [22]	87.35	89.19	88.26	7 500	
Cruz-Roa *et al* [23]	88.15	88.14	88.14	15 000	
Narayanan *et al* [32]					93.50
Proposed method	**88.15**	**88.14**	**88.14**	**88.13**	**94.12**

Figure 8.7. Confusion matrix of the proposed method.

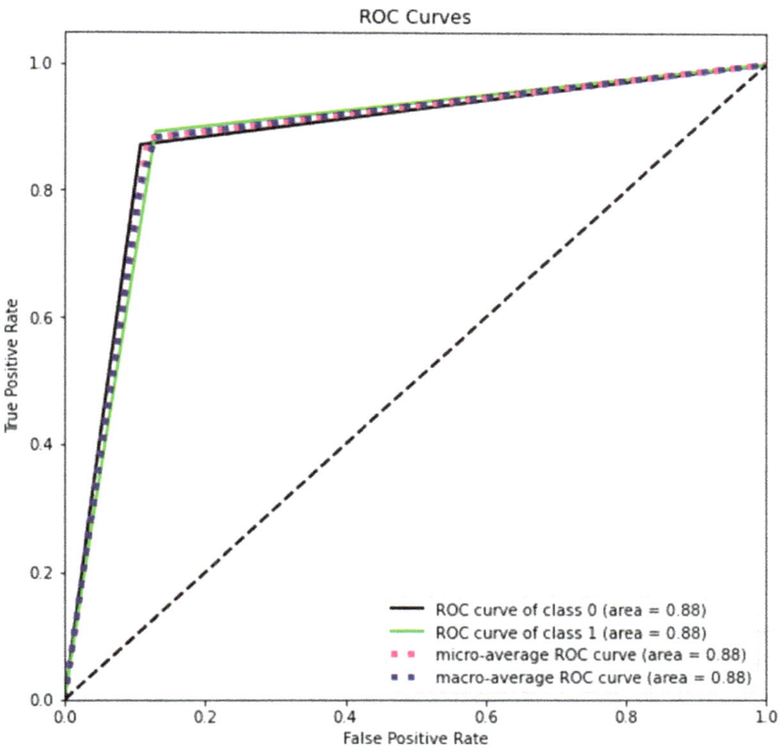

Figure 8.8. ROC curve of the suggested approach.

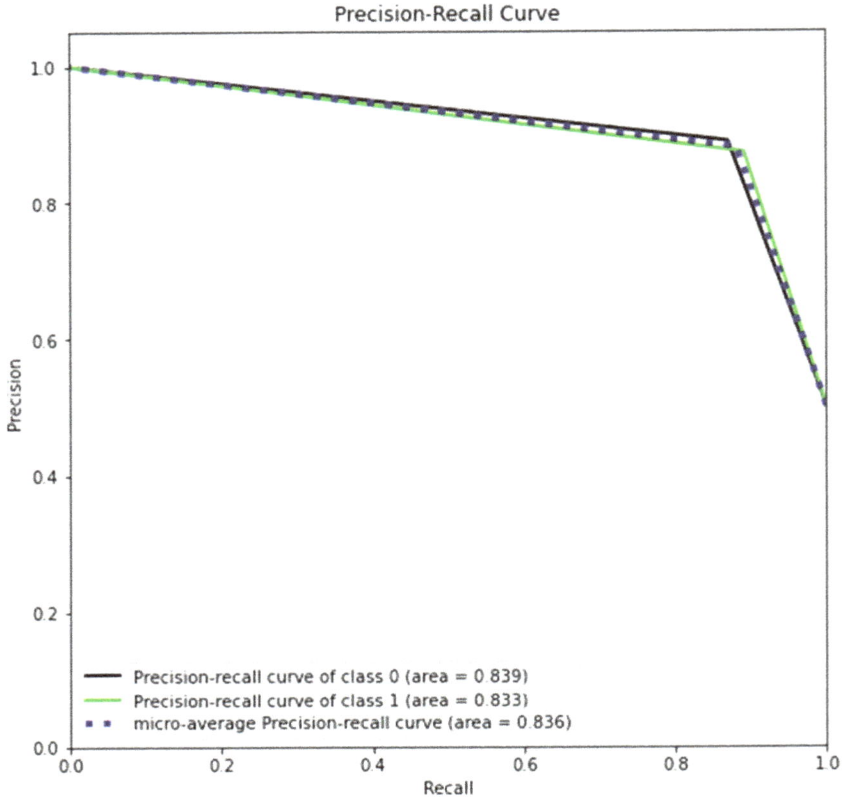

Figure 8.9. Precision–recall curve of the suggested approach.

which was satisfactory. The number of trainable parameters was 717 377, and the total number of parameters was also 717 377, which means that all the parameters were trainable in the proposed method.

Figures 8.8 and 8.9 represent loss and accuracy over the number of epochs during the training and validation phase of the suggested approach.

8.5 Conclusion

Nowadays, breast cancer is considered a hazard to women all over the world and is to blame for the rise in the death rate of women. This study of identifying breast cancer contributes to the growing body of research on deep learning for medical image analysis and provides insights into the future of these strategies for improving breast cancer diagnosis. The study has demonstrated the effectiveness of a deep learning approach for detecting breast cancer from histopathological image data. By preprocessing the imaging data, extracting features using the specific method, and training and evaluating different classification models, a higher-accuracy AUC score has been achieved. Specifically, this model outperformed many existing models, and it is simpler and less time consuming. The findings suggest that deep learning models, particularly CNNs, are effective for analyzing medical images and that earlier and more accurate diagnosis of breast cancer may result from this. These results provide important insights for researchers and health care professionals working towards more accurate and efficient breast cancer diagnosis and treatment.

Acknowledgments

I express my appreciation to all the faculty members of Department of Information and Communication Technology, Bangladesh University of Professionals for their assistance and cooperation.

Bibliography

[1] Alwan N A 2016 Breast cancer among Iraqi women: preliminary findings from a regional comparative breast cancer research project *J. Global Oncol.* **2** 255

[2] Arnold M *et al* 2022 Current and future burden of breast cancer: global statistics for 2020 and 2040 *Breast* **66** 15–23

[3] *Breast Cancer Statistics* (www.cdc.gov/cancer/breast/statistics/index.htm) (accessed 16 May 2023) updated June 8, 2023

[4] *About Breast Cancer* American Cancer Society (www.cancer.org/cancer/types/breast-cancer/about/how-common-is-breast-cancer.html) (accessed 19 June 2023) updated September 14, 2023

[5] *Breast Cancer* (www.who.int/news-room/fact-sheets/detail/breast-cancer) (accessed 26 May 2023)

[6] Hulka B S and Moorman P G 2008 Reprint of breast cancer: hormones and other risk factors *Maturitas* **61** 203–13

[7] Mariotto A B, Yabroff R, Shao K, Feuer Y, Brown E J and M L 2011 Projections of the cost of cancer care in the United States: 2010–2020 *J. Natl. Cancer Inst.* **103** 117–28

[8] Mathers C D, Lopez A D and Murray C J L *et al* 2006 The burden of disease and mortality by condition: data, methods, and results for 2001 *Global Burden of Disease and Risk Factors* ed A D Lopez, C D Mathers and M Ezzati (New York: Oxford University Press) 45 10–1596

[9] Anderson B O, Yip C H, Smith R A, Shyyan R, Sener S F, Eniu A, Carlson R W, Azavedo E and Harford J 2008 Guideline implementation for breast healthcare in low-income and middle-income countries: overview of the breast health global initiative global summit 2007 *Cancer* **113** 2221–43

[10] Lima Z S, Ebadi M R, Amjad G and Younesi L 2019 Application of imaging technologies in breast cancer detection: a review article *Macedonian J. Med. Sci.* **7** 838

[11] Heywang-Köbrunner S, Viehweg P, Heinig A and Küchler C 1997 Contrast-enhanced mri of the breast: accuracy, value, controversies, solutions *Eur. J. Radiol.* **24** 94–108

[12] Allugunti V R 2022 Breast cancer detection based on thermographic images using machine learning and deep learning algorithms *Int. J. Eng. Comput. Science* **4** 49–56

[13] Greenspan H, Ginneken V, Summers B and R M 2016 Guest editorial deep learning in medical imaging: overview and future promise of an exciting new technique *IEEE Trans. Med. Imaging* **35** 1153–9

[14] Shafi A M A, Maruf M M H and Das S 2022 Pneumonia detection from chest x-ray images using transfer learning by fusing the features of pre-trained xception and vgg16 networks

2022 25th Int. Conf. on Computer and Information Technology (ICCIT) (Piscataway, NJ) (IEEE) pp 593–8
[15] Demir C and Yener B 2005 *Automated cancer diagnosis based on histopathological images: a systematic survey* Rensselaer Polytechnic Institute TR–05–09 Rensselaer Polytechnic Institute
[16] Nguyen T T, Nguyen C M, Nguyen D T, Nguyen D T and Nahavandi S 2019 Deep learning for deepfakes creation and detection arXiv:1909.11573
[17] Spanhol F A, Oliveira L S, Petitjean C and Heutte L 2015 A dataset for breast cancer histopathological image classification *IEEE Trans. Biomed. Eng.* **63** 1455–62
[18] Zuluaga-Gomez J, Masry A Z, Benaggoune K, Meraghni S and Zerhouni N 2021 A CNN-based methodology for breast cancer diagnosis using thermal images *Comp. Methods Biomech. Biomed. Eng.: Imag. Vis.* **9** 131–45
[19] Laine R F, Arganda-Carreras I, Henriques R and Jacquemet G 2021 Avoiding a replication crisis in deep-learning-based bioimage analysis *Nat. Methods* **18** 1136–44
[20] Rahman M J U, Sultan R I, Mahmud F, Al Ahsan S and Matin A 2018 Automatic system for detecting invasive ductal carcinoma using convolutional neural networks *TENCON 2018 IEEE Region 10 Conf. (Piscataway, NJ)* (IEEE) pp 673–8
[21] Romano A M and Hernandez A A 2019 Enhanced deep learning approach for predicting invasive ductal carcinoma from histopathology images *2019 2nd Int. Conf. on Artificial Intelligence and Big Data (ICAIBD) (Piscataway, NJ)* (IEEE) pp 142–8
[22] Alghodhaifi H, Alghodhaifi A and Alghodhaifi M 2019 Predicting invasive ductal carcinoma in breast histology images using convolutional neural network *2019 IEEE National Aerospace and Electronics Conf. (NAECON) (Piscataway, NJ)* (IEEE) pp 374–8
[23] Cruz-Roa A, Basavanhally A, Gonz´alez F, Gilmore H, Feldman M, Ganesan S, Shih N, Tomaszewski J and Madabhushi A 2014 Automatic detection of invasive ductal carcinoma in whole slide images with convolutional neural networks *Medical Imaging 2014: Digital Pathology* vol 9041 (Bellingham, WA: SPIE) 904103
[24] Arevalo J, Gonz´alez F A, Ramos-Poll´an R, Oliveira J L and Lopez M A G 2016 Representation learning for mammography mass lesion classification with convolutional neural networks *Comput. Methods Programs Biomed.* **127** 248–57
[25] Kaushiki R, Banik D, Bhattacharjee D and Nasipuri M 2019 Patch-based system for classification of breast histology images using deep learning *Comput. Med. Imaging Graph.* **71** 90–103
[26] Shen L, Margolies L R, Rothstein J H, Fluder E, McBride R and Sieh W 2019 Deep learning to improve breast cancer detection on screening mammography *Scientific Reports* **9** 12495
[27] Lingqiao L, Pan X, Yang H, Liu Z, He Y, Li Z, Fan Y, Cao Z and Zhang L 2020 Multi-task deep learning for fine-grained classification and grading in breast cancer histopathological images *Multimed. Tools Appl.* **79** 14509–28
[28] Wang J, Kato F, Oyama-Manabe N, Li R, Cui Y, Tha K K, Yamashita H, Kudo K and Shirato H 2015 Identifying triple-negative breast cancer using background parenchymal enhancement heterogeneity on dynamic contrast-enhanced MRI: a pilot radiomics study *PLoS One* **10** e0143308
[29] Pesapane F *et al* 2023 How radiomics can improve breast cancer diagnosis and treatment *J. Clin. Med.* **12** 1372

[30] Wang Y, Zeng J, Wu W, Xie S, Yu H, Li G, Zhu T, Li F, Lu J and Wang G Y *et al* 2019 Nicotinamide n-methyltransferase enhances chemoresistance in breast cancer through sirt1 protein stabilization *Breast Cancer Res.* **21** 1–17

[31] Lyu L, Yao J, Wang M, Zheng Y, Xu P, Wang S, Zhang D, Deng Y, Wu Y and Yang S *et al* 2020 Overexpressed pseudogene hla-dpb2 promotes tumor immune infiltrates by regulating hla-dpb1 and indicates a better prognosis in breast cancer *Front. Oncol.* **10** 1245

[32] Narayanan B N, Krishnaraja V and Ali R 2019 Convolutional neural network for classification of histopathology images for breast cancer detection *2019 IEEE National Aerospace and Electronics Conf. (NAECON) (Piscataway, NJ)* (IEEE) pp 291–5

[33] *Breast Histopathology Images* (https://www.kaggle.com/datasets/paultimothymooney/breast-histopathology-images) (accessed 6 July 2023)

[34] Prodhan M M A and Yousuf M A 2023 Combination of the features of pre-trained xception and vgg16 models to identify childhood pneumonia from chest x-ray images *2023 Int. Conf. on Electrical, Computer and Communication Engineering (ECCE) (Piscataway, NJ)* (IEEE) pp 1–6

[35] Arel I, Rose D C and Karnowski T P 2010 Deep machine learning-a new frontier in artificial intelligence research [research frontier] *IEEE Comput. Intell. Mag.* **5** 13–8

[36] Weerakody P B, Wong K W, Wang G and Ela W 2021 A review of irregular time series data handling with gated recurrent neural networks *Neurocomputing* **441** 161–78

[37] Anthimopoulos M, Christodoulidis S, Ebner L, Christe A and Mougiakakou S 2016 Lung pattern classification for interstitial lung diseases using a deep convolutional neural network *IEEE Trans. Med. Imaging* **35** 1207–16

[38] Volpi M and Tuia D 2016 Dense semantic labeling of subdecimeter resolution images with convolutional neural networks *IEEE Trans. Geosci. Remote Sens.* **55** 881–93

[39] Yamashita R, Nishio M, Do R K G and Togashi K 2018 Convolutional neural networks: an overview and application in radiology *Insights Imaging* **9** 611–29

IOP Publishing

Signal Processing with Python
A practical approach
Irshad Ahmad Ansari and Varun Bajaj

Chapter 9

Maximum power point tracking for partially shaded photovoltaic system using advanced signal processing

K P Madhavan

A photovoltaic (PV) array always needs to be operated in maximum power point (MPP) to extract maximum power from it. However, in a partially shaded condition (PSC) there are multiple peaks and traditional MPP methods that fail to capture global maxima and get stuck in local maxima, which results in significant power loss, thereby reducing the efficiency of the PV system. Various techniques are proposed in tracking global maxima in PSC so that the PV system always operates in global maxima. However, every technique has their own advantages and disadvantages. Hence, this chapter aims to provide an explanation of major MPP techniques for PSC conditions along with their comparative advantages and limitations. This chapter has compiled over 106 MPP techniques for PSCs that have been proposed in the literature in the last 11 years (2010–20). This chapter has further classified the works into four techniques: soft computing, exact algorithm, meta-heuristic, and hardware configuration. They are compared and evaluated on various parameters. A brief analysis, along with supported Python codes, is also provided for many methods. This work is intended to give researchers an understanding of existing methods and to serve as a base for their future research.

9.1 Introduction

The depleting reserves of conventional energy, along with the environmental pollution associated with them, have generated interest in renewable energy that provides a cleaner and more viable form of energy (Liu *et al* 2012). Among renewable energy sources, electricity generation by a PV system has been generating a lot of interest due to its environment friendliness, low maintenance cost (for the PV system), absence of sound due to no moving parts, and longevity of the panels. However, one of the major setbacks of a PV array is the efficiency, which typically ranges from 15% to 20%. Hence, it becomes pertinent to operate a PV system in MPP to extract maximum output power from the PV panel. Traditional methods include perturb and observe, incremental conductance, and hill climbing to track MPP under normal conditions (Rezk and Eltamaly 2015, Alsumiri 2019). However, these traditional methods fail to track global peak in PSC and make PV array operate in local peak, which causes considerable power loss. PSC is a condition that occurs when the PV system consisting of PV modules connected in series and parallel receive different irradiance due to clouds, dust, or shadow of buildings. PSCs can have a serious effect on PV arrays, like a hotspot, and can completely damage the PV module. To prevent this from happening, PV modules are generally fitted with a bypass diode in parallel. These bypass diodes under normal conditions operate in reverse bias and if, due to shading, the PV module produces less current than the rest of system, it operates in forward bias in order to short-circuit the PV module; thus, it avoids local hotspots and also avoids causing the PV module to act as load. However, this results in multiple steps in I–V characteristics and multiple power peaks in a respective P–V curve. Traditional MPP tracking (MPPT) methods cannot track them and get stuck in local maxima. Hence, to track MPP in PSCs, many new methods are proposed in order to not fail under partial shading. Soft computing techniques are quite popular due to their ability to handle imprecision, while advantages of meta-heuristic algorithms are that these are system independent. Also, these meta-heuristic optimization algorithms treat PSCs as optimization problems and finding the maxima is set as the objective.

Other techniques involve methods based on exact algorithms; these rely on the exactness of an algorithm as compared to soft computing where exactness is not required. Some other techniques also involve the use of power electronic circuits and array reconfiguration techniques to control the MPPT. These techniques use decentralized controllers and use simple algorithms to find MPP.

Due to a large amount of research into MPPT in recent years for solving the problem of PSCs, this work has been carried out to compile, classify, and compare various MPP techniques during shaded conditions done in the last 11 years (2010–20). Mohapatra *et al* (2017), Kumar *et al* (2020), and Yang *et al* (2020) have extensively reviewed all MPPT methods. However, a comprehensive review of all MPP methods for PSCs, along with the inherent advantages and shortcomings of these methods, with the competitive edge that each method possesses relative to one another, has not been discussed until now. This work is intended to fill that gap and help researchers understand the significant edge each method possesses. However, it is also true that an absolute comparison of each method becomes difficult due to the different environmental conditions and different PV systems that each work covers; this work has attempted to compare them based on practical and logical criteria. The paper is arranged in the following order: section 9.2 deals with modeling and characteristics of PV systems, section 9.3 deals with the need for MPP techniques and traditional MPP techniques, section 9.4 deals with challenges of traditional MPPT systems, section 9.5 deals with soft computing methods, section 9.6 deals with meta-heuristic techniques, section 9.7 deals with exact algorithm-based methods, section 9.8 deals with hardware-based techniques, section 9.9 deals with the analysis and comparison of various methods, section 9.10 deals with the challenge and future scope of MPPT during PSCs in PV systems, and section 9.11 concludes the paper. Also, some sample codes are provided to better understand the concept. This work is intended to help researchers help get a good understanding of existing systems and help them with future work in this domain.

9.2 Modeling and characteristics of PV systems

9.2.1 Modeling of PV cells

A PV cell is the building block of PV modules. Generally a PV module consists of N-cells connected in series. Hence, a modeling of PV cells is important for understanding the characteristics of a PV module. Ideal modeling of a PV cell involves a current source with current I_{ph} flowing out of it. I_{ph} represents the current produced through a solar cell proportional to temperature and irradiance. Since a solar cell is a type of diode, p–n junctions are represented by a diode, and the current through a p–n junction is labeled as i_d (shown in figure 9.1); however, a real diode has losses and is represented by both R_s and R_{sh}. R_s indicates loss due to metallic contacts and semiconductor material; R_{sh} represents the leakage current of a diode (figure 9.2). A two-diode model has also been proposed, which takes into account both diffusion and recombination (instead of just a p–n junction current as in a one-diode model), using two diodes to represent both the currents separately. Even though a two-diode model (shown in figure 9.3) accurately mimics the characteristics of a solar cell, the complexity of simulating it is higher compared to the one-diode model. Hence, a single diode model is generally preferred due to its lower complexity compared to a two-diode model.

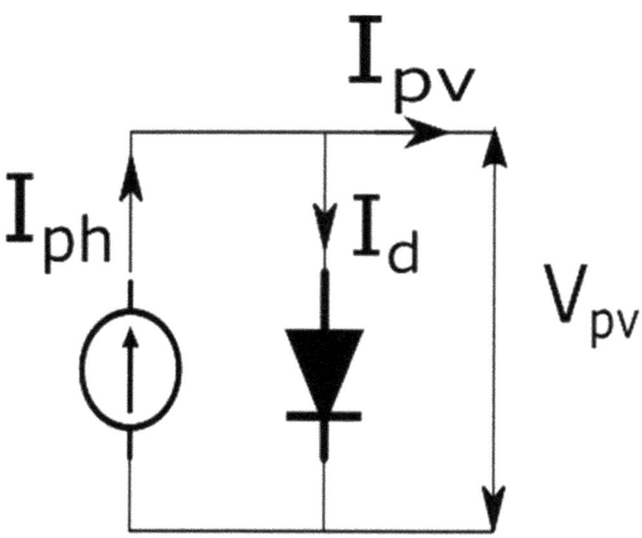

Figure 9.1. Ideal model of a PV cell.

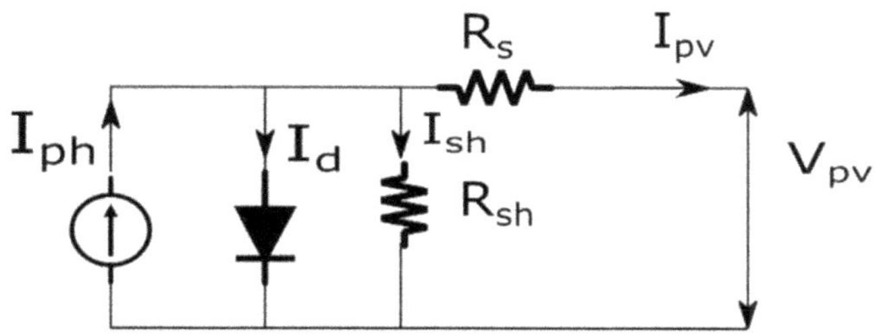

Figure 9.2. Single diode model of a PV cell.

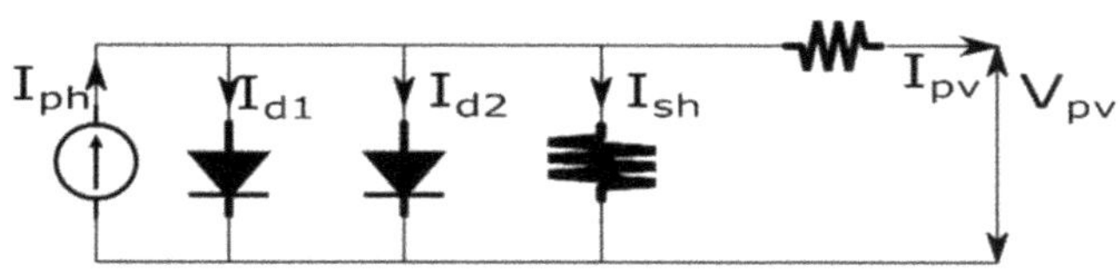

Figure 9.3. Two diode model of a PV cell.

For the single diode model represented in figure 9.2, the terminal current I_{pv} is represented by:

$$I_{\text{pv}} = I_{\text{ph}} - I_d - I_{sh} \tag{9.1}$$

where,

$$I_d = I_o\left(e^{\frac{q\left(V_{\text{pv}}+i_{\text{pv}}R_s\right)}{\eta KT}} - 1\right) \tag{9.2}$$

$$I_{\text{sh}} = \frac{V_{\text{pv}} + I_{\text{pv}}R_s}{R\text{sh}}. \tag{9.3}$$

V_{pv} and I_{pv} are the terminal voltage and terminal currents, I_{o} represents the reverse saturation current, η indicates the ideality factor, T is the cell temperature, K is the Boltzmann constant, R_s is the series resistance, and R_{sh} is the shunt resistance.

9.2.2 Modeling of PV modules

A PV module generally consists of n number of series connected PV cells, hence the PV module current output is n times I_{pv} (however, some panels do contain series and parallel combinations of cells). PV arrays generally consist of series and parallel (S–P) configurations of PV modules (figure 9.4) according to voltage and current levels required.

9.2.3 Characteristics of PV modules

The I–V and P–V characteristics of a PV module is shown in figure 9.5. I–V characteristics of a PV module operate in the 1st quadrant with short circuit currents

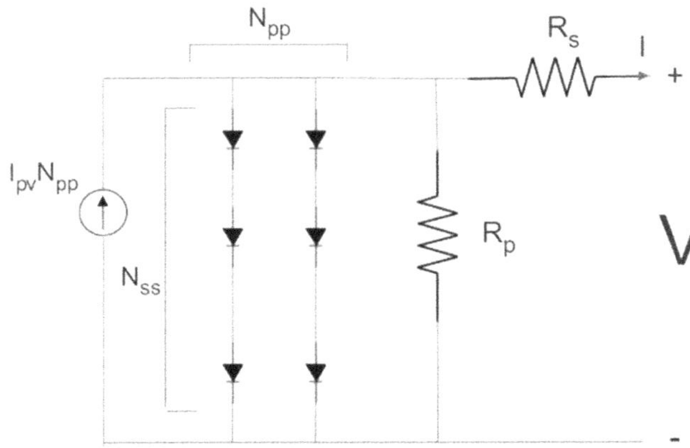

Figure 9.4. S–P combination of a PV array.

represented by I_{sc} and open circuit voltage represented by V_{oc}. The I–V characteristics of a PV module depend on two parameters, namely temperature and irradiance. Variation of I–V and P–V with these two parameters is shown in figures 9.6–9.9. As can be observed, there is a single peak point in the P–V characteristics (figure 9.5(b)) of a PV system, and that peak is referred to as MPP. Our aim is to make the PV panel always operate in that peak point to extract maximum power from the PV system.

9.2.4 PV system schematics

Figure 9.10 shows a PV system connected to a grid. The schematic consists of both a power and a control circuit where the PV system is assumed to be a fixed DC source, and a DC–DC converter (chopper) and a DC–AC (inverter) are control circuits used for control and conversion of the power source. An MPPT controller-aided DC–DC converter is essential for maximum power transfer where the duty cycle of a DC–DC converter is altered by an MPP controller to extract maximum power from the PV system. An inverter is required for the conversion of power from DC–AC, which is subsequently fed to the AC grid.

9.3 MPPT: concept and traditional techniques

9.3.1 Need and concept of MPPT

The requirement of MPPT is explained in this subsection. For simplicity, assume a load is connected across a PV panel. The diagram (figure 9.11) shows a PV panel connected to a load. Load resistance is R_o and voltage is V_o and current flowing through the load is I_o. R_o can vary from 0 (in a short circuit) to infinity (in an open circuit).

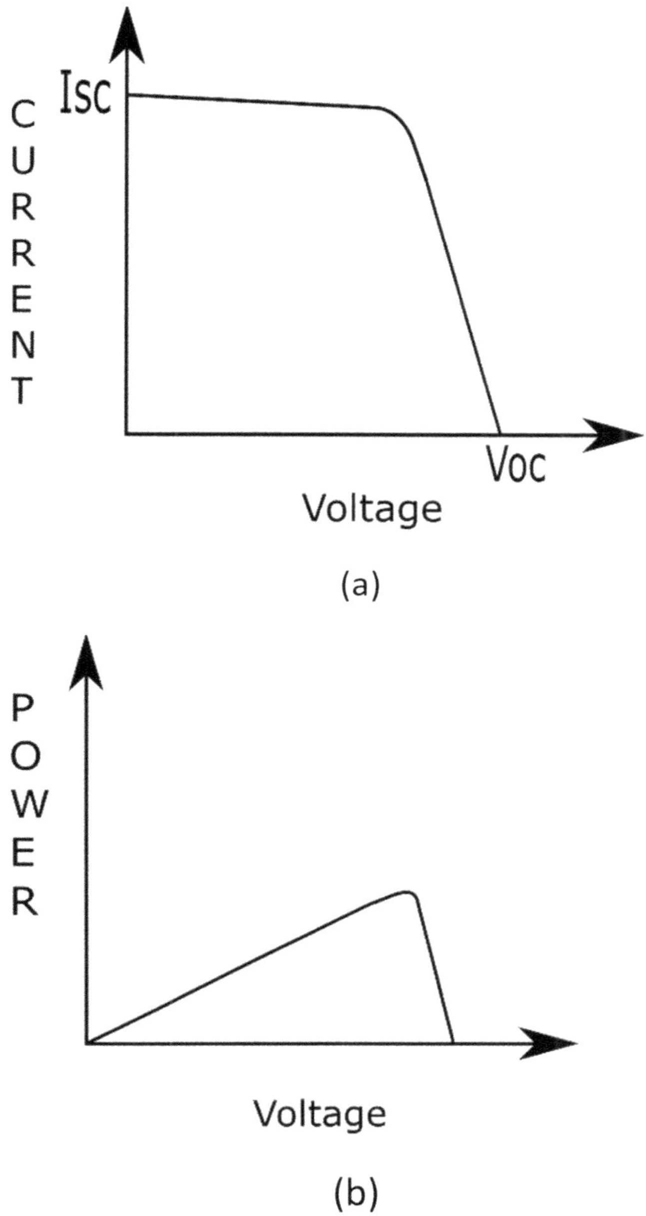

Figure 9.5. (a) I–V characteristics of a PV module; (b) P–V characteristics of a PV module.

As shown in figure 9.12, the point where the load line, having a slope of $1/R_o$ $(= I_o/V_o)$, meets the I–V curve of the PV panel is the operating point of the PV panel. When R_o equals infinity, the load line aligns with the x-axis, and if R_o equals 0, the load line aligns with the y-axis. The aim of MPP is to have the load line meet the characteristic curve at the peak

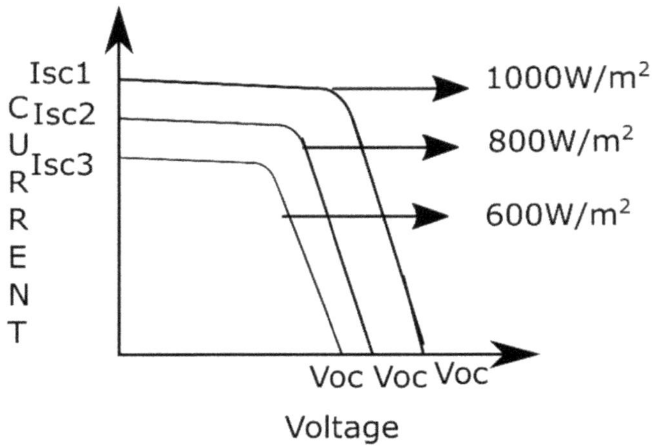

Figure 9.6. *I–V* characteristics for various irradiance.

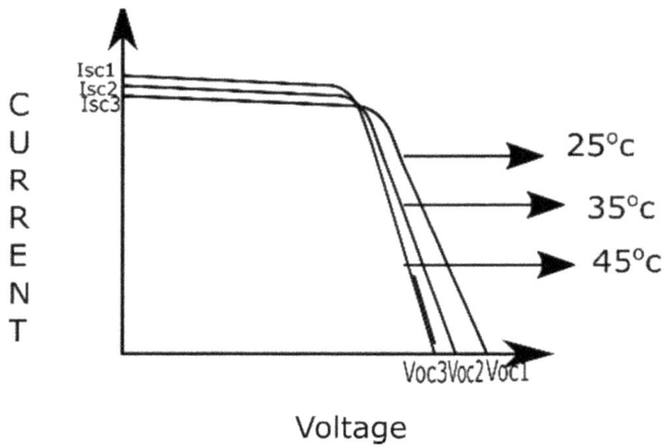

Figure 9.7. *I–V* characteristics for various temperatures.

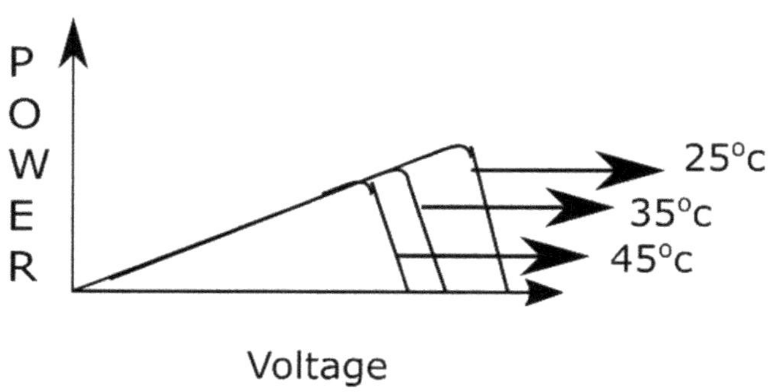

Figure 9.8. *P–V* characteristics for various temperatures.

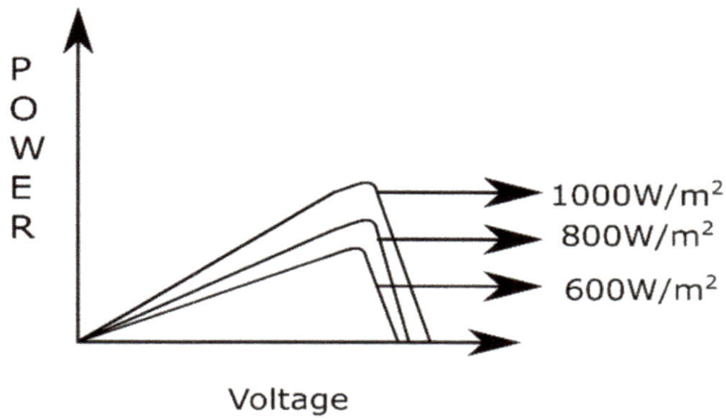

Figure 9.9. *P–V* characteristics for various irradiance.

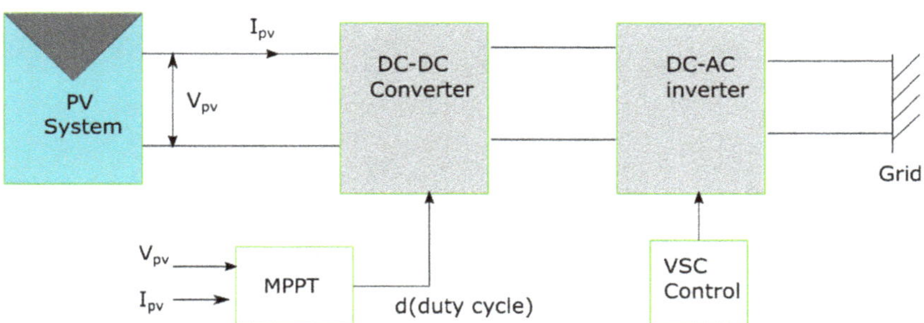

Figure 9.10. PV system connected to a grid.

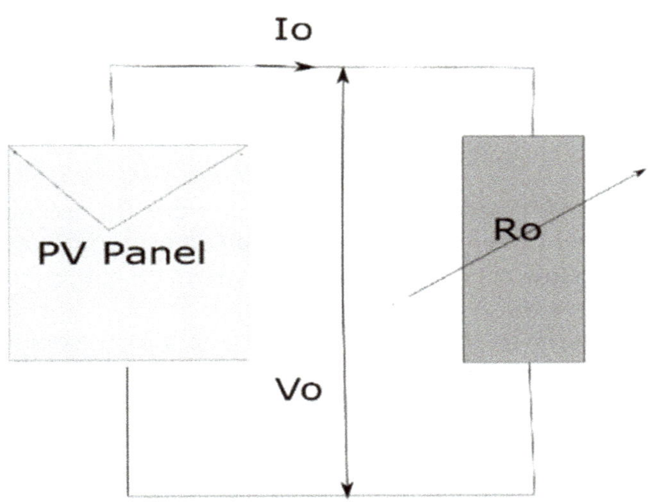

Figure 9.11. A PV panel connected directly to a load.

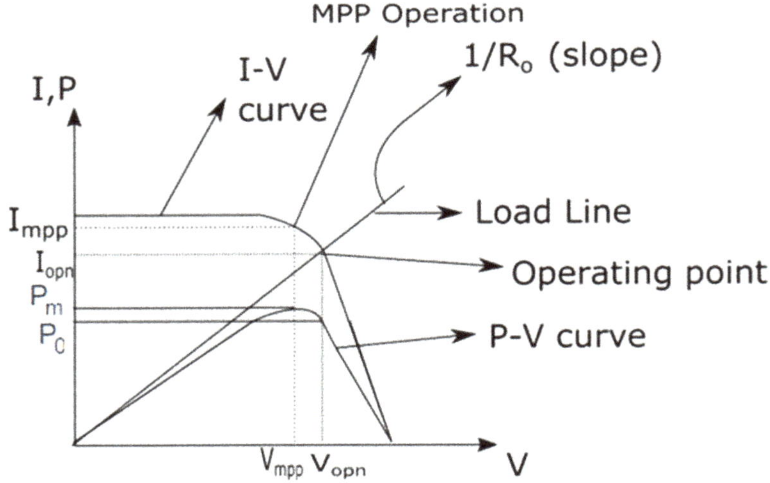

Figure 9.12. PV system not operating at MPP.

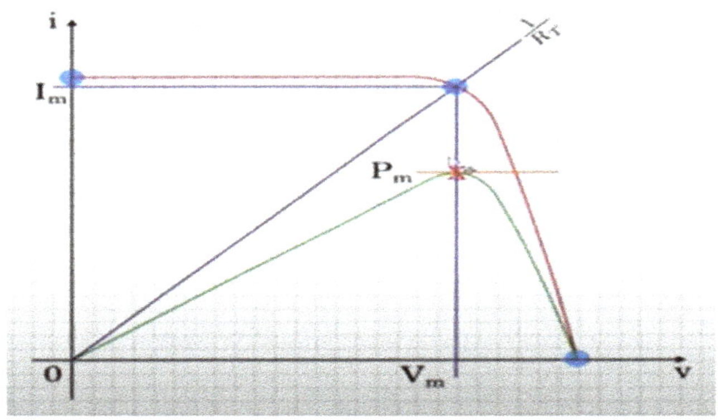

Figure 9.13. A PV system operating at MPP.

point (represented as MPP operation in figure 9.12) so that the operating point (P_o) is always at P_m (in figure 9.12 the I–V curve and P–V curve are superimposed to show power points at operating points). In figure 9.12, the PV system is not operating at MPP and hence experiences a loss of $P_m - P_o$.

Figure 9.13 shows the result of using an MPP controller wherein the PV panel is always operating at peak point regardless of change in load.

Now, to make a PV panel always operate at MPP (P_m), a DC–DC power interface with duty cycle as control input is added between the solar PV panel and load so that regardless of change in load resistance, the input impedance R_t (of figures 9.14 and 9.15) is such that the load line ($1/R_t$) always meets the I–V curve at MPP (shown in figure 9.13).

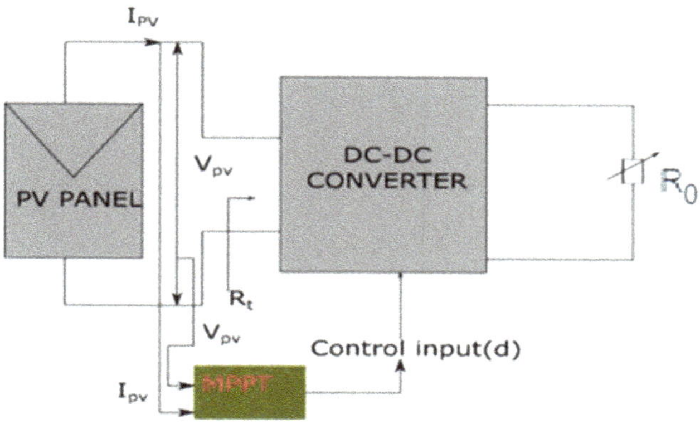

Figure 9.14. MPPT setup.

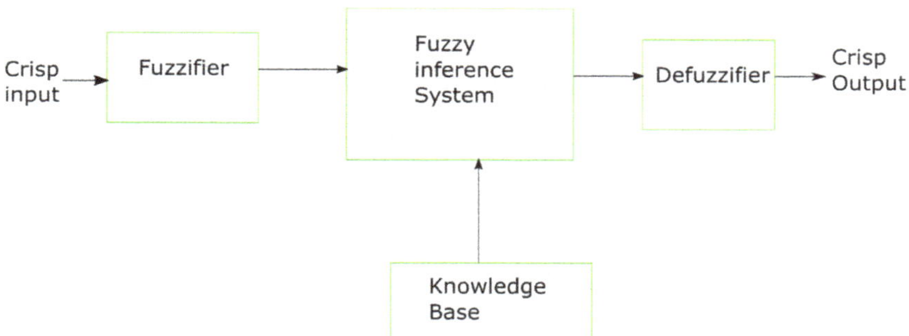

Figure 9.15. Fuzzy system.

9.3.2 Traditional MPPT concepts

Traditional MPP techniques include hill climbing (Xiao and Dunford 2004, Gules *et al* 2008), incremental conductance (Kobayashi *et al* 2006), and perturb and observe (Femia *et al* 2007); there have been many modifications (Chiang *et al* 2002, Wu *et al* Wu 2003) made to these systems for faster tracking.

9.3.2.1 Incremental conductance technique

Incremental conductance is popular due to its high tracking accuracy and low oscillations at a steady state coupled with a high adaptability to rapidly changing atmospheric conditions. The incremental conductance is based on comparing incremental conductance (difference in present conductance with respect to previous conductance) to instantaneous conductance (Eltamaly 2018).

The concept of incremental conductance is shown below.

For a PV panel, power at any instant is defined as

$$P = V \times I. \tag{9.4}$$

The differentiating equation (9.4) with respect to V is

$$\frac{dP}{dV} = V\frac{dI}{dV} + I. \tag{9.5}$$

As we can see in figure 9.5(a), the slope of power with respect to voltage at MPP, for example, is 0:

$$\frac{dP}{dV} = 0 \text{ (@MPP)}$$

$$V\frac{dI}{dV} + I = 0 \tag{9.6}$$

$$\frac{dI}{dV} = -\frac{I}{V} \tag{9.7}$$

$$\frac{I(t) - I(t-1)}{V(t) - V(t-1)} = -\left(\frac{I(t)}{V(t)}\right) \tag{9.8}$$

$$e = \frac{I(t) - I(t-1)}{V(t) - V(t-1)} + \left(\frac{I(t)}{V(t)}\right). \tag{9.9}$$

The error (e) in equation (9.9) keeps reducing as the operating point reaches MPP and at the MPP becomes 0.

Below is a sample Python code for implementation of incremental conductance:

```
i = 0  # Initialize the iteration counter

def MPPT(V, i, D):
  global i  # Use the global variable i

  i += 1  # Increment the iteration counter

  if i == 1:
    V1 = 0
    I1 = 0
  else:
    V1 = V
    I1 = i

  de = (i - I1) / (V - V1) + (i / V)

  if de > 0:
    D += 0.05
  elif de < 0:
    D -= 0.05

  V1 = V
  I1 = i

  return D  # Return the updated duty cycle
```

The above code can be tuned for given input voltage and current so that it senses for better tracking.

9.3.2.2 Perturb and observe

Figure 9.16 shows the flowchart of the algorithm of perturb and observe.

The perturb and observe (P&O) technique is based on perturbing the system by increasing its voltage by ΔV and observing its impact on the array output power. If the power difference is positive, then the voltage is increased, and if the power difference is negative, then the voltage is reduced. Because of constant ΔV, the system faces high oscillation even after achieving a steady state. A lot of modification has been proposed to reduce oscillations and lessen them around the steady state. However, the advantage to P&O is its lower complexity compared to incremental conductance, but it is less robust than incremental conductance.

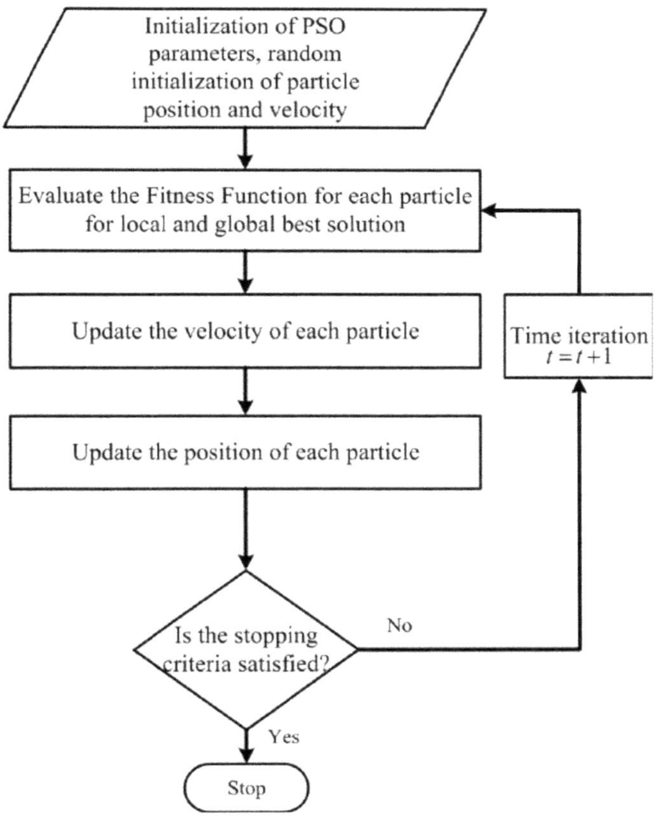

Figure 9.16. Flow chart of particle swarm optimization (PSO).

Sample code for P&O:

```
i = 0  # Initialize the iteration counter

def MPPT(V, i, D):
    global i  # Use the global variable i

    i += 1  # Increment the iteration counter

    p = V * i  # Calculate power

    if i == 1:
        P1 = 0
    else:
        P1 = V1 * I1  # Calculate previous power

    de = p - P1

    if p > P1:
        if V > V1:
            V -= 0.05
        else:
            V += 0.05
    elif p < P1:
        if V > V1:
            V -= 0.05
        else:
            V += 0.05

    V1 = V
    I1 = i

    return V  # Return the updated voltage
```

9.3.2.3 Hill climbing

Hill climbing is among the most widely used MPPT techniques due to its simplicity. It uses a direct duty control–based algorithm where if the power increases, then the duty cycle keeps being reduced, and if the power reduces then the duty cycle is increased. It is similar to the concept of P&O, but direct duty cycle is given as output, compared to P&O where indirect control strategy is used (wherein the voltage response is obtained and then converted to duty cycle).

Python code for hill climbing:

```
i = 0  # Initialize the iteration counter

def MPPT(V, i, D):
    global i, P1  # Use the global variable i and P1

    i += 1  # Increment the iteration counter

    p = V * i  # Calculate power

    if i == 1:
        P1 = 0

    de = p - P1

    if p > P1:
        D -= 0.05
    elif p < P1:
        D += 0.05

    P1 = p

    return D  # Return the updated duty cycle
```

9.4 Challenges of MPPT

Usage of an MPP tracker would make a PV system operate in MPP regardless of fluctuation in load. However, the power produced by a PV system is also greatly influenced by two environmental conditions, namely temperature and irradiance (as already discussed in the preceding section). It is desirable for these environmental conditions to be predictable for efficient and optimum harnessing of solar energy by PV array; however, environmental conditions keep fluctuating throughout the course of day and hence these parameters are also not fixed, continuing to change throughout the day. This makes it essential that an MPPT controller be designed to track MPP continuously to account for environmental changes likely to happen periodically.

Three main challenges associated with MPPT:

(i) changes in environmental temperature
(ii) dynamic changes in irradiance level
(iii) partial shading of the PV panel

9.4.1 Dynamic changes in temperature and irradiance

The dynamic and sudden changes in environmental temperature and irradiance (challenges (i) and (ii)) are quite a regular feature on an everyday basis. The temperature level may be constantly changing throughout the course of a day. IR-radiation changes are also quite frequent in the course of a day, along with the high probability of IR-radiation being received by an entire PV system that changes periodically or even continuously. These change the maximum power operating point. Traditional MPP controllers, due to their continuous tracking of the PV system voltage and PV system current, will be able to successfully track MPP during challenges (i) and (ii). However, they may require extensive oscillation to find a new MPP. New modifications made for the traditional method, like modified incremental conductance, modified P&O, etc, have provided significant reduction in oscillation time.

9.4.2 PSC and its challenges

The third challenge is PSC, which will be discussed in this subsection. Unlike uniform irradiance change due to clouds, here, S–P-connected PV modules receive non-uniform radiation of sunlight either due to shadows of neighboring buildings, trees, or dust accumulation on a PV panel, or even as a result of clouds. Due to these, there might be shading in some PV modules which makes them produce less current compared to other PV panels. Hence there is a mismatch of current in PV modules. In series-connected PV modules, this poses a hazard since all PV panels act as a current source and connecting current sources of different values in a series can create a mismatch in the panel current, which increases heating and creates a hotspot on the shaded panel (the panel receiving relatively less irradiance than the rest of the PV system). So to safeguard against this, all PV panels are attached with a bypass diode in parallel to them, which is generally open circuit in normal conditions but during shaded conditions, acts as a short circuit across the panel, producing less current compared to the rest of the PV array, thereby preventing the hotspot and PV panel acting as a load. While this solves a problem, it brings another challenge where n series–connected PV modules with a possibility of each panel receiving different irradiance may potentially exhibit n stairs in the I–V curve; subsequently, there is a chance of n multiple MPPs with only one being global maxima.

In these PSCs, the global maximum peak may be at any point; thus, full scanning of the I–V curve is required. However, all the traditional techniques mostly get stuck in local maxima and this may result in a large loss of power.

9.4.3 MPPT requirements

An MPPT tracker should ideally have the following capabilities.

(a) It should have mechanisms to differentiate between uniform/dynamic and PSC systems.
(b) It should be able to track MPP instantaneously under uniform and dynamically changing weather conditions.

(c) It should search global maxima under all kinds of partial shading patterns.
(d) It should have robust tracking speed.
(e) It should catch global maxima within an acceptable range of error.
(f) Hardware requirements should be feasible.
(g) It should be system independent.
(h) The oscillations while tracking MPP needs to be minimal, and steady state oscillations should be within a tolerable limit.

Traditional methods with modification are able to track dynamic changing environmental conditions with reasonable accuracy; however, failure to track MPP can occur in partial shading. Hence, in recent times, many non-conventional methods have been proposed to tackle the problem of partial shading for PV systems. The next few sections discuss various MPPT techniques proposed in recent years to solve the problem of PSCs.

9.5 Soft computing methods

Soft computing is a concept that utilizes the concept of imprecise input, uncertainty, and reasoning based on partial truth to arrive at an output compared to hard computing, which has less tolerance for imprecision and works on exact algorithms. Soft computing techniques try to mimic a human brain in decision making in imprecise and uncertain environmental conditions (Dileep and Singh 2017). These techniques have elicited a lot of interest among researchers due to their ability to handle imprecision, non-linearity, and work with limited data to produce satisfactory results.

9.5.1 Fuzzy logic

Fuzzy logic has the advantages of working with non-linear function and imprecise input. Fuzzy logic control can be divided into four types: fuzzification, fuzzy inference, aggregation, and defuzzification. Fuzzification deals with mapping crisp input into fuzzy subsets using a membership function. Membership function is descriptive of the degree of belongingness of a particular value to a linguistic variable. The aim of fuzzification is to convert values to linguist variables.

The next step involves the inference system. Inference contains an if-else rule representing various linguistic variables. Based on the truth or falsity of these rules/statement, an output of a fuzzy set is created. Multiple rules are incorporated to ensure conversion efficiency. System knowledge is required for framing these rules. The next step of aggregation involves the merger of all outputs of a fuzzy set from all rules to form one fuzzy set. Finally, the defuzzification stage is where the fuzzy set is converted to a crisp output.

A fuzzy system has been used to track MPP during dynamically changing and PSCs. Al-Majidi *et al* (2018) has modified fuzzy logic (FL)-MPPT to track MPP during both slow and fast changing irradiance by separately assigning a different membership function for fast and slow changing irradiance. Results show superior tracking compared to normal FL-MPPT and avoidance of the drift problem (commonly associated with fuzzy systems). Macaulay and Zhou (2018) and Loukil *et al* (2020) have incorporated traditional MPPT techniques into the structure of fuzzy systems. The results show better tracking speed compared to the respective traditional technique. Verma *et al* (2020) has proposed and implemented an asymmetrical interval type-2 FL system. The proposed system is compared with type-1 FL in terms of the EN50530 MPPT test and the partial shading test. The proposed system provides higher tracking ability due to optimized membership function. Works like Syafaruddin *et al* (2009) have combined FL control (FLC) to an artificial neural network (ANN), where the ANN is used to train FLC in capturing the global maximum power point (GMPP) during paring partial shading; however, the huge dataset required makes the system non-viable for application. Even though hybrid FLC is able to track GMPP during partial shading, the system's specific design and the requirement of expert knowledge of FLC are the limitations in tracking GMPP during partial shading.

9.5.2 Artificial neural network

An ANN is a soft computing method based on human nervous system. The uniqueness of an ANN is its ability to learn from its experience. It has become popular due to its ability to handle non-linear systems and its ability to be trained. An ANN basically consist of three units: an input layer, a hidden layer, and an output layer. The ANN structure is shown in figure 9.21. First, input datasets are sent iteratively to the ANN to train it. Training of an ANN involves adapting the weights of neurons in such a way that the input produces the desired known output. The iterative process called epoch (epoch refers to the number of times the entire dataset has been sent for training) is continuously carried out until the neurons adapt the weight in such a way that the error between the produced output and the desired known output is low. Then on-line testing occurs, where the ANN is added into the system to test the efficiency of the ANN. Generally, input to the ANN consists of either voltage and current or temperature and irradiance, along with bias, which is given to every hidden layer, and the output would normally be either reference voltage or desired duty cycle. More numbers of hidden layers can increase accuracy with which results can be obtained, whereas having fewer layers may speed up calculations and could produce output at a much faster rate. Although a higher number of ANNs is able to increase accuracy and reduce steady state performance (Asiful Islam and Ashfanoor Kabir 2011), in rapidly changing weather conditions its computational time is higher than those with fewer hidden layers. Hence, optimization of an ANN is required for good performance of the ANN. Optimization of an ANN refers to deciding the:

(a) optimum number of layers,
(b) the activation function,

(c) the training algorithm,
(d) and the learning rate.

There have been many papers which have incorporated an ANN in finding the MPP. In Duman *et al* (2018), hybrid PSO and gravitational search algorithm based on FL (FPSOGSA) is used for optimization of an ANN. FPSOGSA consists of hybrid PSO and gravitational search algorithm (GSA) where the ability of PSO to capture global maxima is combined with ability of GSA to track local best. FL is used to limit maximum velocity (change in duty cycle, in this case) to an upper threshold. This method is tested for various temperature and IR-radiation conditions, and it provides good results. In Messalti *et al* (2017), an ANN is trained using a dataset obtained through P&O, and in Al-Gizi *et al* (2017), the dataset is obtained using I–V and P–V datasets, and this dataset is used to train FLC.

In El-Helw *et al* (2017), ANN is used in combination with one of the conventional MPP methods. Here ANN is used in the detection of an MPP region for partial shading by estimating the boundary voltages (El-Helw *et al* 2017), and a conventional technique is used for the accurate detection of global maxima. Around 17 000 datasets are used for the training of neural network, which are obtained through conventional techniques. The neural network designed has input layers consisting of four neurons, a hidden layer consisting of 180 neurons, and output layers that have two neurons. The algorithm shows good tracking capability of various PSCs. Results show better tracking efficiency than normal conventional algorithms.

In Kota and Bhukya (2019), ANN is used to find shading patterns, and a lookup table is used for accurate GMPP location corresponding to the shading pattern, whereas in Chen and Wang (2019), the authors have implemented a sequential Monte Carlo (SMC) method in incremental conductance, and an ANN is designed for better guidance of the sequential Monte Carlo incremental conductance (SMC-IC) during partial shading.

There also have been works where an ANN is not used as a main controller and instead is used to optimize main control logic like fuzzy, P&O to produce the desired output (El-Helw *et al* 2017).

Even though an ANN is an excellent MPP controller for sudden weather fluctuation due to its ability to handle non-linear characteristics of PV system, the training time may take months or years for the ANN to become completely trained. Even though some researchers have claimed the successful tracking of MPP even under partially shading, there is no inherent mechanism in a neural network to differentiate partial shading and uniformly changing weather conditions; also, partial shading is a dynamically induced phenomenon which an ANN cannot learn easily. Also, since an ANN is trained for a particular type of system, it is system dependent. Hybrid methods of an ANN used to calculate partial shading are also system dependent (i.e. if the same system is upscaled in the future, the ANN will have to be trained again). Hence the main drawbacks—which are: not being able to differentiate partial shading and uniform weather fluctuation, system dependency, and high cost of hardware implementation—make an ANN less preferable to be used for MPPT during partial shading.

9.5.3 Adaptive neural fuzzy inference system

An adaptive neural fuzzy inference system (ANFIS) is a controller that is formed by combining the advantages of a fuzzy inference system, which has a rule-based system, and a neural network, which has the ability to learn from experience. Thus, an ANFIS provides a fuzzy system with learning abilities wherein an ANN is used to train the fuzzy system on various rules. The biggest drawback of a fuzzy system is the requirement of a person with an expert knowledge in framing rules handled by an ANN. Due to this, the ANFIS acts as a robust controller. Hence, ANFIS is also a popular contender for MPPT.

Kharb *et al* (2014) was one of the early incorporations of an ANFIS in MPPT tracking where temperature and irradiance was taken as input for training with hybrid optimization of the least square method, and back propagation was used as the optimization strategy. Amara *et al* (2018) uses information from various PV arrays to train the system, and results show superior tracking speeds compared to the P&O method. Priyadarshi *et al* (2020) uses an ANFIS model where the ANFIS is trained and optimized by PSO due to the simple structure and robustness of PSO. The maximum power extracted from a PV array is then sent to the grid via a space vector modulation (SVM) and hysteresis current controller (HCC) (combining the advantages of SVM and HCC). PSO provides better optimization compared to the least square and back propagation methods proposed in Kharb *et al* (2014). The inverted AC output is tested against EE519 standard and found to be reliable. Pachaivannan *et al* (2021) has implemented a hybrid of ANFIS and crowded plant height optimization (CPHO). ANFIS is used for the estimation of GMPP and the second stage, with CHPO, is used for the accurate tracking of GMPP.

An ANFIS is found to have a high tracking speed of MPP during rapidly changing IR-radiation levels, but it also faces the same problem of an ANN in being system dependent and not being able to differentiate between rapidly changing weather conditions and PSCs. Some hybrid ANFIS systems have been proposed to solve the problem of partial shading; however, they are yet to be experimentally verified.

9.5.4 Extreme learning mechanism

Extreme learning is a feedforward ANN used for clustering, regression, and enhanced feature learning. Behera and Saikia (2020) implemented extreme learning based on variable gradient ascent and PI-fractional order integrator for MPPT. PI-FOI is optimized by the squirrel search algorithm (SSA). The result shows superior tracking speed compared to PI-P&O and P&O.

9.5.5 Extremum seeking control

Extremum seeking control (ESC) is based on adaptive control (where system knowledge is not required) and searching, which is looking for an extreme point based on random features. Bizon (2016) uses a traditional ESC method that requires higher iterations for convergence; a modified ESC-based method, like in Elnosh *et al* (2014), provides a higher convergence speed for tracking MPP.

9.5.6 Reinforcement learning

Reinforcement is a subsection of machine learning that essentially deals with sequential decision making. Reinforcement learning (RL) is based on the principle that an agent can be trained in order to acquire new traits. This training is based on rewards given to agents for better results and penalties for poor performance. RL is an unsupervised learning process, and MPPT is labeled as a Markov decision process. The three functions of RL are state space, action space, and reward function.

The challenges of RL are:

(a) the ability to work only in a discretized environment,
(b) the loss of data due to discretization,
(c) and the greater time requirement.

9.5.6.1 Deep RL

Deep RL combines the advantages of deep neural networks and RL. Chou *et al* (2020) has implemented a deep Q network based on MPPT for PSCs. However, the deep Q optimization also has limitations when working in discrete systems.

Avila *et al* (2020) has proposed a deep RL method in a continuous state action domain. Results show a fast and accurate tracking ability.

9.5.6.2 Memetic RL

Memetic RL combines the searching ability of RL in a memetic structure for enhanced searching. Zhang *et al* (2019) implemented memetic RL in MPPT, and the results show better tracking ability compared to many other meta-heuristic techniques.

9.5.7 Bayesian network

A Bayesian network is a probabilistic graphical model which maps variables based on their conditional dependency (Friedman *et al* 1997). It could be considered unsupervised learning. In Keyrouz and Georges (2011), a Bayesian network is used for mapping two popular network MPP methods, namely incremental conductance (INC) and PSO. It segregates the system containing 20 PV panels connected in series as two sets of ten panels each. For the first ten panels, INC is used for finding MPP, and the remaining ten panels use PSO for computing MPPT. All attributes of the two sets are conditionally independent. The Bayesian network is then used to find the similarity between the two sets of panels. For example, if one of the panels in the first ten has the same output as any of the panels in the other set, then, based on dependency, they are fused. The fusion value of one is given for the panels that have a similar output, and zero is given for dissimilar output. Based on the conditional probability, the node corresponding to optimal value is determined (table 9.1).

9.6 Meta-heuristic techniques

Meta-heuristic techniques are algorithms that try to optimize a given solution iteratively, even though sometimes an optimal solution may not be achieved. These techniques are used when analytic techniques fail to converge to a solution or when exact algorithms to find solutions do not exist (Parejo *et al* 2012). These techniques are generally system independent (i.e. knowledge of the system is not known). Meta-heuristic techniques used in PV systems for finding MPP during partial shading have four categories:

(a) Evolutionary algorithm
(b) Swarm intelligence
(c) Mathematics- and physics-based
(d) Sociology-based

9.6.1 Swarm intelligence

Swarm intelligence is inspired from nature, like the flocking of birds or insects and birds searching for the best source of food, etc (Kar 2016). The underlying aspect of this search is the ability to use swarm intelligence. Swarm intelligence refers to the concept that herd or group intelligence is more than the sum of the individual intelligence. This is achieved through random exploration of all individual insects or birds for the best solution combined with the exchange of information between one another regarding their experience, which helps the group to identify the best solution. Bio-inspired meta-heuristic techniques try to mimic these naturally occurring phenomena and implement them as algorithms to solve optimization problems of a multi-objective function. These techniques start off with random initializing of particles/agents with each particle representing a bird or insect in the herd. Every particle represents potential solution candidates. Then, iteratively, these particles try to optimize the objective function through swarm intelligence. These have a higher probability of not getting stuck in local maxima and finding global maxima. These optimization techniques can be used to solve the partial shading problem of PV arrays. These treat a PSC as an optimization problem, and they use many particles as a means

Table 9.1. Comparative analysis of various soft computing techniques discussed in this paper.

No.	Soft computing method	System dependency	PV partial shading detection	System response	Hardware cost	Algorithm complexity	Digital controller required
1.	FL	Yes	Standalone: no Hybrid: yes	Fast	Moderate	Moderate	Reduced instruction set computer (RISC) microcontroller (PIC 16F872, PIC 16C74, etc)
2.	ANN	Yes	Standalone: no Hybrid: yes	Very fast	Costly	Complex	High-end DSP controllers
3.	ANFIS	Yes	Standalone: no Hybrid: yes	Very fast	Costly	Complex	Digital signal processing (DSP) controllers (dSpace)
4.	Bayesian network	No	Yes	Very fast	Costly	Complex	Multiple DSP are required
5.	Extreme learning mechanism	Yes	Standalone: no Hybrid: yes	Very fast	Costly	Complex	High end DSP controllers
6.	ESC	No	Yes	Fast	Yet to implement in hardware	Complex	Controller not specified
7.	RL (deep and memetic)	No	Yes	Very fast	Yet to implement in hardware	Complex	Controller not specified

to finding the maximum PV array power as a fitness function, and the voltage/duty cycle is an optimization variable. Finding the maximum power is set as the objective. The particles scan the entire *I–V* curve randomly with swarm intelligence to search for maximum power. After a few iterations, when the convergence criteria are satisfied, the global maxima is assumed to be found.

9.6.1.1 Particle swarm optimization

PSO is a swarm intelligence–based meta-heuristic algorithm inspired by the movement of flocks of birds or by birds and insects searching for food (Eberhart and Kennedy 1995). At the start of the algorithm, a swarm of particles (each particle is represented by a coordinate in search space) is initialized randomly. Each particle tries to search for the global maxima of the fitness function. Here, the fitness function is PV panel power and the optimization variable is duty cycle. Initially, particles move in random directions. In subsequent iterations, the particle velocity is influenced by its own personal best position, called cognitive factor, as well as the global best of the entire swarm (social factor)

$$V_i^{k+1} = w \times V_i^k + C_1 r_1 (P_{\text{besti}} - x_i^k) + C_2 r_2 (G_{\text{best}} - x_i^k) \tag{9.10}$$

$$x_i^{k+1} = x_i^k + V_i^{k+1}. \tag{9.11}$$

Equations (9.10) and (9.11) represent the velocity and position updation of particles in ith dimension (represented by subscript i) in the kth iteration (represented by superscript k). w refers to inertia weight, C_1 and C_2 acceleration constant, and $r1$ and $r2$ a random value in [0,1]. P_{besti} represents personal best value of the particles, and G_{best} refers to global best. In the *P–V* curve, the search is limited to one dimension. During the start, the optimization variable of the duty cycle is sent to the power converter, and the respective PV power is obtained. Now the best value of power among the swarm is set as the global best, and the power corresponding to each duty cycle is stored as the personal best. In subsequent iterations, values keep getting updated. After a few iterations, the particles converge towards global best. Sample code is presented below to explain the PSO algorithm to find the global maxima of a 2D objective function.

Sample Python code for PSO:

```
import numpy as np

# Define the fitness function to be optimized (modify this for your specific problem)
def fitness_function(x):
  return x**2  # Example: Minimize the square of the input

# Define PSO parameters
num_particles = 30
num_dimensions = 1
max_iterations = 100
c1 = 2.0  # Cognitive parameter
c2 = 2.0  # Social parameter
w = 0.7   # Inertia weight

# Initialize particle positions and velocities
positions = np.random.uniform(-10, 10, (num_particles, num_dimensions))
velocities = np.random.uniform(-1, 1, (num_particles, num_dimensions))
personal_best_positions = positions.copy()
personal_best_values = np.array([fitness_function(x) for x in positions])
global_best_position = positions[np.argmin(personal_best_values)]
global_best_value = np.min(personal_best_values)

# PSO main loop
for iteration in range(max_iterations):
  for i in range(num_particles):
    # Update velocity
    r1, r2 = np.random.rand(), np.random.rand()
    cognitive_component = c1 * r1 * (personal_best_positions[i] - positions[i])
    social_component = c2 * r2 * (global_best_position - positions[i])
    velocities[i] = w * velocities[i] + cognitive_component + social_component

    # Update position
    positions[i] += velocities[i]

    # Update personal best
    current_value = fitness_function(positions[i])

    if current_value < personal_best_values[i]:
      personal_best_values[i] = current_value
      personal_best_positions[i] = positions[i]

    # Update global best
    if current_value < global_best_value:
      global_best_value = current_value
      global_best_position = positions[i]

# Print the best solution found
print("Global Best Solution:", global_best_position)
print("Global Best Value:", global_best_value)
```

Convergence analysis of PSO:

There have been many studies on the effect of parameters on the convergence of PSO; these are collectively referred to as convergence analysis of PSO (Shi and Eberhart 1998, Eberhart and Shi 2000, Clerc and Kennedy 2002, Trelea 2003, Zheng *et al* 2003, Mendes *et al* 2004, Zhan *et al* 2009, Hu *et al* 2019). The main observation of these studies is the range with which parameters need to be operated for convergence. Let $C1 + C2 = \phi$ and inertia weight be represented by w. Studies prove the best convergence is obtained when the values of inertia weight are in the range [0.4,0.9] and the corresponding values of ϕ are in the range [3,4.1]. For the values of inertia, the corresponding values of ϕ can be found from the graph (shown in figure 9.25). Hu *et al* (2019) shows that when parameter values are near the convergent boundary, convergence is reduced. Hence it is not advisable to operate near the convergent boundary. Influence of w on convergence is high. Shi and Eberhart (1998) have linearly decreased the inertial weight from 0.9 to 0.4 and have proven a higher efficiency wherein, during the initial exploration, area randomness is increased to increase the exploration area, whereas in the later exploitation state, the w is almost set to 0.4 (by linear decrease). In Zhan *et al* (2009), an adaptive change in weights of ϕ and w is proposed where the weights are adaptively changed considering the state in which searching is present. The disadvantages of normal PSO are

(i) excessive oscillations during tracking
(ii) and inability to track subsequent oscillations.

Problem (i) is mainly due to the fact that during the exploration state, all random particles containing duty cycle are sent to the converter to find power. Since random duty cycles are sent there in oscillation, as particles converge, the oscillation reduces. Hence it is general practice to have PSO triggered only during partial shading or dynamically varying weather conditions and resorting back to any other hybrid system (generally any of the traditional techniques) after convergence is achieved; this should reduce oscillation into a steady state. Problem (ii) is because once global maxima in partial shading is found, then the velocity of all particles reach a low value (almost zero), and for subsequent partial shading, particle dispersion is not present, thereby leading to non-exploration of a new global maxima. Hence, many modifications are proposed to PSO to make it more viable to track MPP in partial shading. Some select works of PSO have been reviewed in this paper. In Ishaque *et al* (2012), an improved PSO for a PV system was proposed where PSO has a well-defined initialization and dispersion of particles. During initial shading, PSO detects the partial shading and initializes duty cycle, and PSO is able to catch the global maxima. For subsequent partial shading, dispersion is carried out using the following two equations:

$$d_{\text{new}} = d_{\text{old}} - \frac{1}{k_1}(P_{\text{old,MPP}} - P_{\text{MPP}}) \tag{9.12}$$

$$d_i{}^k = \{d_1 - k_2,\, d_2,\, d_3 + k_2\}. \tag{9.13}$$

Equation (9.12) changes the duty cycle according to change in irradiance and equation (9.13) perturbs the duty cycle apart from the current global best (the current global best is represented by d_2). The proposed system is implemented in hardware to test its efficiency. The tracking efficiency is 99% under partial shading and dynamically changing weather. However, the computation of k_1 requires system knowledge. In Ishaque and Salam (2013), a modified PSO is presented which reduces the influence of random parameters ϕ. This makes the system more deterministic with only w to be tuned. The proposed system is able to track the MPP during PSCs accurately. In Hu *et al* (2019), the control parameter w is further fine-tuned compared to Ishaque and Salam (2013). In Ishaque and Salam (2013), the inertia weight is constant throughout the program, whereas in Hu *et al* (2019) it is linearly decreased from 0.9 at exploration (to aid in a large area to be explored) to 0.4 at exploitation (where convergence to the global peak needs to be achieved). Ji *et al* (2020) is based on a particle jump where each particle is assigned a particular search space, and if the search space is ended, then the particle goes to that space which has the GMPP. Alshareef *et al* (2019) has proposed an accelerated PSO where the PSO is accelerated using the output gained from the P&O method. Hayder *et al* (2020) has implemented an improved PSO method based on optimal parameter selection, producing better results compared to normal PSO.

Hybrid PSO:

Hybrid PSOs are also gaining popularity due to the complementary working fashion of some algorithms which, when combined, can aid each other in smoothly searching for the GMPP. Hybrid algorithms typically involve either two algorithms merging to find the global peak or a hybrid, which may involve PSO to search during partial shading and after the convergence of any other techniques used in the fine-tuning of the found global maxima. Some hybrid algorithms also involve a particular algorithm used to optimize the searching of PSO, like Juang (2004). In Kermadi *et al* (2019), a popular method called skip search judge (SSJ) is combined with PSO. The SSJ method is a modified incremental conductance method which, based on judgment, skips some search space in I–V and searches the remaining I–V curve. SSJ has the ability to not get stuck in local maxima. PSO is used in fusion with SSJ where the search space of all particles is fixed and search happens. If the particle, while searching, goes into the predefined search area of another particle, both the particles are merged. The results show that this has a better speed than normal SSJ. In Ji *et al* (2020) a hybrid of Gaussian PSO and novel simulated annealing is used for tracking MPP. The result shows superior tracking compared with basic PSO. Li *et al* (2019) has implemented a hybrid of PSO and overall distribution, where the overall distribution is used for limiting search space and PSO is used for searching for the GMPP. Farh *et al* (2018) is another hybrid PSO-FLC where PSO is made to handle partial or dynamic shading, and after convergence the control baton is handed over to FLC to fine-tune the global maximum point in steady state. A summary of all the PSO methods discussed in this section is summarized in table 9.2.

Table 9.2. Summary of PSO for MPPT methods discussed in this paper.

Authors and year of publication	Reference	Control variable	Parameters	Remarks
Kashif Ishaque *et al* (2012) [IEEE]	Ishaque *et al* (2012)	Duty cycle	$C_1 = 1.2$, $C_2 = 1.6$, $w = 0.4$	An improved PSO-based MPP tracker where novel particle dispersion under partial and dynamically changing weather is proposed. Tracking is robust. Eliminated the need of PI controller. However, the determination of 'k' during particle dispersion needs to be studied with PV panels under various irradiance level.
Kashif Ishaque *et al* (2013) [IEEE]	Ishaque and Salam (2013)	Duty cycle	$C_1 = 1$, $C_2 = 1$, $w = 0.4$	A deterministic PSO where the dependency of PSO on C_1 and C_2 is eliminated. The controller needs to tune only w, which makes control easy. Proposed system implemented by hardware shows high efficiency and speed in tracking global maxima.
Mostefa Kermadie *et al* (2019) [IEEE]	Kermadi *et al* (2019)	Reference voltage	(Not mentioned, so general values are taken) $C_1 + C_2 = 4$, $w = [0.4, 0.9]$	A hybrid PSO where SSJ is combined with PSO. The paper explores the problem of multiple peak clusters and the challenge of a controller scanning the entire I–V curve for peak point. The proposed system is tested against four different patterns and is able to track GMPP under all conditions with greater speed than SSJ and a modified incremental conductance.
Keyong Hu *et al* (2019) [IEEE]	Hu *et al* (2019)	Duty cycle	$C_1 = 1$, $C_2 = 1$, w linearly decreased [0.4,0.9]	A further fine-tuning of inertia weight with respect to iteration. A high value of inertia weight in exploration state and linear decrement and low value in exploitation state. $w^k = 0.9 - \frac{k}{k_{\max}} \times 0.5$k-current iteration, $k_{\max}$ -max iteration
Hassan M H Farh *et al* (2018) [PLOS]	Farh *et al* (2018)	Duty cycle	(Not mentioned, so general values are taken) $C_1 + C_2 = 4$, $w = [0.4, 0.9]$	A hybrid PSO-FLC is proposed where PSO is used for catching global maximum under partial shading and FLC is used for fine-tuning. A PSO re-initialization on subsequent partial shading is found to be superior to PSO re-initialization on a prefixed time.

9.6.1.2 Ant colony optimization

Ant colony optimization (ACO) is based on the swarm intelligence of ants in search of the best source of food. Initially ants move in a random manner in search of food. As they move, ants leave behind a chemical trail made of pheromones. Pheromones evaporate over time, so the shortest path from the colony to the best source of food has the highest secretion of pheromones, and all ants unilaterally follow that path. ACO mimics this concept with algorithms. A solution archive is used to represent a solution. A pheromone update equation makes sure that particles do not get stuck in local maxima. Thus, pheromones are essentially a communication tool among ants in search of the best solution. Pheromone secretion is inversely proportional to the distance covered by ants. The pheromone secretion equation is given in equations (9.14) and (9.15)

$$\Delta\tau_{i,j}{}^{k} = \frac{1}{L_k}\ k^{\text{th}} \text{ ant travels on edge } i, j \tag{9.14}$$

$$=0, \text{ otherwise} \tag{9.15}$$

$$\tau_{i,j}{}^{k} = (1-\rho)\tau_{i,j} + \sum_{k=1}^{m} \Delta\tau_{i,j}{}^{k}. \tag{9.16}$$

The pheromone updation is shown in equation (9.16). To account for evaporation, another term $(1-\rho)$ is added to equation (9.16), where $\rho = 0$ accounts for no evaporation and $\rho = 1$ accounts for full evaporation. The value of ρ is selected between [0,1]. The probabilistic route for ants to travel in the next iteration is given in (9.17). In MPPT, ants are assumed to be agents, and finding the maximum PV power is set as an objective; in their trail, a pheromone is left and, based on the value of power found, the quantity of the pheromone trail left behind is decided. Sahoo *et al* (2017) implements ACO to implement MPPT, and the pheromone is updated in subsequent iterations

$$P_{i,j} = \frac{(\tau_{i,j})^{\alpha}.(\eta_{i,j})^{\beta}}{\Sigma\left((\tau_{i,j})^{\alpha}.(\eta_{i,j})^{\beta}\right)} \text{ where, } \eta_{i,j} = \frac{1}{L_{i,j}}, \alpha, \beta\text{–influence factor.} \tag{9.17}$$

Titri *et al* (2017) proposes a novel pheromone updation strategy in ACO; the results show a higher efficiency of tracking MPP in partial shading than Sahoo *et al* (2017). ACO is also used in optimization of other controllers. For example, in Besheer and Adly (2012), ACO is used to optimize PI controllers for MPPT. Here, ACO tunes the values of PI controllers for the correct tuning of K_P and K_I values.

9.6.1.3 Grey wolf optimization

Grey wolf optimization (GWO) is based on a pack of wolves hunting prey. Wolves generally live as a pack and follow social hierarchy (Mirjalili *et al* 2014). The leader of the pack is referred to as α. Its subordinates are labeled as β. These are followed by a lesser ranked δ, followed by least dominant w. w is essentially a subordinate to

α, β, and δ. GWO implements this in optimization with α as the best solution. The attacking behavior of the wolves is expressed by the following equations:

$$\vec{D} = | \vec{C} \cdot x_P(t) - \vec{x_P}(t) | \tag{9.18}$$

$$\vec{x}(t+1) = \vec{x_P}(t) - \vec{A}.\vec{D}. \tag{9.19}$$

t-current iteration, D, A, C-coefficient vector, x_p-position vector of the prey, x-position vector of the grey wolf, and vectors $\vec{A}$ and $\vec{C}$ are calculated using the following equations:

$$\vec{A} = 2\vec{a} \cdot \vec{r_1} - \vec{a} \tag{9.20}$$

$$\vec{C} = 2 \cdot \vec{r_2} \tag{9.21}$$

where a-linear decrement is from 2 to 0, and $r1$,$r2$-random vector is between [0,1]. In Mohanty *et al* (2016), MPP control using GWO is designed. The proposed system shows a faster convergence towards MPP, an accurate catching of MPP, and less oscillation around MPP compared to Ishaque *et al* (2012). Mohanty *et al* (2017) has further modified MPP control of Mohanty *et al* (2016). The proposed system is a hybrid of GWO and P&O where GWO is used as offline to track GMPP and P&O is used as online to track GMPP. The results show that the modified method is faster at convergence compared to Mohanty *et al* (2016). Tjahjono *et al* (2020) has modified GWO using modified alpha updation, resulting in a higher tracking time. Guo *et al* (2020) has proposed and implemented a GWO where convergence is based on the non-linear factor of tangent trigonometric function. The proposed system was implemented using a boost full bridge isolated converter. The algorithm was compared with P&O, artificial bee colony (ABC), weighed Salp swarm algorithm (WSSA), SSA-PSO, and GWO. The algorithm shows a higher tracking speed compared to the rest of algorithms.

9.6.1.4 Cuckoo search

Cuckoo search is based on brood parasitism. A cuckoo tries to lay its eggs in another bird's nest. However, the host bird could find out the egg is from a different bird, and if that situation arises, the host bird could potentially throw out the egg or abandon the nest altogether. Therefore, to minimize this possibility, it is theorized that the cuckoo has adapted so that its egg mimics the host egg in shape, size, and color. Cuckoo search is based on the above logic. There are three rules to cuckoo search:

(i) random selection of the nest for laying an egg by Lévy flight,
(ii) best nest in terms of the egg quality carrying over to the next iteration,
(iii) and the probability of an egg being discovered and discarded by the host bird, which is P_a. P_a belongs in the range of [0,1].

The random selection of the nest is determined by Lévy flight. Lévy distribution is given by

$$\text{Levy}(\lambda) = t^{-\lambda} 1 < \lambda < 3. \tag{9.22}$$

The updation equation is

$$X_i^{t+1} = X_i^t + a \oplus \text{levy}(\lambda) \tag{9.23}$$

where X_i^t is the current location, $a \oplus$levy (λ) is the transitional probability, and $\oplus$ represents entry-wise multiplication. Mosaad *et al* (2019) has implemented MPPT by using cuckoo search and then compared it with an incremental conductance method and a neural network method. The proposed system has a higher tracking ability than both the incremental conductance and neural network methods. The tracking speed is comparable to a neural network. Peng *et al* (2018) has proposed a modified cuckoo search where the random Lévy flight in a conventional cuckoo search is eliminated. This method uses only three particles for searching, and it shows a better tracking speed compared to the conventional cuckoo search and Mohanty *et al* (2016). Nugraha *et al* (2019) is a hybrid of cuckoo search and golden section search (GSS). GSS is a traditional MPPT used to track MPP in uniform IR-radiation. GSS initially sets the boundary area in which it tries to track MPP iteratively, and the boundary area is shrunk to find MPP.

9.6.1.5 Bat algorithm

The bat algorithm (BA) was conceptualized by Yang in 2010, inspired by movement of bats in the dark by a property known as echolocation. Echolocation is a property used by bats to fly and hunt in the dark. With echolocation, bats emit pulses which are at frequencies ranging from 10 kHz (low pitch) to 200 KHz (high pitch). These pulses, upon hitting the prey or other object, get reflected back to the bats, and the bats analyze these codes to differentiate between prey or an obstacle. A BA tries to mimic this feature for the purpose of optimization. During initialization, the number of bats is initialized along with pulse frequency, loudness, and number of iterations. After each iteration, the pulse frequency is modified (assuming that the wavelength is constant). The velocity and position are also updated every iteration. The updation equation is given by:

$$f_i = f_{\min} + (f_{\max} - f_{\min})\beta \tag{9.24}$$

$$v_i^t - v_i^{t-1} + (x_i^t - x) \times f_i \tag{9.25}$$

$$x_i^t = x_i^{t-1} + v_i^t \tag{9.26}$$

$$\beta \in [0, 1].$$

The global best solution is stored in x. After the selection of a global best in the current iteration, a new best is selected by using a random walk, as illustrated below:

$$x_{\text{new}} = x_{\text{old}} + \varepsilon A^t \tag{9.27}$$

where $\varepsilon \in [-1,1]$. As the bats start converging to the global best, the pulse rate (r_i) is increased and the loudness (A_i) is decreased

$$A_i^{t+1} = \alpha A_i^t \tag{9.28}$$

$$r_i^{t+1} = r_i^t \times [1 - e^{-\gamma t}]. \tag{9.29}$$

In Seyedmahmoudian *et al* (2018), the authors have implemented a BA-based MPPT on a single-ended primary-inductor converter (SEPIC). The result shows a highly accurate result and faster tracking speed compared to the dividing rectangle (DIRECT) method. In Eltamaly *et al* (2020), the authors proposed a modified BA where bat initialization is based on anticipated values. The proposed algorithm is able to track MPP with a higher convergence than GWO, PSO, and a BA based on random initialization. da Rocha *et al* (2020) provided a comparison based on various BAs.

9.6.1.6 Squirrel search algorithm

The SSA is based on squirrels searching for food, where in summer they eat acorns for energy and then in winter they store hickory nuts for winter. Fares *et al* (2021) has implemented MPPT based on SSA and compared it with a genetic algorithm (GA) and PSO. Results show superior tracking with the SSA compared with a GA and PSO.

9.6.1.7 Cat swarm optimization

Cat swarm optimization (CSO) is based on a group of cats resting or in action, which are represented by seeking and tracing modes. Seeking mode is defined as when a cat is in rest mode but constantly tracking the environment, and tracing mode is when the cat is attacking its prey/target. Guo *et al* (2018) has modified CSO to where it has a higher tracking speed than standard CSO.

9.6.1.8 Moth flame optimization

This model is inspired by the spiral motion of a moth around artificial light. Bouakkaz *et al* (2020) has implemented a moth flame optimization model coupled with fractional proportional integral differential and a sliding mode controller. The proposed model exhibits a higher tracking ability compared with conventional approaches.

9.6.1.9 ABC optimization

The ABC algorithm is inspired by the foraging behavior of bees. It provides advantages like less parameter control, faster convergence, and independence of initial initialization. Benyoucef *et al* (2015) implemented an improved ABC by having just two parameters to control. The proposed system is compared with PSO-based MPPT methods, and the results show superior tracking with the proposed method. The updation equation of the duty cycle to be applied to DC–DC converter is calculated below:

$$d_e = d_{\min} + \text{rand}[0,\ 1](d_{\max} - d_{\min}) \tag{9.30}$$

$$(d_e)_{\text{new}} = d_e + \psi_e(d_e - d_k) \tag{9.31}$$

where d_e is the current duty cycle, $d_{\min}$ is the minimum possible duty cycle, $d_{\max}$ is the maximum duty cycle possible, and ψ_e is the mean constant in the interval $[-1,1]$. Pilakkat and Kanthalakshmi (2019) have improved the ABC algorithm based on a hybrid ABC–P&O method. The method shows a reduced tracking time, lessened by 30%.

9.6.1.10 Butterfly optimization algorithm

The butterfly optimization algorithm (BOA) is inspired by the mating and food searching ability of butterflies. Butterflies emit a fragrance that is used by other butterflies in searching for a mating partner. The BOA uses this phenomenon for the optimization of PV MPPT. Aygül *et al* (2019) has applied the BOA in PV MPPT and these results show that BOA has better tracking efficiency compared with GWO.

9.6.1.11 Salp swarm optimization

Salp swarm optimization (SSO) is based on the behavior of a swarm of salps in search of food and better locomotion ability (Mirjalili *et al* 2017). Mao *et al* (2020) has proposed a dynamic non-linear *w* factor salp swarm algorithm with a bi-directional Cuk converter system. The salp swarm algorithm was used for the estimation of MPP. Mirza *et al* (2020) has implemented MPPT for PSCs based on novel SSO. The proposed system is compared with five different popular MPP techniques, and the proposed system provides superior performance. Premkumar *et al* (2019) has proposed SSO based on lesser search agents and lesser iteration compared with normal SSO. The results show better tracking results compared to normal SSO. Wan *et al* (2019) has proposed a hybrid of SSO and GWO; the leader group of SSO is modified and reconstructed using GWO. The algorithm is compared with P&O, PSO, and the salp swarm algorithm. The algorithm shows superior tracking speed compared with the other two algorithms.

9.6.1.12 Whale optimization

Whale optimization (WO) is inspired by humpback whales hunting for food (Mirjalili and Lewis 2016). The hunting method of whales can be represented by an upward spiral method of the three stages whales follow in searching for prey, using a bubble net method for attacking, and encircling the prey. The WO algorithm is popular due to requiring the fine-tuning of just one parameter and also having a robust structure. Each humpback whale is considered a potential solution candidate, and the prey is the objective function. Kumar and Rao (2016) proposed and implemented MPPT based on WO. The proposed system shows a higher tracking speed compared to GWO and PSO.

9.6.1.13 Harris hawk optimization

Harris hawk optimization is based on the coordinated hunting of prey by the Harris hawk; they use a technique called 'seven kill' where the hawks attack the prey from different angles before converging on the prey (Heidari *et al* 2019). Mansoor *et al* (2020) has proposed MPPT based on the Harris hawk, and results show a better tracking efficiency compared to GWO and PSO.

9.6.1.14 Firefly optimization

Fireflies are insects that use a bioluminescent light in their body at night to catch prey or for the mating process. Fireflies are a unisex organism, so the males and females will be attracted to each other. The brightness of their light is dependent on the distance between each other. The attractiveness can dependent on the brightness of light emitted. Sundareswaran *et al* (2014a) has implemented firefly optimization in a PV MPPT system for tracking global peak.

9.6.1.15 Glow worm optimizer

The glow worm optimization is based on the natural phenomenon of glow worms converging to the global best. A glow worm optimizer is based on the behavior of a glow worm, which uses luciferin to produce its glow. The brightness depends on the value of the objective function and the position of the glow worm. A literature survey from Jin *et al* (2017) implemented MPPT of a PV system during PSCs using a glow worm optimizer. The results showed a superior tracking of the proposed method compared to P&O and fractional open circuit voltage technique methods.

9.6.2 Evolutionary algorithms

Evolutionary algorithms (EAs) are a classification of meta-heuristic techniques inspired from natural evolution as proposed by Darwin. EAs mimic biological evolution, which involves mutation, recombination, and natural selection (Câmara 2015). An EA is based on the principle of survival of the fittest. The properties that give an organism an advantage are preserved and passed on to the next generation, and weaker features are eliminated (Vikhar 2017). An EA tries to mimic this is in computations to improve fitness of subsequent iterations.

9.6.2.1 Genetic algorithm

A GA is a subset of EAs that uses crossover and mutation as two parameters for producing offspring. In GAs, the inputs are labeled as chromosomes (the length of the chromosome needs to be specified at the beginning). The GA performs on two operators, crossover and mutation, and the more fit function is carried over to the next iteration. Generally a GA is not preferred for finding GMPP due to its slow convergence; rather, it is used in the optimization of other methods. In Daraban *et al* (2014), a GA is used to optimize P&O, where the P&O algorithm is incorporated into the structure of the GA. The proposed system is able to negate a major drawback of P&O in which it gets stuck in local maxima. The system also shows better tracking speed compared to normal GAs. In Messai *et al* (2011), a GA is used for the optimization of fuzzy rules and its membership function.

9.6.2.2 Differential evolution

In differential evolution (DE), the update occurs by two operators: mutation and crossover. However, it differs from GAs in the fact that a GA utilizes crossover for convergence whereas DE converges by mutation. Like any other evolutionary technique, DE starts with the initialization of random population generation. The solution is iteratively improved using mutation followed by crossover. Iteratively, the solution converges to optimum value. After each iteration where mutation and crossover are performed, the offspring is compared to its parent vector for fitness, and, in the case that the offspring shows better fitness, it is carried on to next iteration; if not, then the parent vector is retained. In this way, the fitness of the function in every iteration is improved until optimum value is found. The working of DE is shown in figure 9.29 and explained as follows. First, a population is initialized; then, from that population, an agent becomes a target vector. A trial vector is obtained by mutating a target vector with two other agents (equation written below is for mutation):

$$\begin{aligned}\text{Trial vector} = \text{Target vector} + \text{scaling factor} \times (&\text{random selected vector1} \\ &-\text{random selected vector2}).\end{aligned} \tag{9.32}$$

In mutation, an agent is chosen as a parent vector. Next, three agents separate from the parent vectors are chosen randomly from the population, and one of them is labeled the target vector. The scaling factor can be randomly chosen, but normally it is set a value of 0.5. By the process of mutation (shown in equation (9.32)), the trial vector is obtained. Now the next step involves crossover, wherein a random recombination of trial vector and target vector is carried out, resulting in offspring. This offspring is checked with the parent vector for fitness of the function. Offspring refers to the vector produced after recombination. The optimization variables that have a better fitness function within the parent vectors and offspring is passed on to next generation (generation meaning iteration) and the weaker fitness function is discarded. In Taheri *et al* (2010), an MPP controller based on DE is proposed and compared with a traditional P&O. The DE-based MPP controller has a higher tracking efficiency. Tey *et al* (2018) has provided an improved DE which could track the GMPP with a higher speed compared to Taheri *et al* (2010). Many hybrid DE methods have also been proposed. The advantage of hybrid DE is that the two algorithms can act in a complementary fashion and, thus, potentially give a superior performance compared to any one algorithm used in isolation. Kumar *et al* (2017a) is a WO and DE hybrid. The excessive spiraling present in WO is reduced by the optimization of DE and thereby helps WO from getting stuck in local maxima. Seyedmahmoudian *et al* (2015) uses another hybrid optimization technique where DE is used to optimize the parameters of PSO. PSO, having a simple structure and updation equation, is considered robust; however, it can get trapped in local maxima due to its random initialization of particles. The random mutation present in DE helps in the exploration process of PSO.

9.6.2.3 Flower pollination

A flower pollination algorithm is a meta-heuristic optimization algorithm inspired from the pollination of flowers. It is a highly popular algorithm due to its fewer parameter requirements and sufficient convergence speed. Shang *et al* (2018) has implemented MPPT for PSCs using a flower pollination algorithm, and the results are compared with PSO and P&O. The comparison shows a higher tracking speed of GMPP in flower pollination algorithms compared with the other two algorithms.

9.6.3 Mathematics- and physics-based methods

These algorithms are based on some popular mathematical/physics models.

9.6.3.1 Sine cosine algorithm

Sine cosine algorithms (SCAs) are based on the trigonometric waves of sine and cosine for optimization. Chandrasekaran *et al* (2020) implemented an improved SCA based on MPPT for PSCs, and the results were compared with SCA and PSO. The improved cosine sine algorithm provided a better tracking time than SCA and PSO. Kumar *et al* (2017b) proposed a hybrid of Cauchy density (CD) along with a Gaussian distribution (GD) function with sine cosine optimization. Here both CD and GD are used in a complementary fashion, where CD is used for exploration and the GD helps in exploitation. The added requirement of the system is the requirement of just one sensor. The method is compared with GWO and Lagrange interpolation with PSO (LIPSO), and the results show a better tracking ability of the proposed system for various conditions.

9.6.3.2 Wind driven optimization

Wind driven optimization (WDO) is based on the movement of the wind following Newton's second law of motion. The air particle is represented as a potential solution candidate. Abdalla *et al* (2019) has implemented a WDO technique for PSCs in a PV system. The work has compared the proposed algorithm with seven other popular algorithms, and the results show a superior tracking in the proposed system during the tracking of GMPP during shaded conditions.

9.6.3.3 Gravitational search algorithms

GSAs are based on Newton's law of gravity (Rashedi *et al* 2009). Saha (2016) implemented MPPT based on a gravitational search for PSCs. Even though results show a better tracking speed compared with PSO, the local searching ability of a GSA is very poor, with a high chance of getting stuck in local maxima. Li *et al* (2018) modified the conventional GSA with a modified gravitational constant change factor, and the proposed algorithm showed a superior tracking performance compared with PSO and GSA.

9.6.4 Sociology-based methods

These optimization methods are inspired by social behaviors/mannerisms of human.

9.6.4.1 Teacher learning–based optimization

The teacher learning–based optimization method is based on teaching methodology handled in school. The learning is divided into two stages:

(i) learning in the classroom where a teacher imparts knowledge to a student,
(ii) and the knowledge shared among the students themselves.

Rezk and Fathy (2017) implemented GMPP during PSCs for a PV system and compared the tracking speed with PSO and FLC. The proposed system exhibited a higher tracking speed than the other two techniques.

9.6.4.2 Human psychology optimization

Human psychology optimization is based on the mental state of an ambitious person. The ambitious person tries to improve and achieve a dream/goal at all costs. The four stages that an ambitious person goes through can be classified as excitement, self-motivation, inspiration, and lesson learning. Kumar *et al* (2017c) has implemented MPPT based on human psychology. The proposed system is compared with a hybrid of P&O and PSO and a hybrid of LIPSO. The results show a superior tracking ability of the proposed system compared with the other two techniques in tracking ten different patterns.

9.6.4.3 Group teaching algorithm

A group teaching algorithm (GTA) is based on the idea that each student has a learning aptitude, and the role of the teacher is to correctly identify the aptitude of each student and pay more attention to the students with a poor learning ability compared to students with a strong learning ability (Zhang and Jin 2020). Zafar *et al* (2020) has implemented MPPT for PSCs with a GTA, and the response is compared with P&O, dragon fly optimization, particle swarm optimization gravitational search, and cuckoo search algorithms. Results show a superior tracking ability of the proposed system compared with the other algorithms.

9.7 Exact algorithms

This section deals with algorithms based on exact algorithms for MPPT of PSCs. Exact algorithms/hard computing–based methods rely on accuracy and the model. These algorithms are subdivided into two categories, namely:

(a) algorithms based on mathematics
(b) and algorithms based on PV system characteristics.

9.7.1 Mathematics-based algorithms

9.7.1.1 Chaotic search

Chaos, even though interchangeably used with randomness, has a fundamental difference from randomness in the fact that it can be generated deterministically, as compared to random/stochastic sequences which are fundamentally non-deterministic (Salam *et al* 2013, Dileep and Singh 2017). Thus, a chaotic search has superior searching capabilities compared to a random search due to its well-defined boundaries. Bifurcation is done by logistics mapping, shown by the two equations below.

$$x_{n+1} = \lambda\, x_n(1 - x_n) \tag{9.33}$$

$$y_{n+1} = \mu \sin\left(\pi y_n\right) \tag{9.34}$$

where λ is the control parameter and x_n is ergodic between [0,1].

The single chaos-based iteration is not efficient; hence, in Zhou *et al* (2011), a dual carrier chaos search is applied. The chaos variable here is defined as PV voltage which is to be optimized.

The two optimization equations are

$$x_i^r = a + x_n(b - a) \tag{9.35}$$

$$y_i^r = a + \frac{1}{4}\left(y_n + 2\right)(b - a). \tag{9.36}$$

a and b are used to define the range of the initial searching zone

Two points, $P\,(Y_2°)$ and $P\,(X_3°)$ (in figure 9.33), are used as end points/boundary points for searching through successive iteration; the boundary of the search zone becomes smaller and smaller until convergence is obtained. Since chaotic searches are within a well-defined boundary, the chance of convergence is higher and the converging time is smaller compared to random initialization.

9.7.1.2 Cubic spline

Cubic spline is a type of spline interpolation technique, and it is popular due to its prevention of Runge's phenomenon and for the smoothness of its interpolation. Ostadrahimi and Mahmoud (2021) implemented MPPT based on cubic spline interpolation. The sampling frequency is optimal without compromising on accuracy. Huang *et al* (2018) proposed and implemented cubic spline guided by a Jaya algorithm. Here, the cubic spline algorithm is incorporated in the Jaya algorithm for better tracking speed. The proposed system is compared with Jaya algorithm and PSO, and the results show a superior tracking of the algorithm compared to the remaining two algorithms.

9.7.1.3 Fibonacci search

The Fibonacci search deals with MPPT tracking based on the Fibonacci sequence. Ramaprabha *et al* (2012) implemented an MPP based on a modified Fibonacci series where the search space in each iteration is reduced based on the Fibonacci series. The results show a better tracking speed compared to a normal Fibonacci-based search.

9.7.1.4 DIRECT algorithm

The DIRECT method is based on two algorithms of area division and subsequent selection of a potential optimal region. In contrast to a meta-heuristic based approach, the DIRECT method employs a deterministic method with just one parameter to be tuned, which aids in accuracy. Nguyen and Low (2010) implemented MPP based on the DIRECT method. Results show superior tracking capabilities compared to P&O.

9.7.1.5 Gold section search

The GSS is a popular algorithm which searches for maxima/minima within a search space by means of repeated iterative searches where the search space is reduced every iteration. The advantage of GSS is the ability to converge to global maxima with less fluctuation.

Kheldoun *et al* (2016) has implemented an MPPT based on GSS where the voltage range is reduced until MPP is achieved. Agrawal and Aware (2012) implemented a modified GSS that has higher convergence rate than normal GSS.

9.7.1.6 Segmentation search method

The segmentation search method (SSM) is a two-stage mechanism for searching for the MPP during partial shading. First, a *I–V* curve is segmented into various regions based on extensive simulation results. The second stage searches for the global MPP in specified regions. The few works in literature that employ this method for tracking MPP related to SSM include Liu *et al* (2014) and Baimel *et al* (2018). This method has the advantage of having a simple structure and a fast tracking time for the MPP.

9.7.1.7 Random search method

The random search method is a type of numerical optimization algorithm where the requirement of optimization of a gradient is not necessary, hence these could be applied to non-continuous non-differentiable functions. It works on the principle of random sampling combined with an updated sampling strategy (Andradóttir 2006). Sundareswaran *et al* (2014b) implemented MPP for PSCs using a random search, and the proposed system is evaluated against PSO and two-stage P&O. The proposed system provides superior performance in terms of accuracy and tracking time.

9.7.1.8 Pattern search

A generalized pattern search is a derivative-free search, which was introduced by Hooke and Jeeves (Audet 2004). The generalized pattern search is divided into two types: searching and polling. The searching phase involves searching for optimum value, and polling refers to updating mesh values (Javed *et al* 2016).

9.7.2 Techniques based on *P V* array characteristics

Wang *et al* (2016) has proposed an SSJ method for tracking MPPT. It is a modified incremental conductance method used to scan the entire *I–V* curve instead of traditional incremental conductance, which gets stuck in local maxima. The flow diagram of SSJ is shown in figure 9.34. The starting and ending points of the searching space are provided to the algorithm, and then the algorithm scans the entire I–V curve; however, it skips some part of the curve due to the judgment of whether MPP is possible in that region.

Ghasemi *et al* (2018) has proposed MPPT in partial shading by approximating the I–V curve to a constant current source. The proposed system divides the I–V curve into various subregions, and the GMPP is found deterministically. It also has an advantage of not needing a large sampling of data compared to other methods. The result shows robust tracking of PSCs. Alik and Jusoh (2018) proposed a modified P&O able to track GMPP. Here the PV system parameters are fed to the controller along with PV voltage and current. Initially, using P&O, the nearest peak is found, and it is then compared to the voltage (say, Vact) that would have been obtained if not for the PSC. If the reference voltage is half than that of the desired Vact, then partial shading is assumed and P&O is started again to track the entire I–V array (instead of settling in local maxima; Sellami *et al* 2018). Javed *et al* (2020) proposed a modified P&O for searching the entire PV array. In Sellami *et al* (2018), the entire curve is tracked by P&O and the GMPP is computed by a tracker, whereas in Javed *et al* (2020) the duty cycle ΔD is regulated according to input to the system. Furtado *et al* (2018) proposed MPPT based on maximum power trapezoidal, which does not include an entire search, thereby reducing the tracking time compared with traditional techniques.

9.8 Hardware configuration–based MPP methods

This section deals with various hardware-based MPPT techniques proposed in literature. This is broadly classified into two methods:

(a) array reconfiguration–based
(b) and power electronics–based.

9.8.1 Array reconfiguration–based methods

This subsection deals with various array reconfiguration techniques used to mitigate the problem of partial shading. An array reconfiguration–based method refers to the strategic placement and connection of PV arrays to minimize/negate the effect of partial shading. The reconfiguration-based MPP is mainly divided into two types: static and dynamic. Static refers to a fixed topology for maximization of power generation, whereas dynamic refers to a change in electrical or physical configuration. Static reconfiguration techniques are generally tedious and complex to implement, whereas dynamic reconfiguration requires a high number of sensors and switches coupled with a complex control algorithm and a large area requirement for installations. Various popular array reconfigurations are S–P, total-cross-tied (TCT), bridge link (BL), and honey comb (HC). Studies show TCT to have the least loss during PSCs.

In Yadav *et al* (2017), a new array reconfiguration called the magical square puzzle configuration was proposed and compared with conventional array reconfiguration techniques like the TCT, S–P, BL, and HC configurations. The algorithm was based on a number placement puzzle where the sum of each row, each column, and each diagonal adds up to same number. The array reconfiguration makes use of this algorithm. The result shows a better tracking result of the GMPP and less power loss faced under a PSC compared to TCT, BL, HC, and S–P.

Satpathy *et al* (2019) presents a static shade dispersion physical array relocation to minimize power loss during partial shading where the PV modules are renumbered and, subsequently, the connection according to the algorithm is made.

Akrami and Pourhossein (2018) proposed a power comparison technique wherein the optimum reconfiguration structure is decided by an algorithm which compares various topologies for the GMPP. Pillai *et al* (2018) proposed a sensorless and fixed reconfigurable scheme where the relocation of the PV panel is done only column-wise. The computational complexity is minimal, and the labor requirement is lower compared with other methods. The proposed system was compared with the dominance square technique, conventional TCT, and S–P schemes. The proposed system is proved superior to the other conventional electrically array reconfiguration and other similar physical relocation schemes.

9.8.2 Power electronics–based methods

Power electronics techniques used to mitigate the problem of partial shading usually involve a converter topology or a combination of converter topologies to efficiently tackle partial shading and the extraction of MPP from a PV system.

Power electronics–based methods for tackling partial shading can be broadly divided into two methods:

(a) voltage/current equalization–based
(b) and distributed MPP–based.

9.8.2.1 Distributed MPP–based method

Distributed MPP–based methods (DMPPs) have been used to nullify the effect of PSCs. In a DMPP, instead of using a centralized controller to track MPP, a decentralized MPP controller is used, which operates on the principle of differential power processing. Here an MPP controller across each PV panel (or in some cases, across each submodule) is setup to make each panel/module operate in maximum peak independently.

(a) *Multiple-input single-output converter–based topology:*

Multiple-input single-output (MISO)–based topology consists of multiple converters used for segmented PV modules (figure 9.36) (Dhople *et al* 2010). Poshtkouhi and Trescases (2011) modified MISO using a single time interleaved inductor, which provided better results compared with a normal MISO converter.

(b) *Microinverter based topology:*

This method eliminates the use of a DC–DC converter. A DC–AC inverter is fitted across each module separately such that the PV module produces MPP and is sent directly to grid. De Abreu Mateus *et al* (2018) has proposed distributed MPPT using a cascaded H-bridge multilevel converter (shown in figure 9.38). This topology has shown good results in terms of efficiency and losses while tracking MPP during partial shading.

(c) *Module integrated converter–based topology:*

This scheme generally employs a DC–DC converter across each module and a DC–AC inverter from all the outputs of the DC–DC converter. Luo *et al* (2016) proposed a distributed MPPT with a buck converter where a DC–DC converter MPP is placed for every module, and the MPP strategy is P&O for MPPT control (figure 9.37). This method provides low losses in power output during mismatching conditions or in a PSC.

(d) *Multilevel inverter based–topology:*

This scheme involves multilevel inverter topology as a power interface between the grid and the PV system (Abdalla *et al* 2013) (figure 9.39). This system experiences less harmonic distortion compared with regular microinverters like in De Abreu Mateus *et al* (2018); however, the cost goes up due to the extra switches used.

9.8.2.2 ***Voltage/current equalization–based methods***

Various converter topologies are proposed to reduce the current mismatch among a series of connected PV panels. These methods are collectively known as voltage equalization/power equalization methods.

These methods reduce the mismatch of current production from different panels and operates the PV system in MPP.

These operate as a typical current compensation technique where the excess current from unshaded PV panels is passed on to the shaded panels to reduce current imbalance. Anjana *et al* (2017) presents a flyback converter in an input-series-output-parallel configuration that is used as a power converter for voltage equalization. A very popular power electronics–based voltage equalization is TEODI. TEODI (Petrone *et al* 2012, Balato *et al* 2015) is a power equalization technique wherein output power equalization from the boost converter connected across each PV panel is achieved by continuous perturbation of the input voltage. The circuit diagram is shown in figure 9.40.

The following conditions are to be met for TEODI:

(a) $I_{a2} = I_{b2}$
(b) $da = db - \Delta d$

where I_{a2} and I_{b2} are representative of boost converter A and B, respectively, and da and db are the duty cycle of the boost converter A and B, $\Delta d = dv/Vs$.

Balato *et al* (2017) proposed a modified TEODI which is able track partial shading within the PV module also. Sharma and Agarwal (2014) implemented current equalization–based topology with a flyback converter as the power interface (figure 9.41); this showed better results compared with an approximate compensation power-dedicated dc-dc converter MPPT method.

There has also been a voltage equalization–based MPPT method like in Pragallapati *et al* (2013) (figure 9.42). In Pragallapati *et al* (2013), a SEPIC with a single switch is used for equalization of the voltage during partial shading, and

Table 9.3. Summary of power electronics–based MPP methods discussed in this paper.

Sl. No.	MPPT technique	Referred work	Remarks
1.	Distributed MPP	Dhople *et al* (2010), Poshtkouhi and Trescases (2011), Abdalla *et al* (2013), Luo *et al* (2016), De Abreu Mateus *et al* (2018)	Uses a decentralized controller strategy where a simple MPP controller is connected across each PV panel to extract local MPP from each panel. Advantages: (i) Simple computational requirement (ii) reliable and robust tracking of MPP Disadvantages: (i) Makes the system bulkier due to the requirement of n switching converters for n PV panels
2.	Voltage/current equalization	Petrone *et al* (2012), Pragallapati *et al* (2013), Sharma and Agarwal (2014), Balato *et al* (2015), Anjana *et al* (2017), Balato *et al* (2017)	Use voltage/current equalization technique to reduce current mismatch between a series of connected PV modules. Advantages: (i) Digital control is not required (ii) Robust tracking Disadvantages: (i) High hardware requirement

results show around a 37% increase in efficiency compared to the normal using of bypass diodes alone (table 9.3).

9.9 Analysis and comparison of various techniques

The preceding sections have compiled and reviewed various MPPT techniques of PV systems proposed in recent times. These techniques have been broadly classified as:

(i) Soft computing methods
(ii) Exact algorithms
(iii) Hardware-based methods
(iv) Meta-heuristic techniques

However, it is obvious from this article that the meta-heuristic techniques for partial shading have generated significant interest, with around 50% of all published works on partial shading MPPT systems falling under meta-heuristic techniques. This should not be surprising considering the falling rates of digital controllers coupled with the gradient free and stochastic nature of meta-heuristic techniques, making them capable of tracking global peak without getting stuck in local maxima with high accuracy. The stochastic nature of meta-heuristics also means there is scant regard for system information. However, the random tracking of meta-heuristics also has a high probability of making the system unstable during the searching phase. The initialization of particles and the number of particles determine both the accuracy and response speed of the controller while tackling PSCs.

The soft computing techniques have generated significant interest in researchers due to its ability to handle imprecision and operate on limited knowledge, making it viable for tracking the global peak of a PV system during partial shading. It has been shown that soft computing techniques generate results with high accuracy and have a fast tracking time. However, the huge requirement of time consumed in learning the patterns (which extends from days to months) along with the high requirement of hardware components (many requiring high end DSP controllers) makes it less preferable. Most of the soft computing techniques proposed are system specific, except RL (i.e. they are trained for a particular system and hence they need to be trained again when incorporated into a different system, or when the system is upscaled in future).

The method of an exact algorithm operates on the simple exactness of the algorithm, thereby requiring complete information of the system. However, the controllers and hardware requirements are very feasible. The study of a PV panel–based method shows there are types of algorithms which have been implemented after studying characteristics of PV systems and exploiting their characteristics for finding global MPP. Some of the methods have also relied on the modification of conventional algorithms to track the global maxima and not get stuck on local maxima. The mathematical modeling base methods have exploited the mathematics to search for global maxima. These, however, suffer from the fact that global peak may not be tracked for every condition. The predefined searching behavior of these algorithms makes it settle after a large amount of time, thereby increasing the convergence time. However, many hybrid methods have been proposed where these methods are combined with any other meta-heuristic technique which results in a substantial reduction of searching time for the global maxima. It should be noted that the computational requirement for implementing a hybrid MPP controller would be higher.

The hardware-based methods essentially comprise of power electronics and array reconfiguration methods. These offer decentralized control with less computational requirement and provide good results while tracking global MPP. Power electronics–based methods offer decentralized control, which provides good results for a small-scale PV system. However, for a large system the power electronics–based system makes it bulkier due to the fact that a power converter is connected across each PV module separately. The usage of switches increases the overall power loss in the system. The array reconfiguration–based methods effectively negate a PSC with less computational requirement; however, the implementation of the array reconfiguration is generally tedious and laborious, along with requiring a high number of sensors.

The comparison of various MPPT techniques during partial shading is a tedious task due to the different PV systems used, different hardware architecture, lack of any benchmark test, and also due to different processing speeds of various controllers. However, it is advantageous to compare the various techniques based on some logical and practical parameters. This comparison is intended to give researchers a basic idea of the relative performance of various techniques and also give an insight about competitive edges and limitations of each technique, with respect to others. Therefore, this comparison is based on five parameters: algorithm complexity, hardware implementation, controller types used, tracking speed, and system dependency. All the comparisons are shown in table 9.4.

9.10 Challenges and future scope

The main challenge of partial shading is the lack of mathematical models to accurately describe its phenomena. Also, the sudden occurrence of partial shading makes it tough to predict global peak beforehand. During partial shading, to protect individual PV panels, a bypass diode is placed in parallel to each PV panel. Although it protects the PV panel from hotspots and reverse current, a multiple peak P–V curve is exhibited.

To solve the problem of partial shading, many power electronics methods, wherein a decentralized MPP controller along with a converter are proposed, can be found in the literature. Even though these methods are very efficient for small systems, they provide considerable power loss for large PV systems due to the usage of switches. The techniques of array reconfiguration have not been used in any power electronics methods to mitigate partial shading. New converter topologies are also being proposed, which use fewer switches to reduce the total converter loss. The main aim of the proposed topology should include fewer switches, less power loss coupled with less harmonic content, and maximum power transferring capabilities.

The MPPT methods based on exact algorithms provide a robust and simple algorithm for tracking GMPP, which could be implemented using a simple controller. The centralized control structure adds to the advantage of the controller. However, due to the exactness of these algorithms, they tend to take a long time to converge to a solution, thereby leading to lower efficiency. Proposed MPPT methods based on mathematical models may not provide accurate GMPP for all cases.

Due to the recent progress in high performance and the low cost of digital processors, the interest in utilization of meta-heuristic techniques for MPPT has also increased. This is also enhanced by the fact that soft computing techniques operate on a centralized controlling scheme. Meta-heuristic techniques, due to their system independency and robust tracking, are quite popular for finding the global peak power. These algorithms do not need system information and treat partial shading as an optimization problem with the objective of finding maximum power. Due to their random search for a solution, they have a higher probability of converging to a solution compared with exact algorithms. However, the excessive oscillation during tracking leads to low efficiency during searching. Swarm intelligence also has the setback of having to tune control parameters, such as the selection of population size

Table 9.4. Comparison of various MPP controllers in PSCs presented in this paper.

No.	Type	Number of works cited in this paper	Algorithm complexity	Hardware implementation	Controller types used	Tracking speed	System dependency
1.	Soft Computing	24	Very complex	Costly	Centralized controller (digital controller)	Very fast	Yes
2.	Exact algorithms	16	Simple	Moderate	Centralized controller (digital controller)	Slow	Yes
3.	Hardware-based	15	Simple	Costly	De-centralized controller (digital/analog controller)	Very fast	Yes
4.	Meta-Heuristic	51	Complex	Moderate	Centralized controller (digital controller)	Fast	Yes

and the initialization of particle coordinates. The random nature of swarm intelligence would require more tuning of control parameters, whereas a deterministic swarm intelligence method reduces the control parameters required and also results in comparatively fewer oscillations during the search, but it increases the converging time. Many hybrid algorithms have also been proposed to reduce the tracking time, but they require a higher processing capability and a complex control structure. The soft computing methods require a huge amount of data for learning, which may increase learning time.

The scope for further research in this domain is proposed as follows.

(a) A new meta-heuristic technique exploiting the advantages of both meta-heuristics and exact algorithms can be combined to provide competitive an MPPT algorithm with an affordable hardware requirement. Hybrid algorithms of two meta-heuristic techniques can also be explored in order to have better exploration and exploitation capabilities during searching. This can be achieved by incorporating the updation equation of one of the meta-heuristic techniques in the structure of the other meta-heuristic technique. However, the tuning of parameters to extract optimal performance from these algorithms still requires extensive study.

(b) Existing meta-heuristic techniques can be combined with any other exact algorithm technique to efficiently track MPP in a smaller search space area based on input from the exact algorithm, which could aid in better tracking speed. EAs have played a major part in finding MPP for PSCs. The latest advances in EAs with various modifications in control parameters can be used for better results.

(c) The various control parameters in meta-heuristic techniques, like particle initialization, particle size, scaling factor, mutation factor, various weighing factors, etc, are very essential for accurate tracking of GMPP during shaded conditions. However, not enough studies have gone into studying the effects of parameters on convergence and tracking time, and these parameters are generally chosen on a trial-and-error basis. More studies on the tuning of these parameters can ensure much better performance while tracking MPP.

(d) Soft computing techniques that modify the existing system can be implemented for PSCs of MPPT of a PV system; the main aim of these proposed systems should be based on higher tracking speed. New AI methods can be explored and used in a complementary fashion with meta-heuristic techniques or exact algorithms to improve the searching ability and make the searching faster and the system independent.

(e) New power converter topology can also be proposed to solve the problem of partial shading. The main disadvantage of existing power electronics systems is the high number of switches involved. The more complex power electronic circuits provide output with less harmonic content and better MPP; however, they suffer from switching losses. Therefore, a power converter balancing these two points can be designed and compared with current power converter topologies on various parameters. These power converter topologies can also be used in combination with array reconfiguration techniques to obtain much better results.

(f) An accurate partial shading detection mechanism is required for accurate operations. However, current literature does not have sufficient research about partial shading detection and triggering conditions. More work can be done on this field to improve partial shading detection algorithms.

(g) Hybrid algorithms that combine the advantages of meta-heuristics and soft computing techniques can also be explored. The stochastic searching ability of meta-heuristic techniques can be used for the purpose of training soft computing techniques, and this could be compared with existing algorithms for a faster tracking time of accurate GMPP.

(h) The existing meta-heuristic algorithms mainly test for simple PSCs. However, with the growth of PV panels becoming a reality, there could be many more peaks, and these proposed systems have not been checked with many peak clusters. This could result in particles getting stuck in local mode for more complex shading patterns. Therefore, the efficiency of algorithms proposed in the future must be evaluated against these cluster MPPs.

9.11 Conclusion

This chapter has compiled, classified, and briefly discussed the 107 MPP techniques in PSCs proposed in the last 11 years. This chapter started with the need of MPPT and the limitation of conventional algorithms to track MPP during shaded conditions. This work has classified the techniques of tracking MPP during PSCs into four groups: soft computing methods, exact algorithms, hardware methods, and meta-heuristic techniques. The various techniques in these classifications have been discussed briefly, and the comparative advantage of those techniques with respect to other techniques have been discussed on various practical and logical parameters. The challenges and future scope of the same techniques have also been elaborated. Various MPPT techniques for shaded conditions come with their own advantages and disadvantages. The selection of an MPPT controller for a specific PV system would depend upon the hardware availability, computational requirements, convergence time, reliability, and adaptability to different system conditions. Considering the above points, it can be concluded that there is lot of research potential for modifying and improving the present set of MPP controllers in order to obtain much better power extraction during partial shading with a faster system response, along with simpler hardware requirements and a fewer number of current sensor requirements. The ability of algorithms to detect partial shading accurately is also a crucial aspect that needs to be investigated. This work is expected to provide researchers working on MPPT of PV systems in a PSC with fundamental ideas about the problem and to motivate them to carry out further research in this domain.

Bibliography

Abdalla I, Corda J and Zhang L 2013 Multilevel DC-link inverter and control algorithm to overcome the PV partial shading *IEEE Trans. Power Electron.* **28** 14–8

Abdalla O, Rezk H and Ahmed E M 2019 Wind driven optimization algorithm based global MPPT for PV system under non-uniform solar irradiance *Sol. Energy* **180** 429–44

Agrawal J and Aware M 2012 *Golden section search (GSS) algorithm for maximum power point tracking in photovoltaic system 2012 IEEE 5th India Int. Conf. on Power Electronics (IICPE) (Delhi, India)*

Akrami M and Pourhossein K 2018 A novel reconfiguration procedure to extract maximum power from partially-shaded photovoltaic arrays *Sol. Energy* **173** 110–9

Al-Gizi A G, Craciunescu A and Al-Chlaihawi S J 2017 *The use of ANN to supervise the PV MPPT based on FLC 2017 10th International Symposium on Advanced Topics in Electrical Engineering (ATEE) (Bucharest, Romania)* 703–8

Al-Majidi S D, Abbod M F and Al-Raweshidy H S 2018 A novel maximum power point tracking technique based on fuzzy logic for photovoltaic systems *Int. J. Hydrogen Energy* **43** 14158–71

Alik R and Jusoh A 2018 An enhanced P&O checking algorithm MPPT for high tracking efficiency of partially shaded PV module. *Sol. Energy* **163** 570–80

Alshareef M, Lin Z, Ma M and Cao W 2019 Accelerated particle swarm optimization for photovoltaic maximum power point tracking under partial shading conditions *Energies* **12** 623

Alsumiri M 2019 Residual incremental conductance based nonparametric mppt control for solar photovoltaic energy conversion system *IEEE Access* **7** 87901–6

Amara K, Fekik A, Hocine D, Bakir M L, Bourennane E B, Malek T A and Malek A 2018 *Improved performance of a PV solar panel with adaptive neuro fuzzy inference system ANFIS based MPPT 2018 7th Int. Conf. on Renewable Energy Research and Applications (ICRERA) (Paris, France)* 1098–101

Andradóttir S 2006 An overview of simulation optimization via random search *Handbooks in Operations Research and Management Science* ed S G Henderson and B L Nelson (New York: Elsevier) 617–31

Anjana K G, Aniruddha Kamath M and Barai M 2017 A differential current compensation technique for PV systems under partially shaded condition *2017 11th IEEE Int. Conf. on Compatibility, Power Electronics and Power Engineering (CPE-POWERENG) (Cadiz, Spain)* 116–20

Asiful Islam M and Ashfanoor Kabir M 2011 Neural network based maximum power point tracking of photovoltaic arrays *TENCON 2011 - 2011 IEEE Region 10 Conf. (Bali, Indonesia)* 79–82

Audet C 2004 Convergence results for generalized pattern search algorithms are tight *Optim. Eng.* **5** 101–22

Avila L, De Paula M, Trimboli M and Carlucho I 2020 Deep reinforcement learning approach for MPPT control of partially shaded PV systems in smart grids *Appl. Soft Comput.* **97** 106711

Aygül K, Cikan M, Demirdelen T and Tumay M 2019 Butterfly optimization algorithm based maximum power point tracking of photovoltaic systems under partial shading condition *Energy Sources, Part A Recover. Util. Environ. Eff.* **45** 8337–55

Baimel D, Tapuchi S, Bronshtein S, Horen Y and Baimel N 2018 Novel segmentation algorithm for maximum power point tracking in PV systems under partial shading conditions *2018*

IEEE 18th Int. Power Electronics and Motion Control Conf. (PEMC) (Budapest, Hungary) 406–10

Balato M, Costanzo L, Marino P, Rubino G, Rubino L and Vitelli M 2017 Modified TEODI MPPT technique: theoretical analysis and experimental validation in uniform and mismatching conditions *IEEE J. Photovolt.* **7** 604–13

Balato M, Costanzo L, Marino P, Rubino L and Vitelli M 2015 Dual implementation of the MPPT technique TEODI: uniform and mismatching operating conditions *2015 Int. Conf. on Clean Electrical Power (ICCEP) (Taormina, Italy)* 422–9

Behera M K and Saikia L C 2020 A new combined extreme learning machine variable steepest gradient ascent MPPT for PV system based on optimized PI-FOI cascade controller under uniform and partial shading conditions *Sustain. Energy Technol. Assess.* **42** 100859

Benyoucef A S, Chouder A, Kara K, Silvestre S and Sahed O A 2015 Artificial bee colony based algorithm for maximum power point tracking (MPPT) for PV systems operating under partial shaded conditions *Appl. Soft Comput. J.* **32** 38–48

Besheer A H and Adly M 2012 Ant colony system based PI maximum power point tracking for stand alone photovoltaic system *2012 IEEE Int. Conf. on Industrial Technology (Athens)* 693–8

Bizon N 2016 Global extremum seeking control of the power generated by a photovoltaic array under partially shaded conditions *Energy Convers. Manag.* **109** 71–85

Bouakkaz M S, Boukadoum A, Boudebbouz O, Fergani N, Boutasseta N, Attoui I, Bouraiou A and Necaibia A 2020 Dynamic performance evaluation and improvement of PV energy generation systems using moth flame optimization with combined fractional order PID and sliding mode controller *Sol. Energy* **199** 411–24

Câmara D 2015 *Bio-Inspired Networking* (New York: Elsevier)

Chandrasekaran K, Sankar S and Banumalar K 2020 Partial shading detection for PV arrays in a maximum power tracking system using the sine-cosine algorithm *Energy Sustain. Dev.* **55** 105–21

Chen L and Wang X 2019 Enhanced MPPT method based on ANN-assisted sequential Monte–Carlo and quickest change detection *IET Smart Grid* **2** 635–44

Chiang M L, Hua C C and Lin J R 2002 Direct power control for distributed PV power system *Proc. of the Power Conversion Conference-Osaka 2002 (Osaka, Japan)* 311–5

Chou K Y, Yang C S and Chen Y P 2020 Deep Q-network based global maximum power point tracking for partially shaded PV system *2020 IEEE Int. Conf. on Consumer Electronics - Taiwan (ICCE-Taiwan) (Taoyuan, Taiwan)* 1–2

Clerc M and Kennedy J 2002 The particle swarm-explosion, stability, and convergence in a multidimensional complex space *IEEE Trans. Evol. Comput.* **6** 58–73

da Rocha M V, Sampaio L P and da Silva S A O 2020 Comparative analysis of MPPT algorithms based on Bat algorithm for PV systems under partial shading condition *Sustain. Energy Technol. Assess.* **40** 100761

Daraban S, Petreus D and Morel C 2014 A novel MPPT (maximum power point tracking) algorithm based on a modified genetic algorithm specialized on tracking the global maximum power point in photovoltaic systems affected by partial shading *Energy* **74** 374–88

De Abreu Mateus T H, Pomilio J A, Godoy R B and Pinto J O P 2018 Distributed MPPT scheme for grid connected operation of photovoltaic system using cascaded H-bridge multilevel converter under partial shading *2017 IEEE Southern Power Electronics Conf. (SPEC) (Puerto Varas, Chile)* 1–6

Dhople S V, Ehlmann J L, Davoudi A and Chapman P L 2010 Multiple-input boost converter to minimize power losses due to partial shading in photovoltaic modules *2010 IEEE Energy Conversion Congress and Exposition (Atlanta, Georgia)* 2633–6

Dileep G and Singh S N 2017 Application of soft computing techniques for maximum power point tracking of SPV system *Sol. Energy* **141** 182–202

Duman S, Yorukeren N and Altas I H 2018 A novel MPPT algorithm based on optimized artificial neural network by using FPSOGSA for standalone photovoltaic energy systems *Neural Comput. Appl.* **29** 257–78

Eberhart R and Kennedy J 1995 New optimizer using particle swarm theory *MHS'95. Proceedings of the Sixth International Symposium on Micro Machine and Human Science (Nagoya, Japan)* 39–43

Eberhart R C and Shi Y 2000 Comparing inertia weights and constriction factors in particle swarm optimization *Proceedings of the 2000 Congress on Evolutionary Computation, CEC00 (La Jolla, California)* 84–8

El-Helw H M, Magdy A and Marei M I 2017 A hybrid maximum power point tracking technique for partially shaded photovoltaic arrays *IEEE Access* **5** 11900–8

Elnosh A, Khadkikar V, Xiao W and Kirtely J L 2014 An improved extremum-seeking based MPPT for grid-connected PV systems with partial shading *2014 IEEE 23rd International Symposium on Industrial Electronics (ISIE) (Istanbul, Turkey)* 2548–53

Eltamaly A M 2018 Performance of MPPT techniques of photovoltaic systems under normal and partial shading conditions *Advances in Renewable Energies and Power Technologies* ed I Yahyaoui (New York: Elsevier) 115–61

Eltamaly A M, Al-Saud M S and Abokhalil A G 2020 A novel bat algorithm strategy for maximum power point tracker of photovoltaic energy systems under dynamic partial shading *IEEE Access* **8** 10048–60

Fares D, Fathi M, Shams I and Mekhilef S 2021 A novel global MPPT technique based on squirrel search algorithm for PV module under partial shading conditions *Energy Convers. Manag.* **230** 113773

Farh H M H, Eltamaly A M and Othman M F 2018 Hybrid PSO-FLC for dynamic global peak extraction of the partially shaded photovoltaic system *PLoS One* **13** e0206171

Femia N, Granozio D, Petrone G, Spagnuolo G and Vitelli M 2007 Predictive & adaptive MPPT perturb and observe method *IEEE Trans. Aerosp. Electron. Syst.* **43** 934–50

Friedman N, Geiger D and Goldszmidt M 1997 Bayesian network classifiers *Mach. Learn.* **29** 131–63

Furtado A M S, Bradaschia F, Limongi L R and Cavalcanti M C 2018 A reduced voltage range global maximum power point tracking algorithm for photovoltaic systems under partial shading conditions *IEEE Trans. Ind. Electron.* **65** 3252–62

Ghasemi M A, Ramyar A and Iman-Eini H 2018 MPPT method for PV systems under partially shaded conditions by approximating *I–V* curve *IEEE Trans. Ind. Electron.* **65** 3966–75

Gules R, Pacheco J D P, Hey H L and Imhoff J 2008 A maximum power point tracking system with parallel connection for PV stand-alone applications *IEEE Trans. Ind. Electron.* **55** 2674–83

Guo K, Cui L, Mao M, Zhou L and Zhang Q 2020 An improved gray wolf optimizer MPPT algorithm for PV system with BFBIC converter under partial shading *IEEE Access* **8** 103476–90

Guo L, Meng Z, Sun Y and Wang L 2018 A modified cat swarm optimization based maximum power point tracking method for photovoltaic system under partially shaded condition *Energy* **144** 501–14

Hayder W, Ogliari E, Dolara A, Abid A, Hamed M, Ben and Sbita L 2020 Improved PSO: a comparative study in MPPT algorithm for PV system control under partial shading conditions *Energies* **13** 2035

Heidari A A, Mirjalili S, Faris H, Aljarah I, Mafarja M and Chen H 2019 Harris hawks optimization: algorithm and applications *Futur. Gener. Comput. Syst.* **97** 849–72

Hu K, Cao S, Li W and Zhu F 2019 An improved particle swarm optimization algorithm suitable for photovoltaic power tracking under partial shading conditions *IEEE Access* **7** 143217–32

Huang C, Wang L, Yeung R S C, Zhang Z, Chung H S H and Bensoussan A 2018 A prediction model-guided jaya algorithm for the PV system maximum power point tracking *IEEE Trans. Sustain. Energy* **9** 45–55

Ishaque K and Salam Z 2013 A deterministic particle swarm optimization maximum power point tracker for photovoltaic system under partial shading condition *IEEE Trans. Ind. Electron.* **60** 3195–206

Ishaque K, Salam Z, Amjad M and Mekhilef S 2012 An improved particle swarm optimization (PSO)-based MPPT for PV with reduced steady-state oscillation *IEEE Trans. Power Electron.* **27** 3627–38

Javed K, Ashfaq H and Singh R 2020 A new simple MPPT algorithm to track MPP under partial shading for solar photovoltaic systems *Int. J. Green Energy* **17** 48–61

Javed M Y, Murtaza A F, Ling Q, Qamar S and Gulzar M M 2016 A novel MPPT design using generalized pattern search for partial shading *Energy Build.* **133** 59–69

Ji B, Hata K, Imura T, Hori Y, Honda S, Shimada S and Kawasaki O 2020 A Novel particle jump particle swarm optimization method for PV MPPT control under partial shading conditions *IEEJ J. Ind. Appl.* **9** 435–43

Jin Y, Hou W, Li G and Chen X 2017 A glowworm swarm optimization-based maximum power point tracking for photovoltaic/thermal systems under non-uniform solar irradiation and temperature distribution *Energies* **10** 541

Juang c f 2004 A hybrid of genetic algorithm and particle swarm optimization for recurrent network design *IEEE Trans. Syst. Man Cybern. Part B Cybern.* **34** 997–1006

Kar A K 2016 Bio inspired computing—a review of algorithms and scope of applications *Expert Syst. Appl.* **59** 20–32

Kermadi M, Salam Z, Ahmed J and Berkouk E M 2019 An effective hybrid maximum power point tracker of photovoltaic arrays for complex partial shading conditions *IEEE Trans. Ind. Electron.* **66** 6990–7000

Keyrouz F and Georges S 2011 Efficient multidimensional maximum power point tracking using bayesian fusion *2011 2nd Int. Conf. on Electric Power and Energy Conversion Systems (EPECS) (Sharjah, United Arab Emirates)* 1–5

Kharb R K, Shimi S L, Chatterji S and Ansari M F 2014 Modeling of solar PV module and maximum power point tracking using ANFIS *Renew. Sustain. Energy Rev.* **33** 602–12

Kheldoun A, Bradai R, Boukenoui R and Mellit A 2016 A new golden section method-based maximum power point tracking algorithm for photovoltaic systems *Energy Convers. Manag.* **111** 125–36

Kobayashi K, Takano I and Sawada Y 2006 A study of a two stage maximum power point tracking control of a photovoltaic system under partially shaded insolation conditions *Sol. Energy Mater. Sol. Cells* **90** 2975–88

Kota V R and Bhukya M N 2019 A novel global MPP tracking scheme based on shading pattern identification using artificial neural networks for photovoltaic power generation during partial shaded condition *IET Renew. Power Gener.* **13** 1647–59

Kumar N, Hussain I, Singh B and Panigrahi B K 2017a MPPT in dynamic condition of partially shaded pv system by using WODE technique *IEEE Trans. Sustain. Energy* **8** 1204–14

Kumar N, Hussain I, Singh B and Panigrahi B K 2017b Single sensor-based MPPT of partially shaded PV system for battery charging by using cauchy and gaussian sine cosine optimization *IEEE Trans. Energy Convers.* **32** 983–92

Kumar N, Hussain I, Singh B and Panigrahi B K 2017c Single sensor based MPPT for partially shaded solar photovoltaic by using human psychology optimisation algorithm *IET Gener. Transm. Distrib.* **11** 2562–74

Kumar N, Nema S, Nema R K and Verma D 2020 A state-of-the-art review on conventional, soft computing, and hybrid techniques for shading mitigation in photovoltaic applications *Int. Trans. Electr. Energy Syst.* **30** e12420

Li H, Yang D, Su W, Lu J and Yu X 2019 An overall distribution particle swarm optimization MPPT algorithm for photovoltaic system under partial shading *IEEE Trans. Ind. Electron.* **66** 265–75

Li L L, Lin G Q, Tseng M L, Tan K and Lim M K 2018 A maximum power point tracking method for PV system with improved gravitational search algorithm *Appl. Soft Comput. J.* **65** 333–48

Liu Y H, Chen J H and Huang J W 2014 Global maximum power point tracking algorithm for PV systems operating under partially shaded conditions using the segmentation search method *Sol. Energy* **103** 350–63

Liu Y H, Huang S C, Huang J W and Liang W C 2012 A particle swarm optimization-based maximum power point tracking algorithm for PV systems operating under partially shaded conditions *IEEE Trans. Energy Convers.* **27** 1027–35

Loukil K, Abbes H, Abid H, Abid M and Toumi A 2020 Design and implementation of reconfigurable MPPT fuzzy controller for photovoltaic systems *Ain Shams Eng. J.* **11** 319–28

Luo H, Wen H and Li X 2016 Distributed MPPT control under partial shading condition *2016 IEEE 8th International Power Electronics and Motion Control Conf. (IPEMC-ECCE Asia) (Hefei, Asia)* 928–32

Macaulay J and Zhou Z 2018 A fuzzy logical-based variable step size P&O MPPT algorithm for photovoltaic system *Energies* **11** 1340

Mansoor M, Mirza A F and Ling Q 2020 Harris hawk optimization-based MPPT control for PV systems under partial shading conditions *J. Clean. Prod.* **274** 122857

Mao M, Zhang L, Yang L, Chong B, Huang H and Zhou L 2020 MPPT using modified salp swarm algorithm for multiple bidirectional PV-Ćuk converter system under partial shading and module mismatching *Sol. Energy* **209** 334–49

Mendes R, Kennedy J and Neves J 2004 The fully informed particle swarm: simpler, maybe better *IEEE Trans. Evol. Comput.* **8** 204–10

Messai A, Mellit A, Guessoum A and Kalogirou S A 2011 Maximum power point tracking using a GA optimized fuzzy logic controller and its FPGA implementation *Sol. Energy* **85** 265–77

Messalti S, Harrag A and Loukriz A 2017 A new variable step size neural networks MPPT controller: review, simulation and hardware implementation *Renew. Sustain. Energy Rev.* **68** 221–33

Mirjalili S, Gandomi A H, Mirjalili S Z, Saremi S, Faris H and Mirjalili S M 2017 Salp swarm algorithm: a bio-inspired optimizer for engineering design problems *Adv. Eng. Softw.* **114** 163–91

Mirjalili S and Lewis A 2016 The whale optimization algorithm *Adv. Eng. Softw.* **95** 51–67

Mirjalili S, Mirjalili S M and Lewis A 2014 Grey wolf optimizer *Adv. Eng. Softw.* **69** 46–61

Mirza A F, Mansoor M, Ling Q, Yin B and Javed M Y 2020 A salp-swarm optimization based MPPT technique for harvesting maximum energy from PV systems under partial shading conditions *Energy Convers. Manag.* **209** 112625

Mohanty S, Subudhi B and Ray P K 2017 A grey wolf-assisted perturb & observe MPPT algorithm for a PV System *IEEE Trans. Energy Convers.* **32** 340–7

Mohanty S, Subudhi B and Ray P K 2016 A new MPPT design using grey Wolf optimization technique for photovoltaic system under partial shading conditions *IEEE Trans. Sustain. Energy* **7** 181–8

Mohapatra A, Nayak B, Das P and Mohanty K B 2017 A review on MPPT techniques of PV system under partial shading condition *Renew. Sustain. Energy Rev.* **80** 854–67

Mosaad M I, O abed el-Raouf M, Al-Ahmar M A and Banakher F A 2019 Maximum power point tracking of PV system based cuckoo search algorithm; review and comparison *Energy Procedia* **162** 117–26

Nguyen T L and Low K S 2010 A global maximum power point tracking scheme employing DIRECT search algorithm for photovoltaic systems *IEEE Trans. Ind. Electron.* **57** 3456–67

Nugraha D A, Lian K L and Suwarno S 2019 A novel mppt method based on cuckoo search algorithm and golden section search algorithm for partially shaded pv system *Can. J. Electr. Comput. Eng.* **42** 173–82

Ostadrahimi A and Mahmoud Y 2021 Novel spline-MPPT technique for photovoltaic systems under uniform irradiance and partial shading conditions *IEEE Trans. Sustain. Energy* **12** 524–32

Pachaivannan N, Subburam R, Padmanaban M and Subramanian A 2021 Certain investigations of ANFIS assisted CPHO algorithm tuned MPPT controller for PV arrays under partial shading conditions *J. Ambient Intell. Humaniz. Comput.* **12** 9923–38

Parejo J A, Ruiz-Cortés A, Lozano S and Fernandez P 2012 Metaheuristic optimization frameworks: a survey and benchmarking *Soft Comput* **16** 527–61

Peng B R, Ho K C and Liu Y H 2018 A novel and fast MPPT method suitable for both fast changing and partially shaded conditions *IEEE Trans. Ind. Electron.* **65** 3240–51

Petrone G, Spagnuolo G and Vitelli M 2012 An analog technique for distributed MPPT PV applications *IEEE Trans. Ind. Electron.* **59** 4713–22

Pilakkat D and Kanthalakshmi S 2019 An improved P&O algorithm integrated with artificial bee colony for photovoltaic systems under partial shading conditions. *Sol. Energy* **178** 37–47

Pillai D S, Prasanth Ram J, Siva Sai Nihanth M and Rajasekar N 2018 A simple, sensorless and fixed reconfiguration scheme for maximum power enhancement in PV systems *Energy Convers. Manag.* **172** 402–17

Poshtkouhi S and Trescases O 2011 Multi-input single-inductor DC–DC converter for MPPT in parallel-connected photovoltaic applications *2011 Twenty-Sixth Annual IEEE Applied Power Electronics Conference and Exposition (APEC) (Fort Worth, Texas)* 41–7

Pragallapati N, Sharma P and Agarwal V 2013 A new voltage equalization based distributed maximum power point extraction from a PV source operating under partially shaded conditions *2013 IEEE 39th Photovoltaic Specialists Conf. (PVSC) (Tampa, Florida)* 2921–6

Premkumar M, Mohamed Ibrahim A, Mohan Kumar R and Sowmya R 2019 Analysis and simulation of bio-inspired intelligent salp swarm MPPT method for the PV systems under partial shaded conditions *Int. J. Comput. Digit. Syst.* **8** 489–96

Priyadarshi N, Padmanaban S, Holm-Nielsen J B, Blaabjerg F and Bhaskar M S 2020 An experimental estimation of hybrid ANFIS-PSO-based MPPT for PV grid integration under fluctuating sun irradiance *IEEE Syst. J.* **14** 1218–29

Ramaprabha R, Balaji M and Mathur B L 2012 Maximum power point tracking of partially shaded solar PV system using modified Fibonacci search method with fuzzy controller *Int. J. Electr. Power Energy Syst.* **43** 754–65

Rashedi E, Nezamabadi-pour H and Saryazdi S 2009 GSA: a gravitational search algorithm *Inf. Sci.* **179** 2232–48

Rezk H and Eltamaly A M 2015 A comprehensive comparison of different MPPT techniques for photovoltaic systems *Sol. Energy* **112** 1–11

Rezk H and Fathy A 2017 Simulation of global MPPT based on teaching–learning-based optimization technique for partially shaded PV system *Electr. Eng.* **99** 847–59

Saha D 2016 A GSA based improved MPPT system for PV generation *2015 IEEE Int. Conf. on Research in Computational Intelligence and Communication Networks (ICRCICN) (Kolkata, India)* 131–6

Sahoo S K, Balamurugan M, Anurag S, Kumar R and Priya V 2017 Maximum power point tracking for PV panels using ant colony optimization *2017 Innovations in Power and Advanced Computing Technologies (i-PACT) (Vellore, India)* 1–4

Salam Z, Ahmed J and Merugu B S 2013 The application of soft computing methods for MPPT of PV system: a technological and status review *Appl. Energy* **107** 135–48

S Kumar C H and S Rao R 2016 A novel global MPP tracking of photovoltaic system based on whale optimization algorithm *Int. J. Renew. Energy Dev.* **5** 225–32

Satpathy P R, Sharma R and Dash S 2019 An efficient SD-PAR technique for maximum power generation from modules of partially shaded PV arrays *Energy* **175** 182–94

Sellami A, Kandoussi K, El Otmani R, Eljouad M, Mesbahi O and Hajjaji A 2018 A novel auto-scaling MPPT algorithm based on perturb and observe method for photovoltaic modules under partial shading conditions *Appl. Sol. Energy* **54** 149–58

Seyedmahmoudian M, Rahmani R, Mekhilef S, Maung Than Oo A, Stojcevski A, Soon T K and Ghandhari A S 2015 Simulation and hardware implementation of new maximum power point tracking technique for partially shaded pv system using hybrid DEPSO method *IEEE Trans. Sustain. Energy* **6** 850–62

Seyedmahmoudian M, Soon T K, Jamei E, Thirunavukkarasu G S, Horan B, Mekhilef S and Stojcevski A 2018 Maximum power point tracking for photovoltaic systems under partial shading conditions using bat algorithm *Sustain* **10** 1347

Shang L, Zhu W, Li P and Guo H 2018 Maximum power point tracking of PV system under partial shading conditions through flower pollination algorithm *Prot. Control Mod. Power Syst.* **3** 38

Sharma P and Agarwal V 2014 Exact maximum power point tracking of grid-connected partially shaded PV source using current compensation concept *IEEE Trans. Power Electron.* **29** 4684–92

Shi Y and Eberhart R 1998 Modified particle swarm optimizer *1998 IEEE Int. Conf. on Evolutionary Computation Proceedings; IEEE World Congress on Computational Intelligence (Anchorage, Alaska)* 69–73

Sundareswaran K, Peddapati S and Palani S 2014a MPPT of PV systems under partial shaded conditions through a colony of flashing fireflies *IEEE Trans. Energy Convers.* **29** 463–72

Sundareswaran K, Peddapati S and Palani S 2014b Application of random search method for maximum power point tracking in partially shaded photovoltaic systems *IET Renew. Power Gener.* **8** 670–8

Syafaruddin , Karatepe E and Hiyama T 2009 Artificial neural network-polar coordinated fuzzy controller based maximum power point tracking control under partially shaded conditions *IET Renew. Power Gener.* **3** 239–53

Taheri H, Salam Z, Ishaque K and Syafaruddin 2010 A novel maximum power point tracking control of photovoltaic system under partial and rapidly fluctuating shadow conditions using differential evolution *2010 IEEE Symposium on Industrial Electronics and Applications (ISIEA) (Penang, Malaysia)* 82–7

Tey K S, Mekhilef S, Seyedmahmoudian M, Horan B, Oo A T and Stojcevski A 2018 Improved differential evolution-based MPPT algorithm using SEPIC for PV systems under partial shading conditions and load variation *IEEE Trans. Ind. Informatics* **14** 4322–33

Titri S, Larbes C, Toumi K Y and Benatchba K 2017 A new MPPT controller based on the Ant colony optimization algorithm for photovoltaic systems under partial shading conditions *Appl. Soft Comput. J.* **58** 465–79

Tjahjono A, Anggriawan D O, Habibi M N and Prasetyono E 2020 Modified grey wolf optimization for maximum power point tracking in photovoltaic system under partial shading conditions *Int. J. Electr. Eng. Informatics* **12** 94–104

Trelea I C 2003 The particle swarm optimization algorithm: convergence analysis and parameter selection *Inf. Process. Lett.* **85** 317–25

Verma P, Garg R and Mahajan P 2020 Asymmetrical interval type-2 fuzzy logic control based MPPT tuning for PV system under partial shading condition *ISA Trans.* **100** 251–63

Vikhar P A 2017 Evolutionary algorithms: a critical review and its future prospects *2016 Int. Conf. on Global Trends in Signal Processing, Information Computing and Communication (ICGTSPICC) (Jalgaon, India)* 261–5

Wan Y, Mao M, Zhou L, Zhang Q, Xi X and Zheng C 2019 A novel nature-inspired maximum power point tracking (MPPT) controller based on SSA-GWO algorithm for partially shaded photovoltaic systems *Electron* **8** 680

Wang Y, Li Y and Ruan X 2016 High-accuracy and fast-speed mppt methods for pv string under partially shaded conditions *IEEE Trans. Ind. Electron.* **63** 235–45

Wu W, Pongratananukul N, Qiu W, Rustom K, Kasparis T and Batarseh I 2003 DSP-based multiple peak power tracking for expandable power system *Eighteenth Annual IEEE Applied Power Electronics Conf. and Exposition, 2003. (Miami Beach, Florida)* 525–30

Xiao W and Dunford W G 2004 A modified adaptive hill climbing MPPT method for photovoltaic power systems *2004 IEEE 35th Annual Power Electronics Specialists Conf. (Aachen, Germany)* 1957–63

Yadav A S, Pachauri R K, Chauhan Y K, Choudhury S and Singh R 2017 Performance enhancement of partially shaded PV array using novel shade dispersion effect on magic-square puzzle configuration *Sol. Energy* **144** 780–97

Yang B, Zhu T, Wang J, Shu H, Yu T, Zhang X, Yao W and Sun L 2020 Comprehensive overview of maximum power point tracking algorithms of PV systems under partial shading condition *J. Clean. Prod.* **268** 121983

Zafar M H, Al-Shahrani T, Khan N M, Mirza A F, Mansoor M, Qadir M U, Khan M I and Naqvi R A 2020 Group teaching optimization algorithm based mppt control of pv systems under partial shading and complex partial shading *Electron* **9** 1962

Zhan Z H, Zhang J, Li Y and Chung H S H 2009 Adaptive particle swarm optimization *IEEE Trans. Syst. Man, Cybern. Part B Cybern* **39** 1362–81

Zhang X, Li S, He T, Yang B, Yu T, Li H, Jiang L and Sun L 2019 Memetic reinforcement learning based maximum power point tracking design for PV systems under partial shading condition *Energy* **174** 1079–90

Zhang Y and Jin Z 2020 Group teaching optimization algorithm: a novel metaheuristic method for solving global optimization problems *Expert Syst. Appl.* **148** 113246

Zheng Y L, Ma L H, Zhang L Y and Qian J X 2003 On the convergence analysis and parameter selection in particle swarm optimization *Proceedings of the 2003 Int. Conf. on Machine Learning and Cybernetics (Xi'an, China)* 1802–7

Zhou L, Chen Y, Guo K and Jia F 2011 New approach for MPPT control of photovoltaic system with mutative-scale dual-carrier chaotic search *IEEE Trans. Power Electron.* **26** 1038–48

IOP Publishing

Signal Processing with Python
A practical approach
Irshad Ahmad Ansari and Varun Bajaj

Chapter 10

Automating Monte Carlo simulation data analysis using Python in Anaconda environment

Abhishek Kumar Singh and Aakash Kumar Jain

This chapter presents an approach of leveraging Python scripts to analyze Monte Carlo simulation data in an Anaconda environment. The Monte Carlo analysis technique is employed in Very Large Scale Integration (VLSI) to evaluate the statistical performance of circuits and systems by randomly varying the tolerances of both its component and model parameters within their specified bounds. To validate the accuracy of the Electronic Design Automation (EDA) tools performing Fast Monte Carlo simulations, the nature of the test cases needs to be verified first. However, analyzing large amounts of data can be a time-consuming and error-prone task. The process can be automated by scripts using various Python libraries to process and visualize the heavy data for advanced node designs. The process deals with the extracting of golden data from Monte Carlo and transient simulations with different process conditions and simulator settings for benchmarking purposes and to validate the generated data with some standard practices. The results show that the proposed method is efficient and greatly simplifies the analysis process and provides a reliable and automated method for analyzing Monte Carlo simulation data for validating new EDA tools.

10.1 Advanced nodes design

The design of integrated circuits (ICs) at advanced nodes, specifically at 16 nm and below, poses significant challenges due to increased variations [1–3]. Traditional approaches to controlling variation, such as using foundry-provided corners characterized for digital circuits, are becoming less accurate in capturing the complexities of these advanced nodes. In this context, Monte Carlo analysis emerges as a crucial technique for accurately capturing statistics on circuit behavior and evaluating performance [4].

doi:10.1088/978-0-7503-5929-0ch10

10.1.1 A simple memory design

Described below in figure 10.1 is a simple Six Transistors (6T) Static Random-Access Memory (SRAM) design explained with a sense amplifier.

In the above diagram, the provided circuit represents a 6T SRAM cell circuit, a fundamental building block used in memory design. It consists of six transistors, including p-type Metal-Oxide-Semiconductor (pMOS) (M1–M4) and n-type Metal-Oxide-Semiconductor (nMOS) (M5–M15) transistors, voltage sources (V1–V9), capacitors (C1 and C2) and VDD and SE as positive voltage supply and sense enable inputs.

At the heart of the circuit, M1–M5 and M2–M6 form a cross-coupled latch, creating a feedback loop that holds the stored data bit (0 or 1) in the SRAM cell. This latch configuration provides data retention. M3 is used for the wordline, enabling read and write operations on the cell. M4 serves as the access transistor for the write operation, allowing data to be written into the SRAM cell.

During the read operation, M5 and M6 function as access transistors, allowing the data to be read from the cell to the output nodes q and q_bar (complementary outputs). The pulse voltage sources (V1, V2, V4, V5, V6, V7, V8, and V9) generate specific voltage signals, simulating different operations such as reading, writing, and setting initial conditions for simulation.

Capacitors C1 and C2 represent capacitances in the circuit for storing charges and affecting the circuit's behavior.

The transistor models (cmosp and cmosn) that are used from TSMC 180 nm library define the property and behavior of the pMOS and nMOS transistors, such as threshold voltage, mobility, and other electrical characteristics.

In figure 10.2, when the write operation is enabled, the capacitance charges and discharges in response to the word signal, resulting in a corresponding change in the

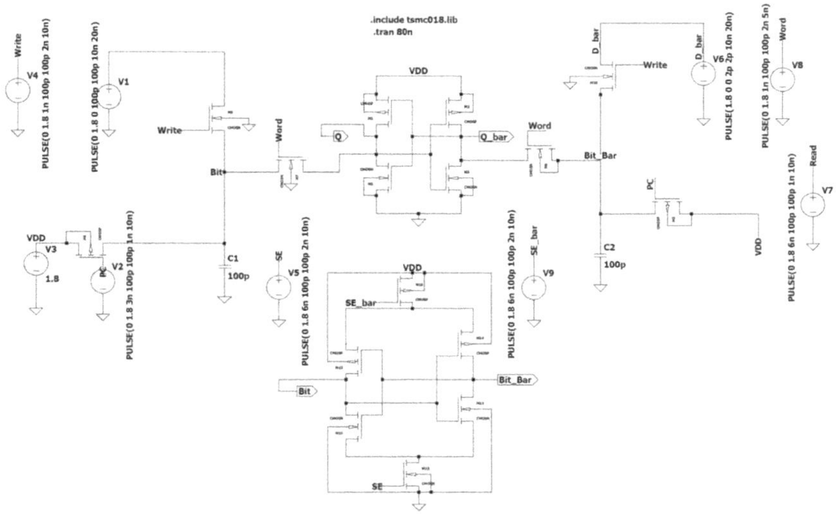

Figure 10.1. A reference SRAM design.

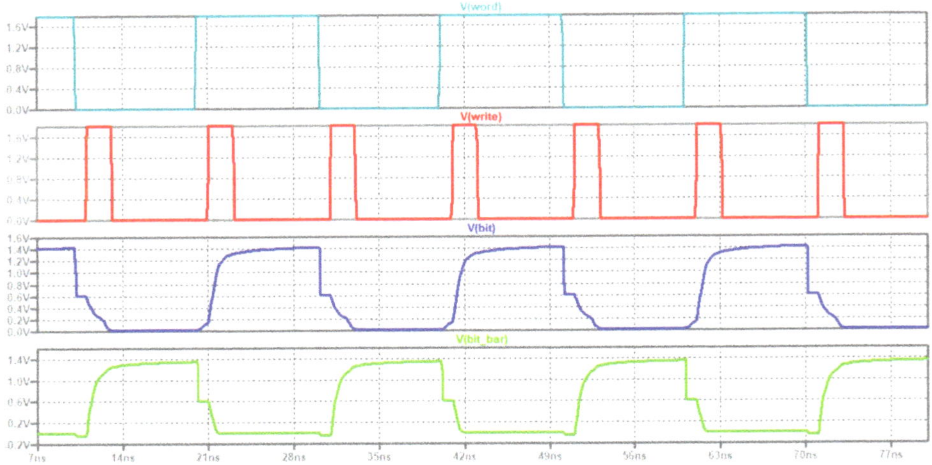

Figure 10.2. Write operation waveform.

bit signal. Similarly, the read operation involves the utilization of sense amplifiers and pre-charge circuits, which have been excluded from the analysis to maintain simplicity and relevance reasons.

Overall, the 6T SRAM SPICE circuit is designed to store and manipulate a single bit of data. The cross-coupled latch configuration ensures data retention, whereas the access transistors enable read and write operations. Through the application of various voltage pulses, the circuit's functionality can be simulated, providing insight into its performance and behavior in different scenarios.

Multiple such cells are connected in a structured manner. These connections form rows and columns, allowing for the storage and retrieval of data. The memory array is organized as a grid, with rows and columns of 6T SRAM cells. Each row represents a wordline, and each column represents a bitline. The wordlines and bitlines are used to access and manipulate the data stored in the memory.

To write data into the memory, the desired wordline is activated, enabling the access transistors of the corresponding row. The data is then written into the cells by controlling the bitlines connected to the desired column. To read data from the memory, the desired wordline is again activated, and the bitlines connected to the desired column are pre-charged. The stored data are then sensed and amplified by the sense amplifiers connected to the bitlines.

10.1.2 Monte Carlo simulations

Monte Carlo analysis is a statistical simulation technique used in IC design to assess the impact of process variations on circuit performance. It involves randomly sampling process parameters within specified ranges and performing repeated simulations to capture the statistical distribution of circuit responses. In IC design, process variations can lead to deviations in circuit performance, affecting important metrics such as timing, power consumption, and yield. Monte Carlo analysis enables designers to quantify the uncertainty associated with these variations and evaluate the likelihood of meeting design specifications.

By simulating the random variations, Monte Carlo analysis provides statistical information, such as mean values, standard deviations, and yield estimates, for different circuit performance parameters. This information helps designers identify potential design issues, optimize circuit performance, and make informed decisions regarding design margins, trade-offs, and reliability.

10.1.3 Challenges in Monte Carlo simulation

However, the traditional Monte Carlo approach has a drawback: it often requires a large number of simulations to provide a more accurate picture of design yield [5, 6]. Conducting a significant volume of simulations can be extremely time-consuming, posing a challenge in meeting tight project deadlines. Designers often face the dilemma of sacrificing accuracy for time or investing a considerable amount of time in running extensive simulations.

10.1.4 Fast Monte Carlo simulations

To address this challenge, researchers and engineers have developed advanced techniques that offer the same level of accuracy as traditional Monte Carlo simulations but with significantly fewer runs. These techniques leverage intelligent sampling methodologies, statistical analysis, and advanced algorithms to efficiently explore the design space and identify critical design parameters [7–9]. By focusing on these parameters, designers can tune them more effectively to improve yield and performance.

Designs that demand high reliability, such as automotive systems, face challenges when relying solely on traditional Monte Carlo analysis due to its time-consuming nature. These designs often require yield estimations at 4–6 sigma levels [10]. However, our algorithm offers a solution by accurately predicting circuit yield at 4–6 sigma without the need for millions or billions of simulations typically associated with traditional Monte Carlo analysis. This is achieved through techniques like scaled-sigma sampling [11], a technique that gradually scales up the variation, enabling yield estimation using larger scaling factors and advanced algorithms like K-Sigma Corners Algorithm [12], Worst Samples Algorithm [1], etc. These techniques have proven their accuracy in handling non-linear behavior and their efficiency in managing a vast array of statistical parameters and specifications. They provide a more efficient and effective approach for yield estimation in complex designs, particularly those that prioritize high reliability.

By adopting these advanced Monte Carlo techniques, designers gain the advantage of reducing simulation time from weeks to just a few hundred runs while maintaining accuracy. This drastic reduction in simulation time empowers designers to meet project deadlines without compromising the thoroughness of their analysis. Moreover, it allows designers to gain confidence in the yield of their designs without spending excessive time on simulations.

10.1.5 Validation of design for tool evaluation

When evaluating a fast Monte Carlo tool for a semiconductor design company, it is crucial to ensure that you consider all types of designs that exhibit significant variations and skewed Monte Carlo outputs. Here are a few reasons why this exhaustive evaluation is necessary:

1. Comprehensive assessment: By including a wide range of varying designs with skewed Monte Carlo outputs, you can thoroughly evaluate the tool's performance across different scenarios. It allows you to assess the tool's accuracy, efficiency, and reliability in handling diverse design characteristics and statistical distributions.
2. Realistic representation: Semiconductor designs often exhibit complex and non-uniform variations, leading to skewed Monte Carlo outputs. By incorporating such designs in your evaluation, you create a more realistic representation of the challenges faced in real-world design scenarios. This ensures that the tool can effectively handle and analyze the intricacies of these designs.
3. Robustness verification: Evaluating the tool's performance on designs with skewed Monte Carlo outputs helps verify its robustness and adaptability. It allows you to determine if the tool can accurately capture and analyze the non-normal statistical behavior and provide reliable results. This verification is crucial to ensure that the tool can handle a variety of design scenarios without compromising accuracy.
4. Decision-making confidence: By exhaustively evaluating the fast Monte Carlo tool on diverse designs, including those with skewed outputs, you can gain more confidence in its capabilities. It enables you to make informed decisions regarding the tool's suitability for your specific design requirements. Evaluating the tool on a broader spectrum of designs reduces the risk of potential limitations or biases that may arise when applied to specific design types.

Apart from this, semiconductor designs are continuously evolving, and new challenges may arise in the future. By considering a wide range of designs during the evaluation, one can assess the tool's potential to handle upcoming design complexities and ensure its long-term relevance.

In summary, evaluating the fast Monte Carlo tool on a comprehensive set of designs, including those with varying characteristics and skewed outputs, provides a more robust assessment of its performance, adaptability, and suitability for semiconductor design applications.

10.2 Simulation setup and tools overview

For this experiment, Synopsys PrimSim XA from PrimeSim Continuum design suite is used. Performing Monte Carlo analysis on PrimeSim XA involves setting up the simulation environment and executing the necessary steps to capture statistical variations. Here is an overview of the process:

1. Simulation setup:
    (a) Design setup: Set up the design files and necessary libraries for simulation. This includes the netlist, models, and any other design-specific files.
    (b) Model setup: Specify the models and parameters required for simulation, including process corners, temperature, and voltage conditions.
    (c) Monte Carlo configuration: Define the Monte Carlo analysis settings, such as the number of iterations or samples to be generated, statistical distributions for random variations, and any specific requirements for data collection.
2. Monte Carlo analysis execution:
    (a) Run Monte Carlo simulation: Execute the Monte Carlo simulation command in PrimeSim XA, providing the necessary parameters and options.
    (b) Iteration generation: PrimeSim XA will generate random variations for the specified design parameters based on the defined statistical distributions. These variations represent the range of possible values within the given statistical constraints.
    (c) Circuit simulation: For each Monte Carlo iteration, PrimeSim XA will simulate the circuit using the randomly generated parameter values. This involves solving the circuit equations and analyzing the behavior of the circuit under these variations.
    (d) Data collection: Collect the desired circuit performance metrics or results for each Monte Carlo iteration. These metrics could include timing delays, power consumption, signal integrity, or any other relevant parameters.
    (e) Statistical analysis: Analyze the collected data using statistical techniques to extract meaningful insights. Calculate statistical measures such as mean, standard deviation, yield, or any other relevant statistical indicators to assess the behavior and variability of the circuit under the considered variations.
3. Results visualization and interpretation:
    (a) Plotting: Visualize the statistical distribution of the circuit performance metrics using histograms, scatter plots, yield curves, or other appropriate visualization methods.
    (b) Interpretation: Interpret the results to understand the impact of variations on the circuit performance. Identify design sensitivities, optimize parameters, or make informed decisions based on the analysis outcomes.

```
.MEASURE TRAN tdelay TRIG V(XITOP:xibankbot:xipage_r:xilio:xilio_mux8:xilio_first:xilio_wa_ll:xilio_top.xibase.xisa_ctrl.s
aenb) VAL=pvddmp*0.1 FALL=1 FROM=1.501u TO=1.56u TARG V(XITOP:xibankbot:xipage_r:xilio:xilio_mux8:xilio_first:xilio_wa_ll:
xilio_top.xibase.rdcomn) VAL=pvddmp*0.1 RISE=1 FROM=1.501u TO=1.56u

.extract label=signal1_fall (xdown (v(XITOP:xibankbot:xipage_r:xilio:xilio_mux8:xilio_first:xilio_wa_ll:xilio_top.xibase.x
isa_ctrl.saenb),pvddmp*0.1 , 1.501u, 1.56u))
.extract label=signal2_rise (xup (v(XITOP:xibankbot:xipage_r:xilio:xilio_mux8:xilio_first:xilio_wa_ll:xilio_top.xibase.rdc
omn),pvddmp*0.1 , 1.501u, 1.56u ))
.extract label=tplh (signal2_rise - signal1_fall)
```

Figure 10.3. Measurement and extract statements in HSPICE syntax.

```
*
* PrimeSim XA linux64 S-2021.09  (built 06:18:22 Aug 26 2021)
* build id: 7232024
* Copyright (C)  2021 Synopsys Inc. All rights reserved.
*

meas variable = tdelay
nominal        = 1.947534e-10
mean           = 1.944671e-10
varian         = 1.162189e-22
stddev(1-sigma)          = 1.078049e-11
avgdev         = 8.654974e-12
min            = 1.590366e-10
max            = 2.31661e-10
median         = 1.946958e-10
run_min        = 1994
run_max        = 956
run_median     = 168, 1757

1.626679e-10    , nb = 12    , freq = 0.006       |
1.699303e-10    , nb = 47    , freq = 0.0235      |*
1.771927e-10    , nb = 152   , freq = 0.076       |***
1.844552e-10    , nb = 340   , freq = 0.17        |********
1.917176e-10    , nb = 496   , freq = 0.248       |************
1.9898e-10      , nb = 496   , freq = 0.248       |************
2.062425e-10    , nb = 309   , freq = 0.1545      |*******
2.135049e-10    , nb = 120   , freq = 0.06        |***
2.207673e-10    , nb = 26    , freq = 0.013       |
2.280298e-10    , nb = 2     , freq = 0.001       |
```

Figure 10.4. Monte Carlo simulation results.

Figure 10.3 illustrates the measure statement written to record a quantity of interest from a memory design in 18 nm and subjected to Monte Carlo analysis for minANmaxAP process corner.

The simulation was run on PrimeSim XA with multi-core (n = 2) options and distributed processing (dp=10). A screenshot of log file can be seen in figure 10.4.

10.3 Analysis of simulation results with Python

To analyze the simulation results stored in distributed locations and assimilated though a script into a Comma-Separated Values (CSV) file, this code presented here is a Python script for performing Monte Carlo data analysis. It imports necessary packages, reads data from the CSV file, computes statistics, and generates a histogram with a fitted probability distribution function (PDF) curve.

Here is a breakdown of the code:

Importing packages: The required packages are imported at the beginning of the script. These include numpy for numerical operations, os for interacting with the operating system, pandas for data manipulation, matplotlib.pyplot for plotting, and scipy.stats.norm for the normal distribution.

```
# Importing Packages

import numpy as np
import os
import pandas as pd
import matplotlib.pyplot as plt
from scipy.stats import norm
```

Setting current working directory (cwd): The os.getcwd function is used to retrieve the cwd. The cwd is then printed to the console.

```
cwd = os.getcwd()
print(cwd)
```

Loading data: The script reads data from the CSV file located at the specified path using pd.read_csv. The file contains two columns named 'index' and 'delay'. The data is stored in a pandas DataFrame called mcdataframe. The script then prints the first five rows of the DataFrame and its shape (number of rows and columns) to the console.

```
coln = ['index', 'delay']
mcdataframe = pd.read_csv(
'/prj/arcad/users/singhab2/simulation/simulationjob3/TESTCASE_MONTE_MC/sigmaResults4/t
file.csv',
    names=coln)
print(mcdataframe.head(5))
print(mcdataframe.shape)
```

Statistics calculation: The 'delay' column of the DataFrame is extracted using mcdataframe.loc[:,'delay'] and assigned to the variable data. The norm.fit function from scipy.stats is used to estimate the parameters of a normal distribution (mean and standard deviation) for the 'delay' data. The calculated mean and standard deviation are stored in the variables mu and std, respectively. The mean and standard deviation are printed to the console.

```
# Statistics start
data = mcdataframe.loc[:, 'delay']
mu, std = norm.fit(mcdataframe.loc[:, 'delay'])
print(mu, std)
```

Plotting: The histogram of the 'delay' data is plotted using plt.hist. The bins parameter determines the number of bins in the histogram. The density parameter is set to True to normalize the histogram. The alpha parameter controls the transparency of the histogram bars, and the color parameter sets the color to red. The PDF curve is plotted using norm.pdf from scipy.stats. The PDF is computed for a range of values (x) spanning the minimum and maximum of the 'delay' data. The computed PDF values are stored in the variable p. Vertical lines are plotted using plt.axvline to indicate sigma values (mean ± standard deviation) and twice sigma values. The title,

x-axis label, and y-axis label for the plot are set using plt.title, plt.xlabel, and plt. ylabel, respectively. Finally, plt.show is used to display the plot.

```
# Plot the histogram

plt.hist(data, bins=100, density=True, alpha=0.7, color='r')

# plot the probability distribution function
xmin, xmax = plt.xlim()
x = np.linspace(xmin, xmax, 300)
p = norm.pdf(x, mu, std)
plt.plot(x, p, linewidth=2, color='b')

# Plotting vertical lines for sigma values
plt.axvline(x=mu-std, linestyle='-.')
plt.axvline(x=mu+std, linestyle='-.')
plt.axvline(x=mu-2*std, linestyle='-.')
plt.axvline(x=mu+2*std, linestyle='-.')

title = "Fit results(2000 samples), BIN=100"
plt.title(title)
plt.xlabel("Delay Values")
plt.ylabel("Number of Occurrences")
# plt.legend(["Gaussian Curve", "Data Bins", "/mu /std values"], loc="upper right")
plt.show()
```

This code allows for the visualization of the distribution of the 'delay' data and the fitting of a Gaussian curve to analyze its characteristics and compare it to the sigma values (figure 10.5).

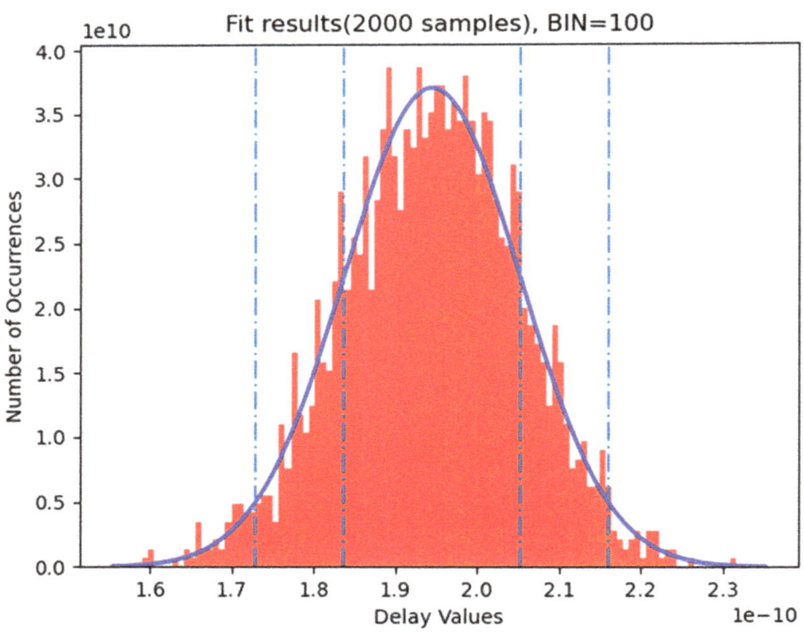

Figure 10.5. Monte Carlo simulation data fitting into Gaussian distribution.

In figure 10.5, for one SRAM design, the simulated data are analyzed against Gaussian distribution. The distribution is slightly left skewed.

10.4 Conclusion

This chapter has provided an overview of the validation of IC designs using Python for fast Monte Carlo tool evaluation. The rapid advancement in semiconductor technology, particularly at advanced nodes, has brought about new challenges in achieving high yield and reliability. Monte Carlo analysis has emerged as a valuable technique for capturing statistics on circuit behavior and assessing design performance under variations.

By leveraging Python in the evaluation process, designers can enhance the efficiency and effectiveness of Monte Carlo simulations. Python's extensive libraries and flexible programming capabilities enable the automation of simulation setups, data analysis, and result visualization. The integration of Python within the Anaconda environment further enhances its capabilities by providing a comprehensive suite of tools for scientific computing. The use of Python in combination with popular EDA tools, such as PrimeSim XA, offers a powerful framework for fast Monte Carlo tool evaluation. The ability to extract relevant data from simulation log files, perform statistical analysis, and visualize the results using Python scripts simplifies the validation process and provides valuable insights into design performance.

10.5 Future scope

Looking ahead, there are several potential future directions for this research. One area of focus could be the development of more advanced algorithms and methodologies for efficient Monte Carlo analysis. This includes exploring techniques such as variance reduction methods, importance sampling, and machine learning approaches to further accelerate simulations and improve accuracy. Leveraging distributed computing frameworks and high-performance computing clusters can significantly reduce simulation time and enable the analysis of larger and more complex designs.

Furthermore, the validation of IC designs using Python can be extended to other aspects of design verification, such as reliability analysis, power optimization, and performance characterization. Python's versatility and wide adoption in the industry make it a suitable platform for addressing various challenges in IC design validation.

In conclusion, the utilization of Python for fast Monte Carlo tool evaluation presents a promising approach to validate IC designs efficiently and accurately. By harnessing the power of Python and integrating it with state-of-the-art EDA tools, designers can overcome the challenges posed by advanced nodes and ensure the reliability and performance of their IC designs.

Bibliography

[1] Zhang L, Liu H and Lewis S 2021 *Monte Carlo Analysis at Advanced Nodes* Cadence https://www.cadence.com/content/dam/cadence-www/global/en_US/documents/tools/custom-ic-analog-rf-design/monte-carlo-analysis-at-advanced-nodes-wp.pdf (accessed 27 August)

[2] Huard V, Parthasarathy C R, Bravaix A, Guerin C and Pion E 2009 CMOS device design-in reliability approach in advanced nodes *IEEE Int. Reliability Physics Symp.* pp 624–33

[3] Huang V, Kim J, Pentapati S, Lim S K and Naeemi A 2020 Modeling and benchmarking back end of the line technologies on circuit designs at advanced nodes *IEEE Int. Interconnect Technology Conf. (IITC)* pp 37–9

[4] McConaghy T and Drennan P 2011 Variation-aware custom IC design: Improving PVT and Monte Carlo analysis for design performance and parametric yield *Solido White Paper*

[5] Watanabe K and Sasaki M 2011 An efficient Monte-Carlo method for calculating free energy in long-range interacting systems *J. Phys. Soc. Jpn.* **80** 093001

[6] Hung H and Adzic V 2006 Monte Carlo simulation of device variations and mismatch in analog integrated circuits *Proc. NCUR*

[7] Bucher C G 1988 Adaptive sampling—an iterative fast Monte Carlo procedure *Struct. Saf.* **5** 119–26

[8] Singhee A and Rutenbar R A 2007 Statistical blockade: a novel method for very fast Monte Carlo simulation of rare circuit events, and its application *Design, Automation & Test in Europe Conf. & Exhibition (Nice, France)* pp 1–6

[9] Mahapatra S, Vaish V, Wasshuber C, Banerjee K and Ionescu A M 2004 Analytical modeling of single electron transistor for hybrid CMOS-SET analog IC design *IEEE Trans. Electron Devices* **51** 1772–82

[10] Balasubramanian S, Peters L, Fisher K, Dyck J, Song Y, Luque M, Higman J and Nigam T 2020 Hierarchical high sigma Monte Carlo simulation of SRAM Vmin shift by parts for automotive grade fail rate projection *IEEE Int. Integrated Reliability Workshop (IIRW)* pp 1–4

[11] Sun S, Li X, Liu H, Luo K and Gu B 2013 Fast statistical analysis of rare circuit failure events via scaled-sigma sampling for high-dimensional variation space *IEEE Trans. on Computer-Aided Design of Integrated Circuits and Systems* **vol 34** pp 1096–109

[12] Canelas A, Martins R, Póvoa R, Lourenço N and Horta N 2016 Yield optimization using k-means clustering algorithm to reduce Monte Carlo simulations *2016 13th Int. Conf. on Synthesis, Modeling, Analysis and Simulation Methods and Applications to Circuit Design* pp 1–4

www.ingramcontent.com/pod-product-compliance
Ingram Content Group UK Ltd.
Pitfield, Milton Keynes, MK11 3LW, UK
UKHW051900100926
8556IPUK00005B/713

9 780750 35927